T0271071

Sustainable Urban Agriculture

In the vibrant discourse of urbanization and climate change, *Sustainable Urban Agriculture: New Frontiers* investigates emerging needs, rising challenges, and opportunities to support urban agriculture. Navigating the dynamic interplay of urbanization and environmental challenges, the book introduces two pivotal agendas for urban sustainability—the "green" agenda, focusing on environmental health, and the "brown" agenda, emphasizing human well-being and social justice. The book embraces a global perspective by confronting geographical biases and advocating for context-specific understanding and early interventions in small and medium cities. This transformative journey guides readers through uncharted territories, fostering profound awareness of urban agriculture's role in shaping a sustainable and resilient future in agriculture.

Features

- Presents information on socio-ecological resilience, shaping a sustainable urban future
- Unveils practical implications, traversing frontiers where urban cultivation extends beyond crops, cultivating a thriving urban ecosystem
- Discusses diverse urban agriculture practices, from traditional methods to cutting-edge technologies

Providing readers with an understanding of the multifaceted layers inherent in urban agriculture, this volume in the NextGen Agriculture: Novel Concepts and Innovative Strategies series is essential for academics, students, practitioners, and experts in urban agriculture and planning, horticulture, landscape architecture, and plant sciences.

NextGen Agriculture: Novel Concepts and Innovative Strategies
Series Editor: Chittaranjan Kole

Allele Mining for Genomic Designing of Fruit Crops
Chittaranjan Kole, Kenta Shirasawa, and Anil Kumar Singh

Allele Mining for Genomic Designing of Cereal Crops
Chittaranjan Kole, Sharat Kumar Pradhan, and Vijay K Tiwari

Allele Mining for Genomic Designing of Vegetable Crops
Chittaranjan Kole, Tusar Kanti Behera, and Prashant Kaushik

Allele Mining for Genomic Designing of Oilseed Crops
Chittaranjan Kole, Manish Kumar Pandey, and Naveen Puppala

Sustainable Urban Agriculture: New Frontiers
Kheir Al-Kodmany, Madhav Govind, Sharmin Khan, and Chittaranjan Kole

For more information, please visit our website: https://www.routledge.com/Nextgen-Agriculture/
book-series/

Sustainable Urban Agriculture
New Frontiers

Edited by
Kheir Al-Kodmany, Madhav Govind, Sharmin Khan, and
Chittaranjan Kole

CRC Press
Taylor & Francis Group
Boca Raton London New York

CRC Press is an imprint of the
Taylor & Francis Group, an **informa** business

First edition published 2025
by CRC Press
2385 NW Executive Center Drive, Suite 320, Boca Raton FL 33431

and by CRC Press
4 Park Square, Milton Park, Abingdon, Oxon, OX14 4RN

CRC Press is an imprint of Taylor & Francis Group, LLC

© 2025 selection and editorial matter, Kheir Al-Kodmany, Madhav Govind, Sharmin Khan, Chittaranjan Kole; XXX individual chapters, the contributors

Reasonable efforts have been made to publish reliable data and information, but the author and publisher cannot assume responsibility for the validity of all materials or the consequences of their use. The authors and publishers have attempted to trace the copyright holders of all material reproduced in this publication and apologize to copyright holders if permission to publish in this form has not been obtained. If any copyright material has not been acknowledged please write and let us know so we may rectify in any future reprint.

Except as permitted under U.S. Copyright Law, no part of this book may be reprinted, reproduced, transmitted, or utilized in any form by any electronic, mechanical, or other means, now known or hereafter invented, including photocopying, microfilming, and recording, or in any information storage or retrieval system, without written permission from the publishers.

For permission to photocopy or use material electronically from this work, access www.copyright.com or contact the Copyright Clearance Center, Inc. (CCC), 222 Rosewood Drive, Danvers, MA 01923, 978-750-8400. For works that are not available on CCC please contact mpkbookspermissions@tandf.co.uk

Trademark notice: Product or corporate names may be trademarks or registered trademarks and are used only for identification and explanation without intent to infringe.

Library of Congress Cataloging-in-Publication Data
Names: Al-Kodmany, Kheir, editor. | Govind, Madhav, editor. |
Khan, Sharmin, editor. | Kole, Chittaranjan, editor.
Title: Sustainable urban agriculture : new frontiers / edited by Kheir
Al-Kodmany, Madhav Govind, Sharmin Khan, and Chittaranjan Kole
Description: First edition | Boca Raton, FL : CRC Press, 2025 |
Includes bibliographical references and index | Summary: "In the vibrant discourse of urbanization and climate change, Sustainable Urban Agriculture: New Frontiers investigates emerging needs, rising challenges, and opportunities to support urban agriculture. Navigating the dynamic interplay of urbanization and environmental challenges, the book introduces two pivotal agendas for urban sustainability-the "green" agenda, focusing on environmental health, and the "brown" agenda, emphasizing human well-being and social justice. The book embraces a global perspective by confronting geographical biases, and advocating for context-specific understanding and early interventions in small and medium cities. This transformative journey guides readers through uncharted territories, fostering profound awareness of urban agriculture's role in shaping a sustainable and resilient future in agriculture"—Provided by publisher
Identifiers: LCCN 2024011938 (print) | LCCN 2024011939 (ebook) | ISBN
9781032417219 (hardback) | ISBN 9781032417165 (paperback) | ISBN 9781003359425 (ebook)
Subjects: LCSH: Urban agriculture. | Sustainable development. | Sustainable agriculture.
Classification: LCC S494.5.U72 S864 2025 (print) | LCC S494.5.U72 (ebook) |
DDC 338.109173/2—dc23/eng/20240711
LC record available at https://lccn.loc.gov/2024011938
LC ebook record available at https://lccn.loc.gov/2024011939

ISBN: 978-1-032-41721-9 (hbk)
ISBN: 978-1-032-41716-5 (pbk)
ISBN: 978-1-003-35942-5 (ebk)

DOI: 10.1201/9781003359425

Typeset in Times
by codeMantra

This book series is dedicated to

Dr. Norman Ernest Borlaug

The Father of Green Revolution, Nobel Peace Prize
Winner and Founder of the World Food Prize

*Who inspired my research and academic works with
his generous appreciation and advice.*

Preface to the Series

Agriculture is going through a phase of transformation now. It has both challenges and opportunities. One of the major challenges it is facing is obviously food security. A population of about 9.7 billion projected by 2050 necessitates an increase of crop production by 60–70%. Another 10–15% increase would be needed to counter the reduction of crop production due to global warming and climate change. Nutrition security is another challenge with more than 800 million people suffering from malnutrition. Depletion of fossil fuel and increasing dependence on biofuel from crop residues are another challenge. Loss of biodiversity and land degradation also add to the list of challenges. The cultivable area is also decreasing fast due to housing, transport, and industrialization. Above all, climate change and global warming have a huge impact on agriculture, health, and environment.

Several novel concepts and strategies are emerging to address the above challenges. For example, the practice of precision farming will lead to precise and timely use of inputs, resulting in higher yield and better quality of produce, on the one hand, and reduction of loss of inputs, on the other. Similarly, urban agriculture and vertical agriculture will add a lot of cultivable areas in the future. Protected farming will facilitate growing crops under controlled conditions, leading to a reduction in biotic and abiotic stresses and precise utilization of inputs.

Agriculture requires maintenance of the philosophy farmers have been following since inception over thousands of years, which are being sold today in new packages with sophisticatedly coined labels such as traditional agriculture, natural agriculture, sustainable agriculture, conservation agriculture, integrated agriculture, regenerative agriculture, and organic agriculture.

In the changing scenarios, next-generation agriculture will need designed crop varieties and breeds produced through designed practices. This book series aims at deliberations on several novel concepts and innovative strategies developed for next-generation agriculture and will hopefully benefit agricultural education, research, and outreach.

Chittaranjan Kole
Kolkata
January 10, 2024

Contents

Preface to the Series...vi
Editors...ix
Contributors xi
Preface...xiii
Abbreviations xv

Chapter 1 Navigating Urban Agriculture and Sustainability: Bridging Green and Brown
Agendas in a Global Context 1

Kheir Al-Kodmany, Madhav Govind, Sharmin Khan, Anita Pinheiro, and
Chittaranjan Kole

Chapter 2 Urban Agriculture: A Historical Perspective 9

Sharmin Khan and Anwar Hussain

Chapter 3 Multifunctional Urban Agriculture and Sustainability: A Global Overview 20

Madhav Govind and Anita Pinheiro

Chapter 4 Multifunctionality of Urban Agriculture in Burkina Faso and Niger:
Environmental, Social, Economic, and Political Perspectives.................................40

Hamid El Bilali

Chapter 5 Urban Home Gardening and Agri-Food System Transitions Toward
Sustainability in Kerala..62

Anita Pinheiro and Madhav Govind

Chapter 6 In Pursuit of a Traditional Livelihood: Insights from Singapore's Farmers 85

Jessica Ann Diehl

Chapter 7 Growing Cities, Changing Demands: Scope of Urban Agriculture as a
Sustainable Agricultural Intensification Strategy in India.......................................113

Sreejith Aravindakshan, Hage Aku, and Michi Tani

Chapter 8 The Social, Economic, and Policy Impacts of Urban and Peri-Urban
Agriculture: Farmer's Experience from Dar es Salaam City and
Morogoro Municipality, Tanzania...134

Betty Mntambo

Chapter 9   In the Face of Climate Change and Food Insecurity in the Middle East
and North African Regions: Are Urban and Peri-Urban Agriculture
Viable Options? 151

Tarek Ben Hassen and Hamid El Bilali

Chapter 10 Urban Agriculture and the Food–Energy–Water Nexus in Qatar
and Drylands: From Permaculture Gardens to Urban Living Labs 164

Anna Grichting Solder

Chapter 11 Cultivating Change in Cities: Exploring and Classifying the Determinants
of Urban Agriculture in India 181

Maitreyi Koduganti and Sheetal Patil

Chapter 12 Impact of Urban Farming on the Built Environment 197

Mohd Khalid Hassan and Sadaf Faridi

Chapter 13 Challenges of Urban Agricultural Production from the Aspects of Plant
Protection and Food Safety: A Case Study of the Republic of Serbia 208

*Dragana Šunjka, Sanja Lazić, Slavica Vuković,
Siniša Berjan, and Hamid El Bilali*

Chapter 14 The Vertical Farm: The Next-Generation Sustainable Urban Agriculture 223

Kheir Al-Kodmany

Index 237

Editors

Dr. Kheir Al-Kodmany, an internationally reputed professor of urban planning at the University of Illinois Chicago (UIC), USA, boasts a prolific career spanning 30 years. His extensive research and teaching portfolio encompasses diverse subjects, including vertical urbanism, sustainable design, geographic information systems (GISs), visualization systems, public participation, crowd management, economic development, and skyscrapers. A prolific author, Dr. Al-Kodmany has published 15 books and over 150 scholarly works, earning widespread acclaim for his comprehensive insights into architecture and urban design. His publications are highly regarded, with over 250,000 reads on ResearchGate. Recognized for his academic excellence, he serves on 20 editorial boards of professional journals. Dr. Al-Kodmany's impactful teaching career spans 30 years at UIC and UIUC, where he secured grants totaling several hundred thousand dollars and developed innovative visualization software. Beyond academia, Dr. Al-Kodmany's expertise has been sought by governments, mayors, and organizations worldwide. Notably, the Saudi government invited him to contribute to the planning of Hajj, earning him international recognition for enhancing safety. His involvement in Chicago's "Taste of Chicago" event and contributions to Mayor Daley's bid for the 2016 Summer Olympics, further demonstrate his impactful civic engagement. Dr. Al-Kodmany delivered over 200 presentations globally at prestigious institutions as a sought-after keynote speaker and trainer. His leadership roles at UIC, including director of graduate studies, associate director of the City Design Center, and co-director of the Urban Data Visualization Laboratory, demonstrate his commitment to innovative tools in participatory planning and design. Affiliated with esteemed professional organizations such as the APA, ACSP, CTBUH, and URISA, Dr. Al-Kodmany continues to contribute significantly to the field of urban planning, carrying forward the legacy of his early architectural training from his father, Dr. Abdul Muhsen Al-Kodmany, a Le Corbusier trainee and École des Beaux-Arts graduate.

Dr. Madhav Govind is a professor and chairperson at the Centre for Studies in Science Policy, Jawaharlal Nehru University, New Delhi, India. After completing his studies in life science, he did his post-graduation in sociology from the University of Lucknow and obtained his first position at the university. Prof. Govind completed his MPhil and PhD from Jawaharlal Nehru University, New Delhi. He has been extensively engaged in teaching, research, and collaborative work for the past 20 years. He has published numerous papers and articles in national and international journals. His broad areas of interest are sociology of science and technology; internationalization of higher education; university–industry relations and transfer of knowledge; waste management and sustainable development; and urbanization and sustainability in the era of globalization.

Dr. Sharmin Khan, an accomplished architecture and construction management professional, holds a PhD in architecture from AMU, Aligarh complemented by a post-graduate degree in construction engineering and management from IIT, Delhi, India. Her academic prowess is evident through the prestigious Merit Scholarship during her undergraduate studies and a GATE Scholarship with an outstanding All India Rank 8. Prof. Khan has contributed significantly to academia, focusing on the correlation between construction management and sustainable development. She served on the faculty at the Institute of Integral Technology, Lucknow and JMI, New Delhi before joining AMU, Aligarh in 2004. Her areas of expertise span construction management, history, and sustainable architecture, with a dedicated commitment to teaching ancient Indian and Western architecture. Dr. Khan's scholarly achievements include receiving the Best Paper Award at the International Conference on "Emerging Trends in Engineering, Technology, Science and Management" in 2017 and first place Best Paper Award and Best Plenary Speaker Award at the International Virtual Conference Workshop on Sustainable Architecture and Urban Design (ICWSAUD2020) in 2020.

She also earned the Best Paper Award at the IIHSG International Conference 2021 on gender security and global politics. Beyond her teaching and research roles, Dr. Khan serves as a respected reviewer for prestigious journals such as the *Journal of Construction Engineering and Management* (ASCE, USA), *Journal of Civil Engineering and Architecture* (Horizon Research Publishing Corporation, USA), and *Frontiers in Engineering and Built Environment* (Emerald Publishing, UK) representing King Khalid University, Saudi Arabia. Her multifaceted contributions underscore Dr. Khan's prominence in academia and her dedication to sustainable architectural practices.

Dr. Chittaranjan Kole is an internationally reputed scientist with an illustrious professional career spanning over 40 years with original contributions in the fields of plant genomics, biotechnology, and molecular breeding, leading to the publication of more than 160 quality research articles and reviews. He has edited over 190 books for leading publishers including Springer Nature, Wiley-Blackwell, and Taylor & Francis Group. His scientific contributions and editing acumen have been appreciated by seven Nobel laureates including Profs. Norman Borlaug, Arthur Kornberg, Werner Arber, Phillip Sharp, Günter Blobel, Lee Hartwell, and Roger Kornberg. He has been honored with many fellowships, honorary fellowships, and national and international awards including the Outstanding Crop Scientist Award conferred by the International Crop Science Society. He has served in many prestigious positions in academia including vice-chancellor of Bidhan Chandra Krishi Viswavidyalaya, project coordinator of the Indo-Russian Center of Biotechnology in India, and director of research of the Institute of Nutraceutical Research of Clemson University in USA. He also worked at Pennsylvania State University and Clemson University as a visiting professor. Recently, he was awarded with the Raja Ramanna Fellowship by the Department of Energy, Government of India. He also heads the International Climate-Resilient Crop Genomics Consortium, International Phytomedomics and Nutriomics Consortium, and Genome India International as their founder president. Recently, he has established the Prof. Chittaranjan Kole Foundation for Science and Society and acting as its chairman.

Contributors

Hage Aku
Agriculture Technology Management Agency
   (ATMA)
Kamle, India

Kheir Al-Kodmany
University of Illinois at Chicago
Chicago, Illinois

Sreejith Aravindakshan
International Maize and Wheat Improvement
   Center (CIMMYT/CGIAR)
Dhaka, Bangladesh

Siniša Berjan
Faculty of Agriculture
University of East Sarajevo
Istocno Sarajevo, Republic of Srpska, Bosnia
   and Herzegovina

Jessica Ann Diehl
National University of Singapore
Singapore, Singapore

Hamid El Bilali
International Centre for Advanced
   Mediterranean Agronomic Studies
   (CIHEAM-Bari)
Valenzano, Italy

Sadaf Faridi
Department of Architecture, Aligarh Muslim
   University
Aligarh, Uttar Pradesh, India

Madhav Govind
Centre for Studies in Science Policy
Jawaharlal Nehru University
New Delhi, India

Mohd Khalid Hassan
Department of Architecture
Aligarh Muslim University
Aligarh, India

Tarek Ben Hassen
Program of Policy, Planning, and Development,
   Department of International Affairs, College
   of Arts and Sciences
Qatar University
Doha, Qatar

Anwar Hussain
University Boys Polytechnic, Faculty of
   Engineering & Technology
Aligarh Muslim University
Aligarh, India

Sharmin Khan
Department of Architecture
Aligarh Muslim University
Aligarh, India

Maitreyi Koduganti
Indian Institute for Human Settlements
Lansing, Michigan, United States

Chittaranjan Kole
Prof. Chittaranjan Kole Foundation for Science
   and Society
Kolkata, India

Sanja Lazić
Faculty of Agriculture
University of Novi Sad
Novi Sad, Serbia

Betty Mntambo
The Open University of Tanzania
Mazimbu, Tanzania

Sheetal Patil
Indian Institute for Human Settlements
IIHS Bengaluru City Campus
Bengaluru, India

Anita Pinheiro
Department of Environmental Studies
Ashoka University
Sonipat, India

Anna Grichting Solder
University of Vermont. Bordermeetings
Geneva, Switzerland

Dragana Šunjka
Faculty of Agriculture
University of Novi Sad
Novi Sad, Serbia

Michi Tani
Division of Fruits & Horticultural Technology
Indian Agricultural Research Institute (IARI)
New Delhi, India

Slavica Vuković
Faculty of Agriculture
University of Novi Sad
Novi Sad, Serbia

Preface

In the bustling landscape of urbanization and climate change discussions, this book emerges as a guiding beacon, directing attention to the nuanced realms of urban agriculture often overshadowed in contemporary discourse. Within the dynamic interplay of urbanization and environmental challenges, our exploration unfolds, introducing two pivotal agendas essential for urban sustainability—the "green" agenda, centering on environmental health, ecosystem protection, and ecological impacts of urban systems, and the "brown" agenda, prioritizing human well-being, social justice, and pro-poor urban development.

As we embark on this intellectual journey, our focus crystallizes on the integral role of urban agriculture. It is a transformative force, seamlessly weaving through the fabric of these agendas, contributing to socio-ecological resilience, and sculpting the contours of a sustainable urban future. Beyond theoretical discussions, this book is a conduit for understanding the practical implications of urban agriculture in fostering inclusive development and environmental equilibrium. Our collective odyssey traverses the new frontiers of sustainable urban agriculture, where cultivation extends beyond crops to cultivate a harmonious, thriving urban ecosystem—marking the path toward a resilient and sustainable urban future.

UNVEILING URBAN AGRICULTURE

Exploring urban agriculture reveals a dynamic and multifaceted landscape, transcending mere cultivation within urban boundaries. This practice assumes diverse forms and dimensions, from subsistence endeavors to sophisticated commercial enterprises. Central to our discourse is the assertion that comprehending urban agriculture necessitates viewing it as an integral component of the urban fabric. In doing so, we unravel its intricate connections to urban life's social, cultural, environmental, and economic tapestry. This book intricately examines the embeddedness of urban agriculture within the broader urban context, shedding light on its profound influence on the very form and function of the urban environment. As we unveil the layers of urban agriculture, our journey extends beyond the mere act of cultivation, delving into its profound impacts on shaping the identity and sustainability of our urban landscapes.

DIVERSE PRACTICES, NEW HORIZONS

Diving into the intricate tapestry of urban agriculture practices, our exploration spans a spectrum that traverses time-honored traditional methods to the cutting-edge frontiers of agricultural technology, ultimately reaching the fascinating domain of vertical farming. Embracing the challenge posed by the contextual variations that defy a one-size-fits-all definition, this book is dedicated to unraveling urban agriculture's intricate nuances. Its purpose is not merely to categorize but to dissect and comprehend the multifaceted layers of urban agriculture's diverse landscape. By undertaking this journey, the book aspires to provide readers with a nuanced and comprehensive understanding of the complexities inherent in urban agriculture. It is a quest that transcends the superficial, inviting readers to delve into the heart of urban agriculture, explore its potential, and unravel its myriad possibilities for sustainable and resilient urban futures.

A GLOBAL PERSPECTIVE

In recognizing the often-neglected geographical disparity within the realm of urban agriculture literature, this book bravely confronts the prevailing bias emanating from publications primarily rooted in the perspectives of the Global North. By acknowledging this imbalance, we embark on a journey to rectify the narrative, offering a more inclusive and representative exploration of urban agriculture practices from a global standpoint. Moreover, we fervently underscore the critical importance of context-specific understanding, advocating for a nuanced comprehension of urban agriculture that transcends geographical boundaries. This emphasis becomes particularly pronounced in our call for early interventions in small and medium cities, which are the focal points for sustainable urbanization efforts. As we delve into the pages of this book, readers are invited to transcend preconceived notions, fostering an appreciation for the varied contexts in which urban agriculture unfolds and recognizing the pivotal role these practices play in shaping a sustainable and resilient urban future.

CHARTING THE COURSE

Embarking on a transformative journey, this book guides readers through the uncharted territories of sustainable urban agriculture. As the curtain rises, it meticulously sets the stage for a comprehensive exploration, unraveling the historical dimensions, diverse practices, and socio-economic implications intricately woven into the fabric of urban agriculture. It boldly confronts intricate policy considerations, paving the way for a holistic understanding of this dynamic field. With intellectual frameworks as our guide, we traverse the diverse landscapes of urban agriculture, unraveling its multifaceted layers. The narrative extends beyond conventional boundaries, giving readers a nuanced understanding of the symbiotic relationship between the city and the harvest. This book aspires to be more than mere documentation; it aims to catalyze transformative thinking, fostering a profound awareness of urban agriculture's pivotal role in shaping a sustainable and resilient future within the intricate tapestry of the global sustainability landscape.

Abbreviations

ACT	Agri-food cluster transformation
AFHVS	Agriculture, food and human values society
AGG	A-Go-Gro
ASFS	Study of Food and Society
AVA	Agri-food and veterinary association
CE	Carbon emissions
CEA	Controlled-environment agriculture
CIWP	Crop irrigation water productivity
CNUP	Crop nitrogen use productivity
CO_2	Carbon dioxide
CO_2e	Carbon dioxide equivalent
CPUL	Continuous productive urban landscapes
CSA	Community supported agriculture
dBA	Decibels
DCEO	Department of commerce and economic opportunity
DWC	Deep water culture
EbA	Ecosystem-based adaptation
EC	Electrical conductivity
EE/EconEff	Environmental efficiency
EES	Ecosystem services
EIGS	Efficiency in implementing government schemes
EPA	Environmental protection agency
EU	European Union
FAO	Food and Agriculture Organization
FB	Facebook-based
FEW	Food energy water
ft	Feet
FYPs	Five year plans
GAM	Greater Amman municipality
GAP	Good agriculture practice
GCC	Gulf cooperation council
GCRF	Global challenges research fund
GDP	Gross domestic product
GHG	Greenhouse gas
GUPAP	Gaza urban and peri-urban agriculture platform
GVO	Gross value output
ha	Hectare
HDB	Housing Development Board
ICT	Information and communications technology
IDP	Internally displaced people
IGAs	Income-generating activities
IoT	Internet of things
IPM	Integrated pest management
IRAS	Inland revenue authority of Singapore
IRB-SBER	Institutional review board for social, behavioral and educational research
i-RTGs	Integrated rooftop greenhouses

IT	Information technology
KCA	Kranji Countryside Association
lb	Pound
LED	Light-emitting diodes
LOD	Limit of detection
LOQ	Limit of quantification
LTA	Land transport authority
m	Meter
MENA	Middle East and North Africa
MLP	Multi-level perspectives
MNEX	Moveable nexus
MOE	Ministry of Education
MOM	Ministry of Manpower
MRLs	Maximum residue levels
MUFPP	Milan urban food policy pact
NbS	Nature-based solutions
NEA	National environment agency
NFT	Nutrient film technique
NGOs	Non-governmental organizations
NParks	National parks board
NSO	National statistical organization
NTU	Nanyang Technological University
NTUC	National Trades Union Congress
NUA	New urban agenda
NUS	National University of Singapore
ODK	Open data kit
PAIA	Priority areas for interdisciplinary action
PCI	Positive cultural impact
pH	Potential of hydrogen
PHI	Pre-harvest interval
POPs	Persistent organic pollutants
PPP	Plant protection products
PRISMA	Preferred reporting items for systematic reviews and meta-analyses
PSI	Positive social impact
PUB	Public utilities board
QNRF	Qatar national research fund
QNRS	Qatar national research strategies
RS	Republic of Serbia
SAFA	Sustainability assessment of food and agriculture systems
SAFEF	Singapore agro-food enterprises federation limited
SAR	Sodium absorption rate
SCDF	Singapore civil defence force
SDGs	Sustainable development goals
SFA	Singapore food agency
SGD	Singapore dollar
SI	Sustainable intensification
SLA	Singapore land authority
SNM	Strategic niche management
SOC	Soil organic carbon
SSFCs	Short supply food chains

SUW	Solid urban wastes
TM	Trademark
TSh	Tanzanian Shilling
UA	Urban agriculture
UAE	United Arab Emirates
UAS	Urban agriculture systems
UEIP	Urban environment improvement project
UGI	Urban green infrastructure
UHI	Urban heat island
UPA	Urban and peri-urban agriculture
UPAGrI	Urban and peri-urban agriculture as green infrastructure
URA	Urban Redevelopment Authority
URDPFI	Urban & Regional Development Plans Formulation & Implementation
UREP	Undergraduate research and entrepreneurship program
USD, $	United states dollar
VDP	Vegetable development program
VIGS	Vertical indoor greenery system
W	Watt
WHO	World Health Organization
WoS	Web of Science

1 Navigating Urban Agriculture and Sustainability
Bridging Green and Brown Agendas in a Global Context

Kheir Al-Kodmany, Madhav Govind, Sharmin Khan,
Anita Pinheiro, and Chittaranjan Kole

Urbanization is a significant global phenomenon, with 55% of the population living in urban areas. By 2050, 68% of the world's population is expected to be urban (United Nations, 2019). This shift from rural to urban areas presents opportunities and challenges, including socio-economic and environmental issues. The shift from rural to urban society leads to irreversible changes in people–nature interactions. Moreover, urbanization is exacerbated by climate change, which is further exacerbated by urbanization. Urban sustainability is a critical topic due to the strategic importance of urban systems and the vast opportunities and obstacles they can contribute to global and local sustainability. However, many urban sustainability interventions overlook the pivotal role of agriculture in urban areas. This raises questions about the lack of attention to food production in discussions on sustainable urbanization and climate change challenges. The "brown" agenda of urban sustainability prioritizes human well-being, social justice, and pro-poor urban development. The "green" agenda centers on environmental health, ecosystem protection, and the ecological impacts of urban systems. The book aims to bridge these two agendas while answering the raised questions within the context of urban agriculture (UA) and sustainability (Allen & You, 2002; Burch et al., 2018; Castán Broto et al., 2019; Hölscher & Frantzeskaki, 2021).

UA, a practice since ancient civilizations, encompasses cultivating food and non-food produce within urban limits. It includes various practices, from subsistence to commercial enterprises, frugal to technologically sophisticated production systems, and individual to community practices. UA is often perceived as an extension of conventional agriculture in the urban context, which can overlook its social, cultural, environmental, and economic outcomes. To fully explore the holistic potential of UA, it is crucial to consider it as a practice integrated into the urban fabric. The embeddedness of UA within the urban fabric significantly influences its form and function. The various urban characteristics influence UA activities, impacting the urban system. This embeddedness plays a significant role in determining and shaping the impacts of UA, making it distinct from rural agriculture. Finding a universally applicable definition for UA is challenging due to its diversity and varying characteristics dependent on local contexts. Different definitions are available in the literature, but most capture only part of the characteristics of UA. Van Veenhuizen and Danso (2007) define UA flexibly and broadly, encompassing all activities related to producing, processing, and marketing food and non-food products from plant and animal sources. However, many studies use different definitions most suited for the specific type of UA they explore. This volume explores varying notions of UA, ranging from traditional forms to more advanced and technologically sophisticated forms of vertical farming (Cook et al., 2015; Korth et al., 2014; Madaleno, 2001; Mougeot, 2000; Nugent, 1999; Smit et al., 2001; Van Veenhuizen, 2006; Van Veenhuizen & Danso, 2007).

DOI: 10.1201/9781003359425-1

Urban sustainability concerns are divided into two main agendas: the "green" agendas, which focus on environmental health and pro-poor urban development, and the "brown" agendas, which focus on human environment and infrastructure development. The brown agenda is primarily a priority in the Global South, particularly in lower-income countries. The green agenda focuses on ecosystem protection and ecological impacts of urban systems at local, regional, and global levels. UA has the potential to bridge these agendas, and it is increasingly being used in research and interventions aimed at increasing urbanization, socio-ecological resilience, and sustainability. However, sustainability is contested, and addressing all three pillars of sustainability with a single solution is overambitious. Different types of UA deliver different outcomes to sustainability, such as vertical farming and urban home gardens. Many urban sustainability interventions focus on infrastructural development instead of the overall development of urban socio-ecological systems, leading to green gentrification and further social injustice. UA cannot be considered a one-size-fits-all solution to urban sustainability challenges. Its potential and limitations depend on the specific type of practice and the local context. The chapters in this volume explore the intricacies of UA and sustainability in different contexts (Allen & You, 2002; Anguelovski et al., 2019; Gould & Lewis, 2016; McGranahan & Satterthwaite, 2000; Shrestha, 2019; Simon, 2016; Simon, 2016; UN Habitat, 2009).

UA is a globally distributed practice with numerous research articles, books, and reports available worldwide. However, there is a significant gap in the literature on UA, with most publications coming from countries and institutions in the Global North. This lack of research from the Global South results in a biased understanding of UA practices, as the intricacies of multifunctionality, sustainability potential, risks, barriers, and other dynamics may not be reflected in studies from the Global North. Context-specific understanding of UA and its local relevance is crucial for policy formulation and interventions. Small and medium cities, with populations of less than 1 million, are the fastest-growing cities and represent most of the global urban population. They are an essential focus for early interventions to make them sustainable and resilient. However, urban planning in developing countries has received less attention than in metropolitan cities. Most studies on UA focus on large cities. Still, it is essential to recognize that small urban areas can embed UA into design and planning to ensure sustainable urbanization. UA may act as a mitigation strategy in large cities, while urban home gardening can contribute to adaptation measures in small urban areas (Pinheiro, 2022; Pinheiro & Govind, 2020; Rao et al., 2022; UN Habitat, 2020; UN Habitat & UN ESCAP, 2015).

BOOK OUTLINE

This book is a comprehensive exploration that delves into various research areas within UA, offering a rich tapestry encompassing historical and contemporary dimensions. The following chapters encapsulate the critical foci of the research:

Insights gained from past experiences can lay the foundation for future achievements. **Chapter 2** provides a historical overview of several green spaces created for diverse purposes at different times. These include the Hanging Gardens of Babylon, green spaces in Mesopotamia, and the Mughal gardens. The creation of these green areas can be attributed to the need for aesthetic representation that embodies art and architecture, serving as a symbol of civilization or cultural philosophy. However, all of them ultimately contribute to their eras' UA. UA has developed over time because of several variables, such as changes in climate, the demand for green spaces, technological advancements, socio-political reasons, and the support of rulers. Gaining a profound understanding of this UA development might also facilitate comprehension of their socio-economic contributions. This chapter aims to comprehend the insights gained from past UA operations and their role in promoting sustainable practices.

The multifunctionality of UA encompasses diverse interpretations and activities, providing numerous concrete and intangible advantages for social, ecological, cultural, and economic sustainability. However, many conversations about UA remain focused on providing food for many people, emphasizing intense production driven by entrepreneurship within urban areas. The limited perspectives on UA greatly diminish the many capabilities of UA, particularly for small-scale, livelihood-focused, or non-commercial projects. Within this framework, **Chapter 3** addresses the deficiencies in existing research by examining the diverse capabilities provided by different forms of UA practices and their potential for long-term viability. The chapter delves into small-scale UA to implement "Nature-Based Solutions" to establish sustainable urban ecology and food systems. The chapter emphasized the significance of tailoring UA to unique contexts to achieve its highest potential for sustainability.

Burkina Faso and Niger are two landlocked countries situated in the Sahel and West Africa regions, characterized by their status as developing nations. Both nations have undergone significant population expansion accompanied by an unparalleled rate of urbanization. Simultaneously, they experience extensive poverty, susceptibility to economic instability, inadequate access to food, and hunger, all of which are intensified by social upheaval and wars (such as internally displaced individuals) as well as climate change. These problems and challenges also impact food systems, including food production (agricultural), conversion, distribution, and consumption. **Chapter 4** aims to examine the ecological advantages of urban and peri-UA, as well as its role in enhancing food security, generating revenue, and improving the livelihoods of residents in Burkina Faso and Niger. The additional objective is to examine the policy position related to UA, considering sector-specific policies (such as agriculture and environment) and local legislation about land management and zoning in urban regions. Although the various advantages of UA, such as its ability to serve multiple functions and provide diverse benefits in terms of the environment, society, and economy, are widely acknowledged, specific measures are required to fully harness its potential in Burkina Faso, Niger, and the broader Sahel region. It is crucial to have a favorable policy and regulatory framework for agriculture in urban and peri-urban areas.

Urban home gardens are complex systems that serve as critical Nature-Based Solutions for urban and agri-food system transitions. They provide several functions, such as social, ecological, and economic benefits. However, governments tend to overlook these private behaviors, resulting in insufficient attention being given to them. In contrast, the government of Kerala, a state in South India, has made significant efforts to prioritize and promote urban and rural home gardening to tackle the risks and challenges linked to food dependence. **Chapter 5** examines the role of urban home gardening in promoting agri-food system transitions and sustainability in Kerala, using the Multi-Level Perspectives of sustainability transition theory. As a state with a rural–urban continuum, the government's interventions prioritize promoting UA through several subsidy schemes across the entire state, regardless of the size and population of urban districts. These initiatives played a vital role in transforming home gardening into a popular movement and facilitating the adoption of specialized innovations. However, the resistance from the regime and the changing views and values of certain state actors are hindering the progress of urban home gardening. This is preventing the adoption of both the "fit and conform" and "stretch and transform" patterns of transitions. Nevertheless, given the robust and ongoing participation of social actors striving to address the deficiencies of government interventions, it is highly improbable that niche innovation will be unsuccessful. This chapter diverges from prior research on agri-food system transitions by specifically examining incremental changes in practices and policy rather than major socio-technical shifts. These changes are rooted in traditional traditions and specific to local contexts. Furthermore, it emphasizes the significance of urban home gardening in the Global South, a topic that has been greatly neglected in the research on UA and sustainability transitions.

The rising need for food due to expanding urban populations and the inherent uncertainties of depending on the global food chain have motivated municipal governments to adopt diverse approaches to incorporate UA at various levels. Singapore is enhancing its national food security because it is a high-income city-state that relies on imports for 90% of its food supply. The Singapore Green Plan 2030 aims to achieve a 30% increase in domestic food production by implementing the "30 by 30" Grow Local initiative. However, can food self-sufficiency be achieved realistically if food production transitions from traditional farming methods to advanced technologies such as rooftop or vertical farms? **Chapter 6** examines UA via the political ecology framework to gain a deeper understanding of food security in Singapore from the viewpoint of conventional commercial farmers. The semi-structured qualitative interviews encompassed four critical areas of inquiry: general farm practices, social networks, advantages and impediments, and livelihood alternatives. The findings were examined within the framework of an emerging group of technologically advanced farmers who are confronting difficulties in meeting customer demand due to the limited availability of land and manpower. This comprehensive case study illustrates that technology, alongside others, influences the sustainability of the local food system in the global city-state of Singapore.

The extensive worries regarding chemical-intensive farming, recent disruptions in agricultural supply chains caused by the pandemic, and compromised public food distribution systems in urban and peri-urban regions of developing nations have created favorable conditions for expanding UA in cities. City residents are acquiring the skills to adjust, innovate, and acquire knowledge to grow their vegetables and fruits for personal use. For instance, in Indian urban areas, the pandemic and subsequent lockdown have led to the emergence of several ideas like rooftop gardening, vegetable kitchen gardens, playhouse cultivation, and vertical farming. Nevertheless, implementing UA in cities has yielded uncertain ecological benefits and has not notably enhanced urban food security, particularly within socioeconomically disadvantaged communities. Observing the swift expansion of urban farmscapes, it is natural to question if this is a long-term plan for sustainable intensification or only a transient trend. UA's origin, sustenance, and sustainability are believed to be influenced by various underlying environmental, economic, social, cultural, and political variables. **Chapter 7** utilizes a mixed-method approach, using primary data from urban farming households in Kerala and Arunachal Pradesh in India. Additionally, secondary data from the national statistical organization and literature reviews augment the analysis. The objective is to define the methods and criteria of UA that promote sustainable intensification.

Like other sub-Saharan African countries, Tanzania is undergoing urbanization, which brings up various food, money, and employment difficulties. Urban and peri-urban agriculture (UPA) is a method urban residents employ to tackle problems related to income, food, and employment. **Chapter 8** analyzes the UPA's social, economic, and policy effects using farmers' experiences in Dar es Salaam City and Morogoro Municipality. The analysis is based on 415 surveys conducted among farmer households, 20 in-depth case studies of individual farmers, and field observations. The key advantages of UPA, as reported by farmers in both regions, are increased income and improved food production. Additional benefits encompass dietary variety, physical activity, personal interests, and environmental considerations. The UPA additionally fostered social connections among farmers and encouraged women. Policies acknowledge the concept of UPA but impose limitations on specific actions, indicating that policies are not effectively incorporated and tailored to the specific needs of individual municipalities. The implementation of UPA is challenging due to variations in policies. Consequently, farmers encounter obstacles such as restricted availability of extension services, uncertain land ownership, exorbitant input costs, and theft. These impediments restrict UPA's capacity to maintain revenue, nourishment, and the urban ecosystem. The chapter finishes by outlining strategies for supporting, promoting, and sustaining Tanzania's UPA.

The Middle East and North Africa (MENA) region encounters many climate challenges, such as rising temperatures and frequent droughts, which harm the already-scarce water resources and threaten agricultural and food security. UPA is a rapidly growing industry recently gaining significant attention in the region. UPA's utilization can potentially address food security and sustainable development concerns in the MENA region. It facilitates the augmentation of food production inside urban areas, enhances the availability of nourishing food for poor and marginalized populations, and advocates for sustainable development. Nevertheless, the UPA encounters numerous obstacles in the MENA region, including insufficient infrastructure and resources to sustain it, such as limited access to water, proper land, and equipment. UPA is less common and sustainable in many regional cities due to the absence of government incentives and challenges associated with land tenure and high land prices. **Chapter 9** examines the status of UPA in the MENA region and the primary obstacles that hinder its development. The chapter is structured into two sections: the first piece addresses the primary concerns of agriculture in the MENA region regarding climate change and food insecurity. In contrast, the second section examines the key elements that either promote or impede the growth of UPA in the region. The chapter integrates bibliographical and topical evaluations of both scholarly and grey literature. It also incorporates case examples from different nations in the MENA region and policy recommendations to enhance UPA. The study determines that the region needs enhanced policy assistance and enabling legislation to stimulate UPA. Consequently, it is imperative to establish zoning restrictions to promote urban food production and provide financial and technical assistance to urban food producers while enhancing marketing strategies.

Chapter 10 explores the incorporation of UA into the curriculum of Qatar University's Architectural and Urban Design program, a subject that gained significance following the blockade of Qatar in 2017. The author highlights the importance of food systems and UA in governmental policies and research objectives and their efforts in establishing a network of stakeholders, partners, and specialists to tackle these matters. To develop and implement projects, events, and publications connected to Food Urbanism and the Food Water Energy nexus, they utilized research funding, financial and in-kind assistance from business companies, and collaborations with government agencies and the Swiss Embassy. The project incorporated experiential learning, design-build elements, and partnerships with Permaculture experts, international landscape architects, green roof specialists, and environmental science center scientists. This allowed students to acquire tools and knowledge in science-based approaches, gain practical experience, and learn design methods. In the Gulf region, there is a lack of Landscape Architecture schools. However, prominent educational institutions are actively involved in developing Landscape Urbanism. This growing field explores productive landscapes, UA, systems design, and the interconnectedness of food, water, and energy. The primary motivations for undertaking this research and deploying a prototype at Qatar University were the scarcity of arable urban land, public consciousness regarding food production, poverty, and unregulated urban expansion. Universities have a vital role in disseminating information, cultivating highly skilled personnel to address recognized economic demands, and incorporating food and water security into several academic disciplines.

The growing metropolitan areas quickly expand, resulting in various socio-economic issues and opportunities. UA has emerged as a possible solution within this complex network of issues. The current body of research on UA primarily highlights its benefits, difficulties, and results. However, only a few studies go beyond providing descriptive evaluations of these outcomes. **Chapter 11**'s investigation commences with the recognition that socio-economic factors have a crucial influence on developing UA practices. The current study aims to understand the impact of UA practices in the Indian cities of Bengaluru and Pune by examining the role of socio-economic factors such as space, time, gardening resources, economic resources, knowledge, skills, and individual choices. To analyze this, we employed a mixed-method strategy involving a thorough online survey conducted in 29 Indian cities between 2021 and 2022 and online in-depth interviews with 50 key informants in Bengaluru and Pune. The online survey's quantitative data was visually shown using frequency tables, graphs, and crosstabs. We employed a thematic analysis methodology to methodically encode the qualitative data. Our investigation revealed that several factors influenced UA practices, which may be categorized into three main groups: demographic factors (such as age and gender), contextual factors (including space, resources, income, and other economic resources), and intrinsic factors (such as attitudes and beliefs toward UA, as well as knowledge about UA. Demographic factors, like age, have facilitated technological integration into UA, while gender has significantly influenced the selection of crops and allocation of work. The scale of UA is influenced by the availability of space and resources, which influences the selection between tiny and large plants and trees. Training, education, peer learning, and social networks act as factors that disseminate knowledge about "optimal user acquisition practices" and promote environmental awareness and sustainability. Essential factors offer a deeper and more individualized understanding of the many UA procedures. Attitudes influence objectives and, as a result, individuals' views of the advantages of UA. Different factors such as demographics, context, and inherent characteristics lead to different outcomes for UA practitioners, often resulting in unequal adaptation and unfair scaling.

The exponential increase in the urban population in emerging countries is exerting significant pressure on the infrastructure of cities. The limited land in these densely populated urban regions is facing significant strain to accommodate the necessary infrastructure for meeting the fundamental demands of the population. Frequently, these advancements undermine the urban infrastructure's ecological, sociological, and environmental factors. Urban farming is an emerging area that tries to reduce the negative impact of urban-built environments on the environment. It offers opportunities to enhance cities' physical, social, and environmental elements, promoting urban sustainability. **Chapter 12** provides a comprehensive analysis of urban farming, explicitly examining the utilization of urban land and the incorporation of building facades for this purpose. It explores the effects of urban farming on the built environment, considering various factors such as thermal dynamics, structural integrity, aesthetic considerations, maintenance requirements, greenhouse gas emissions, energy expenses, and water quality. The analysis is based on existing literature on urban farming.

Modern agricultural practices face significant difficulty meeting the rising need for food production while addressing environmental issues. Developed nations have achieved the optimal equilibrium regarding resource utilization (water and soil) and preserving ecological sustainability and biodiversity. The dwindling expanse of cultivable agricultural land and the surging need for food have intensified the focus on UA. In urban settings, agriculture cultivates crops in small-scale gardens for personal consumption or nearby markets. At the same time, more intense farming practices are focused on the outskirts of cities. **Chapter 13** emphasizes the safety of the food generated inside UA systems. This study will provide an overview of the current state of UA in the Republic of Serbia, focusing on chosen case studies. It aims to enhance our understanding of the significance of proper plant protection and the reduction of synthetic pesticides in urban settings, particularly concerning food safety. Moreover, it is widely recognized that locally sourced food from small-scale producers enjoys a superior reputation among consumers while often having less stringent regulations than those from more extensive retail operations. Nevertheless, food generated through UA must adhere to the same high food safety standards as other sources. Pesticides, heavy metals, persistent organic pollutants, nitrate and nitrite, and mycotoxins in agricultural products, soil, and water primarily stem from agricultural activities, specifically plant protection measures. Urban horticulture can contaminate agricultural goods through polluted soil, water, and air. Hence, it is imperative to impose stringent regulations and minimize the usage of artificial pesticides and other agrochemicals to manage a diverse array of pests in UA. Application of biological measures, biopesticides, and ecologically sound synthetic pesticides is necessary.

Chapter 14 examines the increasing demand for vertical farming as a response to food insecurity, exacerbated by climate change, growing urbanization, limited availability of fertile land, and excessive carbon emissions. Several urban planning and agricultural specialists have suggested that cities should produce food to meet the increasing demands of their populations and mitigate the negative consequences of long-distance food transportation, such as traffic congestion, air pollution, and elevated food prices. The vertical farm is optimal for densely populated urban regions with limited space and water resources and high demand for fresh produce. The chapter also covers the main vertical farming techniques, such as aquaponics, aeroponics, and hydroponics, as well as the latest developments in greenhouse technologies, such as sensors and automation, machine learning and data analytics, artificial intelligence, IoT, autonomous robotics, and agritech analytics. These innovative solutions embody a novel agriculture and food production approach that can optimize yield. The chapter presents several project prototypes that demonstrate the possibilities of vertical farming. Nevertheless, it also underscores the potential drawbacks of adopting the vertical farm, including substantial upfront and ongoing expenses, restricted crop diversity, emissions of greenhouse gases, use of energy, production of waste, and adverse effects on rural communities reliant on agriculture.

Embarking on the journey of sustainable UA, we find ourselves standing at the crossroads of innovation and tradition. The unfolding narrative of this intricate tapestry beckons us to peer into the horizon of possibilities that the future promises. In this ever-evolving landscape, technology emerges as a driving force poised to usher in a new era of agricultural practices within urban spaces.

Smart farming techniques, fueled by data analytics, promise to revolutionize the efficiency and precision of cultivation. The integration of precision agriculture and optimizing resource utilization through cutting-edge technologies stands as a beacon of transformative change. Simultaneously, the pulse of urban life beats in tandem with shifting lifestyle patterns. A heightened emphasis on sustainability, an embrace of health-conscious choices, and a burgeoning desire for locally sourced produce reshape not only our dietary preferences but also redefine the very essence of how our cities cultivate food. This envisioned direction casts a visionary light on the trajectory ahead, where the intersection of technology and lifestyle converges to influence the fabric of UA profoundly. It foretells a future where cities embrace innovative approaches, cultivating a sustainable and resilient ecosystem that harmonizes with the evolving needs and aspirations of urban dwellers.

REFERENCES

Allen, A., & You, N. (2002). *Sustainable Urbanisation: Bridging the Green and Brown Agendas*. UN-HABITAT.

Anguelovski, I., Connolly, J. J., Garcia-Lamarca, M., Cole, H., & Pearsall, H. (2019). New scholarly pathways on green gentrification: What does the urban 'green turn' mean and where is it going? *Progress in Human Geography*, *43*(6), 1064–1086. https://doi.org/10.1177/0309132518803799

Burch, S., Hughes, S., Romero-Lankao, P., & Schroeder, H. (2018). Governing Urban sustainability transformations: the new politics of collaboration and contestation. In T. Elmqvist, X. Bai, N. Frantzeskaki, et al. (Eds.), *Urban Planet Knowledge towards Sustainable Cities* (pp. 303–326). Cambridge University Press. https://doi.org/10.1017/9781316647554.017

Castán Broto, V., Trencher, G., Iwaszuk, E., & Westman, L. (2019). Transformative capacity and local action for urban sustainability. *Ambio*, *48*(5), 449–462. https://doi.org/10.1007/s13280-018-1086-z

Cook, J., Oviatt, K., Main, D. S., Kaur, H., & Brett, J. (2015). Re-conceptualizing urban agriculture: an exploration of farming along the banks of the Yamuna River in Delhi, India. *Agriculture and Human Values*, *32*(2), 265–279.

Gould, K., & Lewis, T. (2016). *Green Gentrification: Urban Sustainability and the Struggle for Environmental Justice*. Routledge. https://www.taylorfrancis.com/books/mono/10.4324/9781315687322/green-gentrification-kenneth-gould-tammy-lewis

Hölscher, K., & Frantzeskaki, N. (2021). Perspectives on urban transformation research: transformations in, of, and by cities. *Urban Transformations*, *3*(1), 1–14.

Korth, M., Stewart, R., Langer, L., Madinga, N., Da Silva, N. R., Zaranyika, H., van Rooyen, C., & de Wet, T. (2014). What are the impacts of urban agriculture programs on food security in low and middle-income countries: a systematic review. *Environmental Evidence*, *3*(1), 1–10.

Madaleno, I. M. (2001). Cities of the future: Urban agriculture in the third millennium. *Food Nutrition and Agriculture*, *29*, 14–21.

McGranahan, G., & Satterthwaite, D. (2000). Environmental health or ecological sustainability? Reconciling the brown and green agendas in urban development. In R. Zetter & R. White (Eds.), *Sustainable Cities in Developing Countries* (pp. 73–90). Earthscan Publications Ltd. https://scholar.google.com/scholar_lookup?title=Environmental+health+or+ecological+sustainability?+Reconciling+the+Brown+and+Green+agendas+in+urban+development&author=McGranahan,+G.&author=Satterwaithe,+D.&publication_year=2000&pages=73%E2%80%9390

Mougeot, L. J. (2000). Urban agriculture: Definition, presence, potentials and risks, and policy challenges. *Cities Feeding People Series*; *Rept. 31*.

Nugent, R. A. (1999). Measuring the sustainability of urban agriculture. In *For Hunger-Proof cities. Sustainable Urban Food Systems* (pp. 95–99).

Pinheiro, A. (2022). *Technology and Policy Landscpe for Urban Agriculture in Kerala: Exploring the Sustainability Implications*. Jawaharlal Nehru University.

Pinheiro, A., & Govind, M. (2020). Emerging global trends in urban agriculture research: a scientometric analysis of peer-reviewed journals. *Journal of Scientometric Research*, *9*(2), 163–173. https://doi.org/10.5530/jscires.9.2.20

Rao, N., Patil, S., Singh, C., Roy, P., Pryor, C., Poonacha, P., & Genes, M. (2022). Cultivating sustainable and healthy cities: A systematic literature review of the outcomes of urban and peri-urban agriculture. *Sustainable Cities and Society*, *85*, 104063.

Shrestha, P. (2019). Mainstreaming the 'Brown' Agenda. *Sustainability*, *11*(23), Article 23. https://doi.org/10.3390/su11236660

Simon, D. (Ed.). (2016). *Rethinking Sustainable Cities: Accessible, Green and Fair*. Policy Press.

Smit, J., Nasr, J., & Ratta, A. (2001). Urban agriculture yesterday and today. In J. Smit, J. Nasr & A. Ratta (Eds.), *Urban Agriculture: Food, Jobs and Sustainable Cities*, pp. 1–31. The Urban Agriculture Network, Inc.

UN Habitat. (2020). *World Cities Report 2020: The Value of Sustainable Urbanization* (p. 418). https://unhabitat.org/sites/default/files/2020/10/wcr_2020_report.pdf

UN Habitat, & UN ESCAP. (2015). *The State of Asian and Pacific Cities 2015 Urban transformations Shifting from quantity to quality* (p. 204). https://www.unescap.org/sites/default/files/The%20State%20of%20Asian%20and%20Pacific%20Cities%202015.pdf

UN Habitat. (2009). *Planning Sustainable Cities: Global Report on Human Settlements*. Earthscan. https://unhabitat.org/planning-sustainable-cities-global-report-on-human-settlements-2009

United Nations. (2019). *World Urbanization Prospects 2018: Highlights*. https://population.un.org/wup/Publications/Files/WUP2018-Highlights.pdf

Van Veenhuizen, R. (2006). Cities farming for the future. *Cities Farming for Future, Urban Agriculture for Green and Productive Cities* (pp. 2–17). RUAF Foundation, IDRC and IIRP, ETC-Urban Agriculture, Leusden.

Van Veenhuizen, R., & Danso, G. (2007). *Profitability and Sustainability of Urban and Periurban Agriculture* (Vol. 19). Food & Agriculture Org.

2 Urban Agriculture
A Historical Perspective

Sharmin Khan and Anwar Hussain

INTRODUCTION

The history of architecture is complete only with a discussion of gardens and agricultural practices. Human beings initiated the development of farming practices to serve the settlements. According to some scholars, the abrupt postglacial environmental changes (after the Pleistocene) drove certain large animals out of their natural habitats, severely disrupting the existing food chain and leading to a shortage of food for hunter-gatherers (Cohen, 1977), which in turn forced them to choose for agricultural practices (Boserup, 1965). However, there is another school of thought that agriculture did not develop as a response to the food shortage caused by an unstable wildlife source. The early people started domesticating plants and animals to increase the variety in diet (Hayden, 2003), even though they had plentiful food supplies, and gradually established communities after settling down in areas conducive to cultivation. Therefore, it is difficult to refer to a single factor that inspired people to start farming in different locations worldwide. For instance, it's believed that climatic shifts during the last phase of the ice age introduced seasonal circumstances that benefited annual plants such as wild grains Near East. The pressure on natural food resources was continuously increasing elsewhere, such as in East Asia, which might have compelled individuals to develop their remedies. The cities and civilizations were established as an outcome of agriculture because human beings learned the domestication of plants and animals. As a result, the global population multiplied from some 5 million people almost 10,000 years ago to approximately 7 billion today (The Development of Agriculture, 2022).

Ancient Indian history also reflects an overview of artificial and natural green space evident from various literary sources. Dasgupta (2016) emphasizes that the ideas and procedures around using natural resources evolved and shifted political and social paradigms. The desire to create shade and shelter from climatic conditions introduced gardens and trees in these gardens, especially in the courtyards of the palaces of Mesopotamia. Some historians believe that fishponds and orchards co-existed, implying that such gardens' scale was large (Dalley, 1993). The gardens and water bodies eventually became an integral part of urban design. Raising animals and growing plants in and near cities can be summed up as urban agriculture (Awasthi, 2013). The development of these green spaces is credited to the need for aesthetic expression reflecting art and architecture and a symbol of civilization or philosophy of culture. Still, they are all ultimately contributing to the urban agriculture of their times.

A deep insight into this urban agriculture development can help understand their environment and socio-economic contribution. It is essential to understand the historic urban agriculture practices concerning their contribution to architecture and examine their adaptive possibilities in current times, in context with the changing scenario and global warming issues. An attempt has been made to present the historical background of different regions and periods across the globe in the following sections. A thorough investigation of historical facts related to early settlements with reference to the selection of a site shall help identify the need for urban agriculture. Further, the beliefs and practices of different civilizations led to the creation of gardens and adorable landscapes. A wide variety of vegetation can be witnessed in history, together with the fact that some of them were imported from distant places for the purpose of fulfilling their needs. The magnificent water

DOI: 10.1201/9781003359425-2

management systems have been employed by people of different periods; those shall help in appreciating the technological advancements and may open new insights into readopting the systems in achieving sustainable development.

THE EARLY SETTLEMENTS

According to the archaeological record, the Pleistocene period from 20,000–16,000 B.C. was the time of hunter-gatherers (Fletcher, 1999; Zeder, 2011; Zohary et al., 2012) before a profound change occurred across Eurasia at the beginning of the Holocene, 11,700/500 Cal B (Zeder, 2011; Zohary et al., 2012). The first agricultural system eventually resulted in the "Domestication Syndrome," which includes both – plants and animals. A population boom created by forming long-lasting, agriculturally oriented permanent communities later spread westward and eastward (Bouquet-Appel, 2011). Humans hunted and caught their favorite mammal, bird, reptile, and fish species, felled trees for construction materials and the production of countless items, and dumped waste materials near their sites. Hunter-gatherers later cleared grounds near their homes for planting as successful, purposeful farming began (Alpert et al., 2000). Foraging societies spread out across the Fertile Crescent during the following several millennia started growing, tending to their goats, sheep, pigs, and cattle in addition to herding them. The first evidence of agriculture dates to 9000 B.C., when settlements of communities of Natufian culture started developing (Fletcher, 1999). The process is termed the "Neolithic Revolution," i.e., when people began employing stone tools for agricultural practices (Childe, 1942).

Between 7500 and 6000 B.C., permanent agricultural villages were established with buildings constructed in mud bricks from Southern Turkey to the Nile Delta and Southwest Asia with time. However, a prototype agricultural economy was developed around 12,000 B.C. in the lower Egypt region. There were three broad zones in Egypt and the northeast region: the Arabian Peninsula in the south, the Mediterranean coastal plain, and Palestine in an arc including the Fertile Crescent and the mountains and plateaus from west to east. The early settlements took place in Egypt owing to the valley of the Nile and rich alluvial soil. In the middle of the 4th millennium B.C., optimum climatic conditions prevailed in western Iran and Mesopotamia. The climate was warmer and more humid, which resulted in settlements. The first complex societies of Southwest Asia evolved in the southern Mesopotamian region. However, the environment was arid and received insufficient rainfall for crop growing without the maintenance of irrigation facilities. However, before the dynasties, Egypt had a stable climate and the advantage of the River Nile. They had a narrow strip of the alluvial plain bordered by desert. The presence of natural resources helped the people of the Fertile Crescent develop the early agricultural economies (Fletcher, 1999). In the Helmand Valley in southwest Afghanistan, west of the Indus, remains of the Helmand civilization from the 3rd millennium B.C. have been discovered (Cortesi et al., 2008). It is assumed that the city civilization was centered on agriculture.

The Indus Valley civilization, before 2600 B.C., was also centered on agriculture (Wright et al., 2008). India may have been the first civilization to practice primitive agriculture (Sharma et al., 2004) (Cahil, 2012). The Rigveda divides Indian peninsula areas into four categories based on their level of fertility: anurvar means infertile, artana means less fertile, apnasvati means fertile, and urvar means the most fertile (Yadav et al., 2008). The Vedic civilization was primarily agricultural, and certain cities were called Janapadas. The agricultural boom led to steady social, economic, political, and technological advancement. Mahajanpadas rose to prominence starting in the early 6th century B.C. due to this multidimensional expansion (Tripathi et al., 2008). However, according to the sacred Indian texts of Atharvaveda, Viṣṇu Purāṇa, and Śrimad Bhāgvad Mahāpurāṇa, a king named Pṛthu was regarded as the inventor of agriculture, who effectively brought the agriculture into practice. Mahajanpadas realized the importance of agriculture as the only means of food security for human beings. It is also emphasized that, regardless of whether they acquire farmland from a king or buy one for themselves, men should work hard and dedicate themselves to farming (Dwivedi, 2017).

Scholars refer to ancient Japanese civilization as "Jomon hunter-gatherers." The period of Jomon is frequently referred to as the "Jomon culture" (Jomon bunka) or sometimes as the "Jomon period" in Japanese-language literature (Jomon jidai). In the 1980s, the focus on "Jomon hunter-gatherers" increased. The presence of lacquer trees (*Toxicodendron verniciflua*) raised concern because they have been regarded as a Chinese import. The historians suggest that the lacquer trees were frequently connected to Jomon graves as they grow in what is known as early successional or secondary forests, which are not regarded to be "natural" woods (Yokoyama, 1984).

Scholars believe that Japanese islands witnessed some cultural changes around 3000 B.C. when immigrants arrived from the Asian continent, identified as the Yayoi people. They brought new technology, the skills of working on iron and bronze, and also the agricultural skills of growing rice. These people were identified later as "the people of Wa" by the Han dynasties, and it is believed that the Iron Age arrived in Japan. They established themselves first in the south, i.e., in northern Kyushu, and spread quickly north-eastward along the Sannin coast and to the Kanto plain. The new tools and technology contributed to more food production than the previous Jomon culture, which was largely dependent on stone tools and hunting and gathering. The settlements began to be permanent, as the food production (rice) was now more stabilized and served a larger population. The extra food supplies also led to the economic growth of the society and were managed by an elite group of people. The Yayoi period lasted about seven centuries, i.e., until the middle of the 3rd century A.D. (Heritage of Japan: The Yayoi years, n.d.). The botanists deduct that the lacquer tree was brought from China to Japan. Southwest Japan had considerable socio-economic changes beginning in the first millennium B.C. The Yayoi era is marked by the growth of a ranked society efficient in metallurgy and substantial engagement with regions outside Japan. Japan established wet rice production, a form of intensive agriculture, but that was not the only component of the agricultural system during the Yayoi era. The Japanese Satsumon culture that followed marked a substantial shift in technology, way of life, and settlement patterns (Yokoyama, 1984). According to Crawford (Crawford, 2011), the Tohoku Yayoi culture, which communicated with civilizations from southwestern Japan, gave rise to the Satsumon culture. The first Japanese state was created as a result of significant socio-political changes in southwestern Japan (Crawford, 2011). Typically, the Okhotsk civilization is described as a marine culture that traveled from Sakhalin to the Hokkaido north shore region (Yamada & Tsubakisaka, The spread of cultivated crops from the mainland. In Final Report on Research Project of the Historical and Cultural Exchange of the North, 1995). The center Ryukyu Islands, located between Kyushu and Taiwan, had a comparatively late and rapid development of agriculture (Takamiya, The transition from foragers to farmers on the island of Okinawa, 2001).

The domestication of the crop was a long process in Africa. It took place outside the Niger River Basin, unlike Vigouroux, which emphasizes climate as the reason agriculture developed in the region mentioned above. Sylvain Ozainne, an archaeologist at the University of Geneva, Switzerland, contends that Saharan pastoralist movements assisted in establishing and spreading a culture of growing crops like pearl millet. "The spread of African agriculture may be best explained over a more complex process, which incorporated socio-economic adjustments, as compared to direct response to abrupt climatic change," he claims (Pennisi, 1).

It is thought that, around 8000 B.C., rice was first grown in terraces along the Yangtze River in central China (Normile, 2003). From 8000 B.C. until 1911 A.D., the agricultural technology system of ancient Chinese evolved on the mobilization of resources such as water and land. In China, agricultural technology development was prolonged and followed an S-shaped trajectory, only tripling in size over 8,000 years. The Neolithic Period saw the predevelopment of primitive millets from the "agricultural crop," the slash-and-burn farming techniques from the "agricultural practices" subsystem, and the method of stone, bone, and wood implementation from the "agricultural engineering" subsystem. These innovations in the Yellow River valley were the foundation for the early agricultural technology system. The Yangtze River region was another location for the "agricultural engineering," "agricultural practices," and "agricultural crop" subsystems during the Neolithic Era. The primary materials used to make farming implements were shell and bone, and similar slash-and-burn techniques were used for growing rice in paddy fields. The adoption of furrowing techniques in the northeastern, Yellow River, and northwestern regions served as an indicator of the acceleration stage. Technological growth was first determined by physical instruments (40% of development) and then by technical theories and practices (more than 50% of growth). Almost 45% of its technologies were developed in the Yangtze River region, which is physically oriented toward the Yellow River (Wu et al., 2019).

BELIEVES AND PRACTICES RELATED TO URBAN AGRICULTURE

Scholars have discussed various beliefs that can be associated with the development of gardens during ancient times. The importance of trees in early Mesopotamia can be justified by their personifying them as "Nin-Gishzida," a God, meaning "trusty tree." It is also believed and evident from some ancient texts that the Assyrians had their tombs in the gardens, especially in the courtyards of their palaces. This implies that they created gardens in memory of the dead. The Babylonians appreciated the natural beauty and used gardens for daily activities like picnics, official meetings, and gatherings to display their assets, like military trophies. Another King Sargon II (722–704 B.C.) built parks and orchards for pleasure to spend time with his family and hunt lions and falconry in the city Dur-Sharrukin, north-east of Nineveh. The royal gardens inside the cities witnessed trees, a hilly terrain, and flowing water, and they became the architectural characteristics of the cities. Temple gardens inside a central courtyard of a rectangular building, having regular rows of trees or shrubs, were also constructed by Sennacherib. Paradeisos, a Persian word, was used by the Greeks to describe the Garden of Eden, paradise, and other enclosed spaces. Famous literature like the Gilgamesh Epic and the stories from the Arabian Nights also describe gardens such as Lebanon's grove of pines or cedars, a holy site guarded by the enormous Humbaba, which the greatest gods had chosen. The giant watches over the Garden of Eden's trees, and the garden is made up of trees, demonstrating the relationship between trees and the Garden of Eden, with "paradise" as the final resting place of great sages (Dalley, 1993).

According to the designs discovered and preserved, the gardens in Egypt, the oldest artificial gardens, were built using a layout similar to Islamic gardens. Egyptians created gardens to provide shade and bring fresh air to their dry land that lacked a wide range of trees and flowers. The manuscripts reveal that the kings and queens imported species of plants from distant places and ordered the making of their pictures on the temple walls. These were used as decorative elements in their private palaces and temples. Egyptian gardens are known for their functional and purposeful design, having rectilinear and geometric gardens enclosed by high walls (The uniqueness of gardens in ancient Egypt, 2020).

As mentioned in the second mandate of Indian ruler Ashoka, a widespread initiative was taken to plant medicinal fruit-bearing trees, roots, and herbs for the advantage of humans and animals, as well as to dig wells and establish common shady trees for long-distance visitors (Hultzsch, 1925; Barua, 1943/1990 reprint). The ruler was concerned about morality or Dhamma, i.e., the "civil religion," as indicated by the repeated use of terms like "harapitani/halapita," especially concerning the medicinal plants that were probably not growing generally in the area concerned (Olivelle et al., 2012). There is an indication of the initiative the ruler took for growing such plants in edicts. The inscription carved in caves of the Kalinga king Khāravela period also refers to the landscaping and gardening adoption for the embellishment of settlements, especially the urban spaces (Neelis, 2011). The inscriptions mention the repair works of all the gardens in the royal city of Kaliṅga (including the walls, gates, houses, and tanks) undertaken and prioritized by the ruler, thus emphasizing that the architectural layouts of royal spaces were inclusive of gardens. Kauṭilya's Arthaśāstra of the Ashokan period discusses the settlements and agricultural land and mentions that the floral areas termed vana, araṇya, and ārāma were also brought under administrative regulation and management (Arthaśāstra 2.1.33). The Arthaśāstra classifies six different types of vana, or middling forests, besides ārāma, i.e., gardens or parks for promenades. It is also mentioned that puspaphalavātān (Arthaśāstra 2.4.25) were the gardens of flowers and orchards having fruits, existed adjacent to ordinary urban quarters (Kangle, 1960–65, Reprint 2010). The administrative texts reflect the fact that there are places for hermitages, areas for wildlife hunting, and forests for goods like timber and elephants. It was also professed that all of these things were kept according to strict guidelines. Interestingly, the Arthaśāstra advised establishing facilities like lower gardens and orchards in the ordinary householders' building places within the capital city and the royal fort in the 2nd century C.E. (Kauṭilya Arthaśāstra: 2.4.25). The Gupta ruler Viśvavarman planned an urban city on the banks of the river Gargarā overstated with exciting features, including leisure ways, wells, tanks, temples, halls of gods, drinking wells, large water bodies, and leisure gardens (Fleet, 1898–99). Evidence from the text of Amarakoṣa belonging to the 5th century C.E. discusses various types of green spaces: the gardens attached to houses for the general well-to-do householder are referred to as Gṛhārāma, the garden house attached to the premises of the counselors, aristocrats, rich ministers, and courtesans were the Vrikṣavaṭikāor, and the recreational parks for the royal members were termed ākriḍa or udyāna. The gardens attached to the Royal Harem were known as pramodavana (Amarakoṣa: Ch. II. Vanauṣadhivarga, 2–7). The Mahābhārata belonging to the post-Maurya Indian period mentions artificial spaces with flowers or gardens of different kinds (Mahābhārata: I. 70. 30). The tapovana is represented as a beautiful and peaceful space with flowers that are carved out from the natural forest, before the urbanization. Thus, it can be concluded that tapovana was not the forest itself; instead, it may be defined as a replacement of the forest that had been altered to meet the ascetic community's basic minimum needs while still retaining a sufficient amount of the natural landscape (Dasgupta, 2016).

The concept of the Garden of Eden is also associated with the Muslim paradise. It is believed that a river flows from Eden and is used for watering the garden, which is further divided into four headwaters. Eden is a garden and a location that God saves for His chosen people and the enjoyment of humanity forever. The aesthetic value of Islamic gardens was great, as it was related to the Garden of Eden. There was an agricultural drive, in addition to the joy of vision and fragrance, that resulted in the intelligent usage of water and the creation of a science of acclimation of novel botanic species (The uniqueness of gardens in ancient Egypt, 2020). Mughal gardens are a combination of landscape beauty and practical requirements. They provided a setting for spatial protocols, social functions, and social communication standards while simultaneously creating poetic, artistic, and individual beauty. A garden was already running alongside the wall on the other three of the city's four sides when the Mughal Emperor Akbar instructed the fortification of the existing settlement in Lahore. Small gardens might be found inside and outside the fort boundaries (Mubin et al., 2013).

TYPE OF VEGETATION

Ancient Mesopotamian gardens' most common evidence is the palm tree and tamarisk. The advantages of these two trees are manifold. The dates from palm trees and strong tamarisk were valuable, besides providing shaded areas for kings and queens to enjoy meals. The dates were surplus in quantity after serving the royal family. The tamarisk was used for furniture, utensils, tools, etc. The cities of Egypt were settled on raised platforms or mounds with fortifications to protect them from floods due to their proximity to the Tigris and Euphrates. The concept of courtyard gardens must have evolved to create pleasant shaded areas and, at the same time, protect them from floods and the reach of wild animals. The Assyrian king Assurnasirpal II planted pines, cypress, almonds, dates, ebony, rosewood, walnut, fir, pomegranate, fig, olive, tamarisk, olive, etc., in the orchard having streams of water (Dalley, 1993).

In the Egyptian gardens, edible plants like figs, dates, and pomegranates were grown alongside herbs, medicinal plants, and fragrant, medicinal, and edible plants. Botanical species aided a supply of raw materials for creating the everyday tools, clothing, and other goods that were a basic necessity. The commonly found trees include date palm (*Phoenix dactilyfera*), sycamore fig (*Ficus sycomorus*), papyrus (*Cyper papyrus*), Egyptian lotus or the blue lotus (*Nymphaea caerulea*), pomegranate (*Punica granatum*), mandrake (*Mandragora officinarum*), dum palm (*Hyphaene thebaica*),

Tamarisk fruits (*Tamarix nilotica-articulata*), Apple trees (*Malus sylvestris*), Egyptian plum (*Cordia myxa*), grapevine (*Vitis vinifera*), etc. Trees like the olive tree (*Olea europea*) were imported from Minor Asia, Syria, and Greece, and almond trees (*Prunus dulcis*) were imported from Palestine (The uniqueness of gardens in ancient Egypt, 2020). Before 2600 B.C., there was evidence of planted wheat, barley, and linseed, according to excavations near cities like Shahr-e-Soktha. Melons and grapes were also grown, requiring careful cultivation and irrigation systems (Habib, 2002).

The Indus people were adept at growing multiple crops, and they raised wheat, barley, numerous legumes, and possibly grapes, jujube, dates, and cotton (Wright et al., 2008). They were aware of irrigation techniques used in science.

Chestnut, persimmon (*Diospyros kaki*), apricot (*Prunus armeniaca*), fig (*Ficus*), prickly ash (*Zanthoxylum*), *Perilla*, ginger (*Zingiber*), and also a variety of other plants have been established in the areas between homes and gardens of Jomon period (Nishida, 1983). Some of these plants were probably purposefully planted, while others may not have been. Nishida observed that certain plants were carefully removed while others were kept. Additionally, he suggested that since chestnuts and walnuts thrive in the sun, they were likely placed in the area of groves during the Jomon period. Their usage followed an outline similar to what he had noticed in the Mukasa region. Remains of wood from the sites of Jomon show an evident fondness for chestnut as a building material (Noshiro & Sasaki, 2007). A substantial part of the new agricultural system in the Yayoji period included a wide variety of plants, such as wheat, millet, barley, and other dry crops during the Yayoi period. Barley, foxtail millet, broomcorn millet, wheat, perilla, soybeans, Japanese red beans, melon (*Cucumis melo*), flax (*Linum usitatissimum*), and some other crops were grown during the Japanese Satsumon culture. At eight Ainu sites, plant remnants have been collected by flotation. Most of the collections are rice, but it is generally believed that the rice was acquired through trading. There were also other crops planted, like barnyard millet. Foxtail, broomcorn millet, and barley are among the crops in the Okhotsk civilization. Chenopod, silvervine, grape, elderberry, and other weedy plants are also typically included in the plant collection, as are some nuts (walnut) (Yamada & Tsubakisaka, The spread of cultivated crops from the mainland. In Final Report on Research Project of the Historical and Cultural Exchange of the North, 1995). The earliest barley from the Okhotsk region is broader and thicker than the Satsumon variety (Crawford, 2011). Most of the plant's remnants are nuts and seeds from fruits like silvervine and grapes. Barley, rice, wheat, and foxtail millet were the main crops during the 8th and 10th centuries A.D.; weeds like sedges and knotweeds show that the crops were locally produced; possible fields have been identified at the Yayoi-Heian era Nazakibaru site (Takamiya, Island prehistory, 2005).

Ethiopia's crop complement includes all stages of plant development, i.e., increased use of wild plants to domestication. Several wild plants were used as food, especially during periods of scarcity of food, i.e., the time between the sowing of seeds and harvesting them. The bulk of these plants were consumed as green vegetables, followed by the plants having tubers, roots, and edible fruits. The grass, *Snowdenia polystachya* (Fresen), is an example of a plant that is harvested and utilized in a manner akin to that of teff. *Avena abyssinica* and *Coccinia abyssinica* are two examples of semi-domesticated plants (Edward, 1991).

The Fertile Crescent of the Middle East is possibly where wheat and other plants were first cultivated. The plant genome investigations reveal that a second early cradle of agriculture was located thousands of kilometers distant, near the Niger River Basin in West Africa. There, the traditional food crops of the continent were developed, including the rice alternative used in Africa and the pearl millet cereal. Yams have been added to the African crops in that same region, domesticated thousands of years ago, according to a paper published in *Science Advances*. According to Yves Vigouroux (a population geneticist at the French Research Institute for Development in Montpellier), the dry climatic conditions might have prompted the transition to farming. In addition, Scarcelli, Vigouroux, and their associates also determined that the contemporary domesticated variety of yam originated from the Niger River Basin, located between western Nigeria and eastern Ghana. Their investigation identified genes that changed along the route but could not determine the exact domestication time. A plant that thrives in direct sunlight likely evolved from a woodland dweller. Gene changes related to starch generation and root growth are probably responsible for tubers' regular shapes and higher starch content. The most significant cereal for dry regions of Asia and Africa with weak soil conditions, i.e., the pearl millet (*Cenchrus americanus*), was also found to have a West African origin. After sequencing and comparing the genomes of 221 domesticated and wild millet species, Vigouroux and his team concluded that all the varieties of pearl millet that were domesticated had the exact origin that grew north of the Niger River in the western Sahara Desert, which includes northern Mali and Mauritania in the present times. According to Fuller, the genetic research published in the journal *Nature Ecology & Evolution* fits well with the 2011 finding of 4,500-year-old pearl millet remnants at an archaeological site in southeast Mali (Pennisi, 1).

Wheat, peas, and barley are the crops whose wild families can be witnessed throughout the Near East. Cereals were first farmed in Syria some 9,000 years ago, and figs were planted earlier. The evidence from seedless fruits of ancient times found in the Valley of Jordan suggests that fig trees were first established 11,300 years ago. At the same time, corn (maize) was planted later, as they had to wait for spontaneous hereditary alterations. The first directly dated corn cob only appears to have existed around 5,500 years ago. However, maize-like plants descended from teosinte appear to have been farmed at least 9,000 years ago. Later, corn was transported to North America, where domesticated sunflowers had already begun to grow around 5,000 years earlier. Moreover, this was the beginning of potato cultivation in the Andes region of South America (The Development of Agriculture, 2022).

Most of the alfalfa, buckwheat, yam, and sesame grew in the Yellow River regions. Later, canola and cotton spread from the northeastern and northwestern regions to the area of Yellow River, where they developed into primary sources of textiles for clothing and oil for human consumption throughout all of China. Moreover, kapok and multi-seasonal rice varieties were accepted, and they were initially planted in the warmer regions of the southeastern before being moved to the area of the Yangtze River. In the southeastern part, foreign crops such as sweet potato, peanut, tobacco, potato, and corn were introduced before becoming widely used throughout China (Wu et al., 2019).

THE WATER MANAGEMENT SYSTEMS

The development of irrigation technologies was crucial to the culture and agricultural practices of the ancient civilizations of Egypt, the Indus Valley, Mesopotamia, and China. It is commonly referred to as "hydraulic civilizations" (Leach, 1959).

North of Cairo had fertile land, and the temperature rarely reached 38 degrees Celsius. However, rainfall was scarce, and agriculture could not occur without irrigation. The appropriate temperature and humidity were helpful for a wide variety of plants. The great cities enjoyed the Mediterranean climate. They had short winters and received adequate rainfall in autumn, winter, and spring. The summers were generally hot and dry. Rivers flowing from the Anatolian plateau provided them with rich alluvial valleys. They had the cultivable plain they used for livelihood, contributing to the size and importance of the communities that settled in these regions (Fletcher, 1999). The Assyrian king Assurnasirpal II built a canal to transport water from the foothills and mountain streams for irrigating the orchards in the city of Nimrud. Sargon's successor, Sennacherib (704–68 B.C.), constructed channels and aqueducts from the beautiful Bavarian region northeast of Nineveh to transport fresh water from a mountain river down to his gardens and parks at Nineveh. The inclusive urban design example is quoted from the Palace at Mari (1800 B.C.), popularly called the Court of Palms. The remains from the Palace at Ugarit (on the coast of Syria) show evidence of a stone basin for water in the courtyard. Some researchers believe that the Hanging Gardens of Babylon were constructed on artificial slopes raised on terraces built of stone, bitumen, and timber. These were then topped with soil, and trees were planted, similar to the construction technique of Greek theatre. It is also believed that they were built upon vaulted stone terraces and were meant to imitate mountainsides with trees.

The system of carrying water to the top of the garden was not visible (Dalley, 1993). The rectangular ponds received water from the Nile's water channels. They were typically surrounded by lotus flowers (*Nymphaea careluea*), as their floating leaves provided shade for the fish, Nile grass (*Cyperus papyrus*). They also had other aquatic plants, a common sight in the middle of gardens. Additionally, the routes from the gardens to the dwellings were marked by covered boats and short pergolas that offered shade. The flower beds and quadrants were created using a cruise's transversal and longitudinal axes. The Egyptians were renowned for being excellent farmers who marveled at making the most of their vast, fertile grounds, which were perpetually covered in mud because of the annual flooding of the river Nile. The people of this civilization respected nature and benefited from all it had to offer while living in harmony with it. Muslims learned the irrigation techniques and also the methods for the extraction and use of water, which were highly advanced in the region at the time, and employed them in the gardens of Al-Andalus in the 7th and 8th centuries after Islam spread across the Egyptian, Persian, and Syrian empires (The uniqueness of gardens in ancient Egypt, 2020).

In subhumid climates, the Indians learned the art and science of managing water. There are still remnants of the understanding of natural resource management throughout rural India. "The first farmers of India" were wheat and barley growers of Mehrgarh, where pieces of evidence of granaries and seeds of cereals belonging to the 7th millennium B.C. are found. The tank systems on the Karnataka plateau, the cascading waterfalls in Chhotanagpur, the community canals in Manbhum, and the groundwater structures in Rajputana all provide glimpses of this traditional wisdom (Ghosh & Sarkar, 2014).

The central theme of Mughal gardens was water, which can be witnessed in Persian, Central Asian, and Indian gardens (Fatma & Fatima, 2012). However, the square or rectangular ground, typically walled with intersecting channels carrying water and cross lined with walkways, tank(s), usually one in the center, and well(s) divided to create the same pattern on varying scales depending on the available area enclosed was the main characteristics features of Mughal gardens (Dickie, 1985). The four primary sources of water employed in the Mughal gardens are wells or step-wells, lakes or tanks, canals, and natural springs (Siddiqui, 1986).

According to previous studies, ancient China had highly sophisticated water supply systems, and all of China's affluent eras featured thorough water supply planning and rational water source selection. In the City of Yangcheng, the water delivery system included a pipeline system and cleaning procedures (of the Eastern Zhou Dynasty) (Du & Chen, 2007). When substantial infrastructure projects for water storage and diversion for paddy fields in the southeastern area and the Yangtze River region were created, the agricultural subsystem became more fragmented (Wu et al., 2019). China boasts "the richest variety, largest distribution, and highest irrigation effectiveness in heritage constructions," according to the evidence discovered at four sites: Xinghua Duotian Irrigation and Drainage System, the Tongjiyan Irrigation System, the Chongyi Shangbao Terraces, and the Songgu Irrigation Scheme. Prehistoric Chinese used the idea of "man as an inherent part of nature" to promote ecological and agricultural development in harmony. With the help of dikes, canals, and sluice gates, the Xinghua Duotian Irrigation and Drainage System distributes water to dry mountains in Xinghua, Jiangsu Province, East China. Throughout times, the residents of Xinghua constructed structures of wood on which they deposited little stacks of mud to create the duotian, converting untamed marshes into arable land and preventing flood damage to crops. The duotian is the ideal example demonstrating the technique of living in peace with nature while making the most of the available resources. The 3,400-hectare-plus Chongyi Shangbao Terraces, located in the mountainous region of Chongyi County, Jiangxi Province, East of China, was initially constructed on sloppy hillsides during the reign of Song Dynasty (960–1279) and developed under the Ming (1368–1644) and Qing (1644–1911) dynasties later. They used cutting-edge irrigation techniques and ecological systems to minimize droughts and flooding. Farmers developed these terrace fields by the natural terrain rather than attempting to level the steep area, using gravity for irrigation to generate fields that yield about 500 kg of rice per 600 m^2 (Reporters, 2022).

CONCLUSION

The issue of global warming is not new and was perhaps the first reason to look for agricultural practices. The early settlements provided human beings an opportunity to adopt agricultural practices, complemented by the invention of tools. This ultimately converted human society from nomadic life to civilized societies of the ancient world. Similar examples may be quoted across the globe, ranging from Mesopotamian to Indian civilizations. They were eventually converted to producers from food gatherers. These settlements evolved near existing water bodies, and hence the availability of water enhanced the agricultural practices and gardens in later times. The civilizations gradually converted into empires ruled by strong rulers who started the fortification of cities, thus strengthening the territories. The evolution of urban societies was inclusive of green spaces, including agricultural land and gardens with shady trees for leisure, hunting, tombs, etc. The climatic and socio-cultural conditions enforced the agricultural practices in urban areas. Irrigation practices were developed to bring water to the gardens and agricultural land, as it would not be possible to grow crops or maintain gardens without water. Religious beliefs also promoted urban green spaces, including agriculture. The demand for different types of vegetation ranging from wood to fruits and medicinal plants led to a variety of agricultural practices. Those were sometimes native species and also imported at times, depending on the need.

It is very clear from historical studies of different periods that agriculture has been an integral part of societies and moved with them to the places of human settlement. The communities were segregated with increasing modernization and turned into rural and urban areas. The distinction between urban and rural development brought disaster to the urban settlements regarding slum development, food security, etc. The urban areas were developed with all infrastructural facilities and technological advancements, whereas the rural areas were confined to agricultural practices only in most countries. The growing trend of urbanization further stressed the existing cities, and people moved to urban areas to attain modern means of life. The distance between the food producer and the consumer affects the ecological and environmental balance.

The building energy can amount to twice its operational energy because it includes transportation energy owing to the building location (Environmental Building News, 2007). There has been an urge for sustainable development over the past couple of years, thus leading to significant changes in building delivery systems. Scholars suggest that, to achieve sustainable land use, land development should be curtailed, especially the undeveloped, natural, or agricultural land (green fields), as it is a limited source (Kibert, 2013).

Urban agriculture's most notable characteristic that sets it apart from agriculture in rural areas is that it is incorporated into the urban system, both ecologically and economically. The practice of urban agriculture is intertwined with urban ecology. Urban agriculture has indirect connections with consumers of urban regions, impacts (positive and negative) directly on the urban ecology, enhances participation in the food system, and uses typical urban waste as resources (organic waste for making compost and wastewater for irrigation). Several elements also influence urban agriculture, and vice versa, namely, people, location, food products, product market, and technology (Awasthi, 2013). Historical evidence shows that the rulers patronized urban agricultural practices. Therefore, government policies regarding regulations, guidelines, rebates, and by-laws can help promote urban agricultural practices.

REFERENCES

Alpert, P., Bone, E., & Holzapfel, C. (2000). Invasiveness, invasibility and the role of environmental stress in the spread of non-native plants. *Perspectives in Plant Ecology, Evolution and Systematics*, *3*, 52–66.

Awasthi, P. (2013). Urban agriculture in India and its challenges. *International Journal of Environmental Science: Development and Monitoring (IJESDM)*, *4*(2), 48–51.

Barua, B.M. (1943/1990 reprint). Inscriptions of Asoka. In *Calcutta: Sanskrit College Research Series* (pp. 2–3.). Calcutta, India: New Age Publisher.

Boserup, E. (1965). *The Condition of Agricultural Growth: The Economics of Agrarian Change Under Population Pressure.* Chicago, IL: Aldine.

Bouquet-Appel, J. (2011). The agricultural demographic transition during and after the agricultural inventions. *Current Anthropology, 52,* S497–S449.

Cahil, M. A. (2012). *Paradise Rediscovered: The Roots of Civilisation.* Brisbane: Glass house books.

Childe, V. (1942). *What Happened in History, Contradictions Over Exact Age (of certain periods).* New York: Penguin Books.

Cohen, M. (1977). *The Food Crisis in Prehistory: Overpopulation and the Origins of Agriculture.* New Haven, CT, USA: Yale University Press.

Cortesi, E., Tosi, M., Lazzari, A., & Vidale, M. (2008). Cultural relationships beyond the Iranian Plateau: The Helmund Civilization, Baluchistan and the Indus Valley in the 3 millennium BCE. *Palorient, 34*(2), 5–35.

Crawford, G. W. (2011). Advances in Understanding Early Agriculture in Japan. *Current Anthropology, 52*(S4), 331–345.

Dalley, S. (1993). *Ancient Mesopotamian Gardens and the Identification of the Hanging Gardens of Babylon Resolved".* Garden History, Vol. 21, 1, pp. 1–13. Retrieved 12 20, 2022, from https://www.jstor.org/stable/1587050.

Dasgupta, N. (2016). Gardens in ancient India: Concepts, practices and imaginations. *Puravritta-Journal of the Directorate of Archaeology and Museums, 1,* 133–151.

Dickie, J. (1985). The Mughal Garden: Gateway to paradise. *Muqarnas, 3*(1), 128–137.

Du, P., & Chen, H. (2007). Water supply of the cities in ancient China. *Water Supply, 7*(1), 173–181.

Dwivedi, D. V. (2017). Development of agriculture in ancient India. *Sanskruti Darpan, 54,* 28–39.

Edward, S. (1991). Crops with wild relatives found in Ethiopia. In *Plant Genetic Resources of Ethiopia* (pp. pp. 42–74). Cambridge: Cambridge University Press.

Environmental Building News. (2007). Retrieved from https://www.buildinggreen.com/blog/buildinggreen-bulletin-december-2007-environmental-building-news

Fatma, S., & Fatima, S. (2012). *Waterworks in Mughal Gardens.* (pp. 1268–1278), Indian History Congress.

Fleet, J. (1898 - 99). Inscriptions at Managoli. *Epigraphia Indica, 5*(9), 31.

Fletcher, S. B. (1999). *20th Edition. A history of Architecture.* New Delhi, India.: CBS Publishers.

Ghosh, A., & Sarkar, A. (2014). Agriculture and irrigation practices in India before Christ. *Indian Journal of Spatial Science, 5*(2), 77–86.

Habib, I. (. (2002). *The Indus Civilization.* New Delhi, India: Tulika Books.

Hayden, B. (2003). Were luxury foods the first domesticates? Ethnoarchaeological perspectives from Southeast Asia. *World Archaeology, 34* (3), 458–469.

Heritage of Japan: The Yayoi years. (n.d.). Retrieved 3 9, 2023, from https://heritageofjapan.wordpress.com/yayoi-era-yields-up-rice/the-advent-of-agriculture-and-the-rice-revolution/

Hultzsch, E. (1925). *Inscriptions of Asoka: Corpus Inscriptionum Indicarum. Vol 1.* Oxford: Clarendon Press.: Clarendon Press.

Kangle, R. P. ((1960–65). (Reprint 2010)). *Kauṭiliya Arthasastra (Translation with Critical and Explanatory Notes). Bombay: University of Bombay Studies.* Delhi: Motilal Banarsidass Publishers Pvt. Ltd.

Kibert, C. J. (2013). *Sustainable Construction- Green building design and Delivery. 3rd Edn.* USA: Wiley Publications.

Leach, E. R. (1959). Hydraulic society in Ceylon. *Past and Present, 15*(1), 2–26.

Mubin, S., Gilani, I. A., & Hasan, W. (2013). Mughal gardens in the city of Lahore - a case study of Shalimar Garden. *Pakistan Journal of Science, 65*(4), 511–522.

Neelis, J. (2011). *Early Buddhist Transmission and Trade Networks: Mobility and Exchange within and beyond the North-western Borderlands of South Asia.* Leiden: Boston: Brill Publisher.

Nishida, M. (1983). The emergence of food production in neolithic Japan. *Journal of Anthropological Archaeology, 2,* 305–322.

Normile, D. (2003). *Earlier start for Japanese Rice cultivation.* (New: Archaeology) Retrieved 3 10, 2023, from https://www.science.org/content/article/earlier-start-japanese-rice-cultivation

Noshiro, S., & Sasaki, Y. (2007). Use of timber resources at the Shimo-yakebe site, Tokyo, and its characteristics in the Kanto plain of the late to latest Jomon periods. *Japanese Journal of Historical Botany, 15,* 19–34.

Olivelle, P., Loeshko, J., & Prabha, H. (2012). Asoka's inscriptions as text and ideology. In Olivelle, P., (Eds.), *Reimagining Asoka, Memory and History* (pp. 157–183). New Delhi: Oxford University Press.

Reporters, G. S. (2022). *Water-control principles used in ancient Chinese irrigation systems remain as useful as ever.* (Global Times) Retrieved 3 12, 2023, from https://www.globaltimes.cn/page/202210/1276996.shtml

Sharma, M., Prasad, V., Saxena, A., & Singh, I. B. (2004). Microscopic Charcoal in Lacustrine Sediments of Lahuradewa, as Evidence of Human Activity. (*Abstract*) *Paper Presented in the Joint Annual Conferences of IAS, ISPQS and IRCS and National Seminar on the Archaeology of the Ganga Plain.*

Siddiqui, I. H. (1986). Water-works and irrigation system in India during Pre-Mughal times. *Journal of the Economic and Social History of the Orient, 29*(1), 52–77.

Takamiya, H. (2001). The transition from foragers to farmers on the island of Okinawa. In P. Bellwood & D. Bowdery (Eds.), *Melaka papers, Special issue, Indo-Pacific Prehistory Association Bulletin 21, 5,* 60–67.

Takamiya, H. (2005). Island *prehistory.* Japan: Borderink.: Naha, Okinawa,

The Development of Agriculture. (2022). (National Geographic Society) Retrieved 3 11, 2023, from https://education.nationalgeographic.org/resource/development-agriculture/

The uniqueness of gardens in ancient Egypt. (2020). (INES) Retrieved 12 23, 2022, from https://medomed.org/2020/the-uniqueness-of-gardens-in-ancient-egypt/

Tripathi, V., (Eds) Gopal, L., & Srivastava, V. C. (2008). Agriculture in the Gangetic Plains during the first millennium BC. In *History of Agriculture in India, up to C. 1200 AD* (p. 361). New Delhi: Concept Publishing Company, New Delhi, p.361.

Wright, R. P., Bryson, R., & Schuldenrein, J. (2008). Water supply and history: Harappa and the Beas regional survey. *Antiquity, 82* (315), 37–48.

Wu, S., Wei, Y., Head, B., Zhao, Y., & Hanna, S. (2019). The development of ancient Chinese agricultural and water technology from 8000 BC to 1911 AD. *Palgrave Communications, 5*(77), 1–16.

Yadav, A. L., (eds) Gopal, L., & Srivastava, V. C. (2008). Some materials for the study of agriculture in Vedic India: Problems and prospective. In *History of agriculture in India up to c 1200 A.D. Vol. V, Part-I.* New Delhi, India: Centre for Studies in Civilizations.

Yamada, G., & Tsubakisaka, Y. (1995). *The spread of cultivated crops from the mainland. In Final Report on Research Project of the Historical and Cultural Exchange of the North.* Hok-kaido Historical Museum. [In Japanese.].

Yokoyama, E. (1984). Hajiki before the use of rokuro in Hokkaido:the formation of the Early Satsumon. *Koukogaku Zasshi [In Japanese], 70,* 52–75.

Zeder, M. (2011). The origins of agriculture in the Near East. *Current Anthropology, 52,* S221–S235.

Zohary, D., Hopf, M., & Weiss, E. (2012). *Domestication of Plants in the Old World: The origin and spread of domesticated plants in Southwest Asia, Europe, and the Mediterranean Basin.* Oxford, UK: Oxford University Press.

3 Multifunctional Urban Agriculture and Sustainability
A Global Overview

Madhav Govind and Anita Pinheiro

INTRODUCTION

Urban agriculture is becoming increasingly popular as it offers a range of benefits to both urban communities and the environment. The sustainability potential of urban agriculture is very much context-specific, and how it is practiced is the decisive factor determining its sustainability potential. The sustainability potential of urban agriculture also varies depending upon the motivating factor for taking up such practices. Therefore, it is highly possible that the same type of urban agriculture can provide differential outcomes between countries and within a country. Therefore, it is highly possible that urban agriculture can produce unjust and unsustainable outcomes if not careful. One key advantage of urban agriculture is its multifunctionality, which means it can serve multiple purposes beyond just producing food. Urban agriculture provides multiple social, ecological, and economic benefits, which may be tangible or intangible. Yet, much of the discussions on urban agriculture are still trapped in the narrative of 'feeding the masses' and emphasizing entrepreneurial-based intensive production in urban landscapes. Such narrowed visions of urban agriculture largely undermine the multifunctionality of urban agriculture, especially of those small-scale, livelihood-oriented, or non-commercial initiatives.

Urban agriculture research publications mostly come from countries and research institutions of the Global North. Publication of original research on urban agriculture from the Global South is relatively less in peer-reviewed journal articles published in the English language (Pinheiro & Govind, 2020). Therefore, the idea of urban agriculture that we get from academic publications is primarily based on studies from the Global North. Hence, the intricacies of the multifunctionality aspects, sustainability potential, risks, barriers, and other dynamics of urban agriculture in the literature might need to be reflected by those from the Global South.

For instance, urban home gardening and livestock keeping are more predominant practices in the Global South than in the Global North. However, these practices have not been captured in the scientometric analysis of urban agriculture literature conducted by Pinheiro and Govind (2020). Instead, the analysis shows the practices of Global North, such as community gardening and allotment gardening, followed by capital-intensive, building-integrated production technologies, as the dominant urban agricultural practices. The dominance of the Global North in urban agriculture research has clearly influenced the focus of research in this field. It is also important to note that urban agriculture research focusing on gender and poverty aspects of developing countries, especially in the African context, has witnessed stagnation over the years. One could say that the driving factors for taking up urban agriculture activities will have conflicting reasons in different contexts. Therefore, it is pertinent to give attention to those practices that have received little attention in the literature.

In this context, this chapter tries to fill the gaps in the literature by analyzing the multiple functions offered by various types of urban agriculture practiced and their sustainability potential. The chapter further explores how small-scale urban agriculture can act as 'nature-based solutions' (NbS) that can contribute to creating sustainable urban ecology and circular urban food systems. The chapter underlines the importance of context-specificity, which is needed to drive the maximum sustainability potential of urban agriculture.

DOI: 10.1201/9781003359425-3

The chapter begins with the analysis of multiple functions offered by urban agriculture for sustainability, followed by an introduction to NbS to urban sustainability. After emphasizing the importance of small-scale forms of urban agriculture, it further focuses on examples of small-scale urban agriculture systems from different parts of the world that can act as NbS to urban sustainability or contribute to circular urban food systems.

NATURE-BASED SOLUTIONS TO URBAN SUSTAINABILITY

NbS is a relatively new approach that is being used to address the sustainability challenges of society. The NbS emphasizes mainstreaming nature in policy planning and interventions (Gómez Martín et al., 2020) or using nature as a means to provide solutions (Nesshöver et al., 2017). Although NbS has originated in the debates on climate change, at present, its potential as an effective approach to sustainability is being explored in multiple areas such as agriculture, urban sustainability, wastewater management, and flood management. The NbS is considered an umbrella concept that incorporates already-existing conservation and sustainability measures such as ecosystem-based adaptation, urban green infrastructure (UGI), and ecosystem services (EES) (Pauleit et al., 2017; Seddon et al., 2019). It is an integrated and systemic approach (Nesshöver et al., 2017) that focuses on multiple dimensions of sustainability by addressing social, ecological, and economic benefits. The NbS is considered an integrated approach that has the potential to contribute to various Sustainable Development Goals (SDGs) (WWF, 2019). However, the long-term effectiveness of NbS is dependent upon the design, implementation, and social-ecological context in which it is applied (Gómez Martín et al., 2020).

Although NbS opposes the idea of capital and technology-intensive solutions (Nesshöver et al., 2017), it is not a complete alternative to technology-based solutions for urban sustainability. Rather, they are considered to be complementary to the human-made technology-based infrastructure (Nesshöver et al., 2017) as they optimize the interrelations between nature, society, and economy (Faivre et al., 2017). Despite the importance of nature in urban sustainability transitions (Muñoz-Erickson et al., 2016), NbS are understudied in the transitions literature (van der Jagt et al., 2020). Various forms of urban agriculture have been considered NbS for sustainability (Artmann & Sartison, 2018; Dorst et al., 2019; Kingsley et al., 2021; Maćkiewicz & Asuero, 2021). The NbS for urban sustainability includes 'urban' and 'peri-urban' forests, urban green roofs, pervious pavements, and restoration of vegetation to improve water systems (Dorst et al., 2019). Urban agriculture can play a crucial role as NbS to facilitate circular metabolism in cities by using the recovered resources in the production of new food and biomass (Canet-Martí et al., 2021). Urban agriculture also acts as UGI that can regulate the 'near-surface temperature' and thereby act as a feasible strategy for urban cooling and improving urban climate (Anderson & Gough, 2022) or addressing global warming by reducing the urban heat island effect (Mancebo, 2018).

The concept of 'edible city' is also considered an innovative NbS (Artmann et al., 2020; Sartison & Artmann, 2020), which in turn is associated with the concept of 'continuously productive urban landscapes' (Scharf et al., 2019). The edible city concept is associated with integrating food production into the urban agenda in all possible ways. It includes innovative designing and planning of urban public landscapes where, instead of ornamental plants, food production can be seen as a normal activity while providing aesthetic beauty. The concept of an edible city is also associated with a circular city in facilitating closed-loop systems of managing wastewater and solid wastes, which eventually helps to create livable, sustainable, and healthy cities (Säumel et al., 2019). Thus, urban agriculture, based on the principle of NbS, performs multiple functions.

MULTIFUNCTIONALITY OF URBAN AGRICULTURE

Due to its multifunctionality, urban agriculture is considered an NbS that can support systemic approaches to address various societal challenges (Artmann & Sartison, 2018; Grigorescu et al., 2022). Though much of the interventions on urban NbS focus on climate change and resilience, there is an emerging attention that focuses on the multifunctionality of urban agriculture that can address various challenges ranging from urban ecology and human wellbeing. Some of the challenges that urban agriculture as NbS can address include climate change, biodiversity and EES, food security, resource efficiency, agricultural intensification, land management, urban renewal and regeneration, public health, economic growth, and social cohesion (Artmann & Sartison, 2018). This argument underlines the fact that the benefits of urban agriculture should be seen from its multifunctionality and not just from the perspective of its ability to support urban food self-sufficiency (Artmann & Sartison, 2018).

However, the functions of urban agriculture as NbS will vary depending upon its type. For instance, certain types of urban agriculture enhance the microclimate while others will enhance the EES, biodiversity, or human wellbeing. Some st0075dies showed that urban gardens could encourage a shift to low-carbon food consumption (Kim, 2017; Puigdueta et al., 2021). Much of the studies on urban agriculture as NbS focus on its functions as UGIs, such as rooftop gardens, edible walls, and green roofs, while urban farmland has been neglected considerably (Evans et al., 2022; Lin & Egerer, 2017; Rolf, 2020). Despite the potential of urban agriculture to function as NbS, there are only limited studies that connect urban agriculture from the perspectives of the NbS (Kingsley et al., 2021).

Urban agriculture also helps in urban habitat management, enhancing the local urban biodiversity, beneficial insects, pollination, and a range of other ecosystem functions (Newell et al., 2022; Orsini et al., 2014; Wilhelm & Smith, 2018). However, it is also important to note that intensive forms of urban agriculture using agrochemicals might have a negative effect on urban biodiversity, and hence it is important to re-iterate that the beneficial outcomes of urban agriculture are completely dependent upon the way it is practiced. Some of the important ways in which NbS–urban agriculture contributes to urban sustainability are shown in Figure 3.1 and discussed in the following paragraphs.

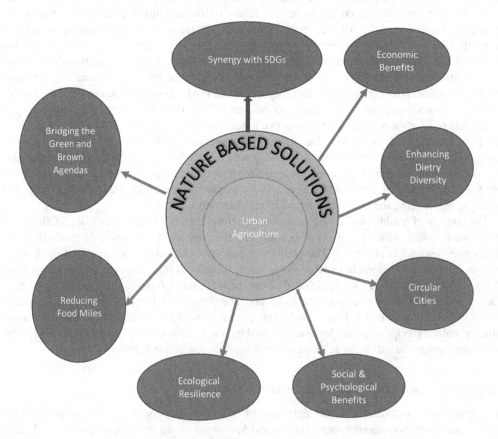

FIGURE 3.1   Multifunctionality of urban agriculture.

BRIDGING GREEN AND BROWN AGENDAS THROUGH URBAN AGRICULTURE

Urban and peri-urban agriculture is one important aspect that could link both green and brown agendas of urban sustainability (Simon, 2016; UN-Habitat, 2009). The green agenda focuses on the ecosystem protection and ecological impacts of urban systems at the local, regional, and global levels (UN-Habitat, 2009). On the contrary, the brown agenda acts to make the city healthy, livable, and functional (Newman, 2009) and engages in improving the water, waste, energy, transport, and building systems of urban areas (McGranahan & Satterthwaite, 2000; Newman, 2009).

In many urban centers, urban agriculture addresses the brown agenda by allowing low-income groups to access the land and have a livelihood from the cultivation of food or non-food crops (McGranahan & Satterthwaite, 2000; Simon, 2016). Urban agriculture also addresses some of the green agenda, such as facilitating local food production, reducing ecological impacts of long-distance transportation of food to feed the urban population (McGranahan & Satterthwaite, 2000; UN-Habitat, 2009), being part of UGI, and facilitating urban greening (Simon, 2016; UN-Habitat, 2009).

Bridging green and brown agendas through urban agriculture, however, requires careful planning and implementation. The use of untreated sewage water for irrigating urban agriculture lands and the use of agrochemical pesticides in urban agriculture are some of the areas that require adequate intervention to bridge both green and brown agendas effectively.

REDUCING FOOD MILES AND FACILITATING SHORT FOOD SUPPLY CHAINS

Food miles indicate the distance that food travels from the site of production to the site of consumption (Al-Kodmany, 2018; Van Passel, 2013). The long-distance transportation associated with the global food system results in increased carbon footprints from fossil fuel emissions and transportation costs. In highly industrialized countries such as the USA, the food miles can be higher.

Food miles are important in determining the carbon footprints of food production (Wakeland et al., 2012). Therefore, the local food production system, including urban agriculture, is getting increased consideration among environmentally conscious people (Gauthier, 2012). Urban agriculture helps reduce the carbon footprints of urban areas by contributing to the reduced food miles through local food production by facilitating reduced fossil fuel usage for transportation. Therefore, it is pertinent to explore all forms of urban agriculture, including rooftop cultivation, for sustainable urban food production so that food miles can be reduced to an extent (Benis et al., 2018). However, it is also important to mention that, even with reduced food miles, the carbon footprint of urban agriculture may be higher if energy- and fertilizer-intensive production practices are adopted (Goldstein et al., 2016; Mok et al., 2014), and, therefore, the technologies and practices of urban agriculture also matter to determine its sustainability outcomes.

Short-supply food chains (SSFCs) are alternative forms of agri-food supply chains that focus on shortening the supply chains by reducing the number of intermediaries and the geographical distance between the producers and consumers (Chiffoleau et al., 2016; Grando et al., 2017; Schmutz et al., 2018; Walthall, 2016). Especially in the Global North, the concept of SSFCs has gained wide-level interest from activists and academia alike. The SSFCs are considered a quality turn in the food system as they are more locally embedded and emphasize sustainable production methods (Goodman, 2003). Direct selling at farmer's markets, farm gate sales, vegetable box schemes, etc. are some forms of short food supply chains in urban areas.

Urban and peri-urban agriculture has the potential to increase the proximity between producers and consumers and, thereby, shorten the food supply chains while helping people reconnect with food production and producers. Being geographically close to the consumers, urban and peri-urban agriculture can reduce the number of intermediaries while enabling the embedding of small-scale or commercial-scale food production into the urban food systems (Chiffoleau et al., 2016; Schmutz et al., 2018; Yacamán Ochoa et al., 2020). An exploratory study by Schmutz et al. (2018) analyzes the social, environmental, and economic sustainability of five types of urban and peri-urban agriculture and/or SSFCs in London, including urban gardening for self-supply, commercial urban gardening, community-supported agriculture (CSA), direct on-farm sale, and direct off-farm sale. All these SSFCs have ranked higher on the social sustainability aspects. Regarding overall sustainability, CSA delivers the highest outcomes in terms of social, environmental, and economic benefits, followed by commercial urban gardening, urban gardening for self-supply, direct off-farm sale, and direct on-farm sale (Schmutz et al., 2018). In SSFCs, urban and peri-urban food production that adopts organic cultivation methods is mostly preferred by consumers (Walthall, 2016; Yacamán Ochoa et al., 2020). This is mainly due to the increased environmental consciousness of consumers regarding food production methods.

ENHANCING FOOD PRODUCTION AND DIETARY DIVERSITY

Urban agriculture contributes to enhanced food production in many instances. Such enhanced food production could be by contributing to food and nutritional security in urban areas or by bridging the disconnection of urban areas with food production.

As urban areas are considered sites of consumption, most food production occurs outside of its peripheries or in distant rural areas. Urban agriculture will enable addressing this issue of the disconnection of urban areas with food production. Edible cities that encompass urban agriculture fields (Palassio & Wilcox, 2005) are also considered important NbS that can contribute to urban sustainability transformations (Sartison & Artmann, 2020). Moreover, urban agriculture can address the issue of 'food deserts', a term that refers to the food-deprived areas within urban areas that will substantially lack retail outlets for fresh food sales (Krishnan et al., 2016).

The contribution of urban agriculture to food security is mainly by enhancing access to food and income generation fields (Mougeot, 2005). Those who engage in urban agriculture or livestock keeping can access more food for their households than those who engage in non-agricultural activities in urban areas (Poulsen et al., 2015). When it comes to home-garden food production, there are chances of growing more variety of crops, thus enhancing the dietary diversity of the families. Economically weaker sections of the urban population benefit from such direct access to food through urban agriculture as they may not be able to afford the food otherwise (Stewart et al., 2013).

Some studies argue that urban agriculture does not have the potential to enhance urban food security or the food security of the poor urban population (Badami & Ramankutty, 2015; Korth et al., 2014). Such studies base their claim on the analysis of agricultural land required to cultivate food for the poor (Badami & Ramankutty, 2015) or by considering urban agriculture as a mere extension of commercial agricultural activities into the urban limits. These studies do not account for the diverse forms of urban farming practices carried out in non-conventional agricultural spaces such as roadsides, backyards, rooftops, and other available spaces. Moreover, comparing urban agriculture with commercial agriculture in rural areas would be naive.

URBAN AGRICULTURE FOR CIRCULAR CITIES

The concept of circular cities aims to circularize the resource flow within the urban systems, and urban agriculture acts as an NbS to bring circularity within the urban systems (Canet-Martí et al., 2021). Such a shift to circularity has the potential to maximize self-sufficiency and resource efficiency in urban systems while also reducing the carbon footprint of urban systems (Langergraber et al., 2020). As food production has the ability to close some loops of energy and material flows, urban agriculture plays a crucial role in bringing circularity to urban systems (Paiho et al., 2021) and helps design regenerative urban food systems (Pascucci, 2020).

Urban agriculture brings about circularity through material and resource cycling and closes the urban nutrient cycle by integrating urban sanitation with food production (Skar et al., 2020). Some urban agriculture systems that embrace a circular economy include hydroponic production linked to district heating and aquaponics (Canet-Martí et al., 2021; Paiho et al., 2021). Significant discussions on urban agriculture in connection with circular cities emphasize circular practices involving nutrient recovery from organic solid and liquid wastes and its use in producing new food and biomass through urban and peri-urban agriculture fields (Mohan et al., 2020; Pascucci, 2020). This helps maximize the resource efficiency within the urban systems while reducing the dependence on external systems for inputs (Canet-Martí et al., 2021; Pascucci, 2020).

Another example of bringing circularity through urban agriculture is by using biomass waste materials from one type of crop grown in urban agriculture practices such as integrated rooftop greenhouses (i-RTGs) as a substrate for growing different kinds of vegetable crops in i-RTG production (Manríquez-Altamirano et al., 2020). It is also important to note that some urban agriculture practices can incorporate circular approaches. In contrast, careful strategic planning and designing are required to integrate circularity in other urban food production methods (Canet-Martí et al., 2021).

ENHANCING SOCIAL-ECOLOGICAL RESILIENCE

Multiple historical and contemporary studies have recognized the importance of urban agriculture for its ability to enhance the adaptive capacity of urban food systems and enhance resilience in urban areas during external shocks (Langemeyer et al., 2021). Urban gardens such as allotment gardens act as 'pockets of social-ecological memories' or 'incubators of social-ecological knowledge' about urban food production and the crisis times in the past so that such memories and knowledge can easily be transferred in some other times of crisis and hence contributing to the resilience (Barthel et al., 2014). Failing to protect such social-ecological memories and the larger trend of 'global generational amnesia' on the importance of urban agriculture as an essential life-support system will have negative consequences for enhancing urban social-ecological resilience (Barthel et al., 2014). Social networking around home gardens and increasing the diversity of resources have critical roles in enabling continuous adaptation to change and building resilience (Buchmann, 2009; Dewaelheyns et al., 2014).

The key examples that highlight the resilience potential of urban agriculture include how urban agriculture helped to respond to crisis times such as the Great Depression (McClintock, 2010), World Wars (Barthel et al., 2014), a Special Period in Cuba (Altieri et al., 1999; Buchmann, 2009), post-Hurricane Katrina (Button, 2013; Kato, 2020; Passidomo, 2014), other natural disasters (Okvat & Zautra, 2014), Economic Crisis in 2008 (Calvet-Mir & March, 2019), COVID-19 lockdown period (Schoen et al., 2021), and poverty (Gallaher, 2017; Malan, 2015).

SOCIAL AND CULTURAL BENEFITS OF URBAN AGRICULTURE

Urban agriculture has the potential to provide many social, cultural, psychological, and health benefits (Glover, 2021; Diekmann et al., 2020; Gallaher, 2017). Urban agriculture strengthens social justice (Reynolds & Cohen, 2016), community development (Slater, 2001), and community self-reliance, helping to protect cultural practices, values, and ethnic identity (Taylor & Lovell, 2015). Urban agriculture also helps to strengthen food justice and food sovereignty (Horst et al., 2017; Pettygrove & Ghose, 2018; London et al., 2021). Many of the community and allotment gardening projects also aim to enhance the connection with nature and act as a therapeutic way to improve the mental health (Rich et al., 2015) of people, especially the elderly and disabled sections of the population.

In the USA and some African countries, specific non-profit organizations try to bring about wellbeing to the deprived sections of society through urban agriculture, mainly community gardening. For instance, in Kibera, one of the largest slums in Kenya, a local version of urban agriculture called sack gardening has helped to regreen the slum while positively changing the social environment in the area and improving household food security (Gallaher, 2017). Similarly, in the USA, urban gardening practiced by women from the colored population in Detroit represents their way of resistance to the consumerist culture and as a way to reassess their cultural roots and reclaim personal power (White, 2011).

Urban agriculture provides differential outcomes to different sections of society. Therefore, in many instances, urban agriculture has exacerbated social inequalities and social injustice (Hoover, 2013; Reynolds, 2015). Hoover (2013) provides an example of such contribution of urban agriculture to social inequality. In Denver, USA, urban agriculture is mostly practiced by middle-class white populations, even in areas dominated by blacks, Latinos, or people other than white (Hoover, 2013). Therefore, when analyzing the sustainability outcomes of urban agriculture, it is also important to investigate who gets the benefits and who is left out in the process. Without considering the different sections of society who get the benefits, urban agriculture cannot offer social sustainability outcomes, and, instead, it intensifies the existing inequalities.

ECONOMIC BENEFITS OF URBAN AGRICULTURE

Not all urban agriculture practices are oriented toward food production for self-provision. Almost all forms of capital-intensive urban agriculture, such as vertical farming, rooftop farming, and rooftop greenhouses, in the Global North (Benke & Tomkins, 2017; Buehler & Junge, 2016; Pfeiffer et al., 2015) and Singapore (Diehl et al., 2020) are innately market-oriented and intended for profits. In the low- and middle-income countries of the Global South, the conventional forms of open-space urban agriculture practiced within the urban periphery (mainly in the Global South) are intended for the market. Urban livestock keeping and aquaculture in the urban periphery in the low- and middle-income countries of the Global South are mostly livelihood options aimed at economic benefits. Among the other types of urban agriculture, CSA and entrepreneurial community gardens are profit-oriented, with social and ecological principles incorporated into them.

Urban agriculture is also a livelihood opportunity for many people, especially in the Global South (Cook et al., 2015; World Bank, 2013). In some cases, it is the primary source of income, while, in most cases, urban agriculture is an additional income source. Although quantitative data on the contribution of urban agriculture to livelihood and food production are mentioned in some literature, there are mostly old data, and, hence, the latest information is not available.

Although urban agriculture can provide economic benefits to the socially weaker sections of society, there exists a gender disparity, especially in the commercial, open-space urban agriculture practices in the Global South (Devi & Buechler, 2009; Hovorka, 2005), which results in men having access to the marketability and profits while women's involvement is limited to the supporting activities only.

SYNERGY WITH THE SUSTAINABLE DEVELOPMENT GOALS

Urban agriculture also contributes to achieving the SDGs proposed by the United Nations General Assembly in 2015. However, the contributions of urban agriculture to sustainability and SDGs completely depend on how it is implemented. According to Nicholls et al. (2020), urban agriculture directly impacts SDGs 1, 2, 3, 8, 11, 12, 13, and 15. Urban agriculture helps to reduce poverty (SDG 1) by facilitating enhanced access to food, increasing income, and reducing expenditure on food. Because of better access to fresh, nutritious, and diverse food, people involved in urban agriculture are observed to eat more food than non-farming households, thereby contributing to reducing hunger (SDG 2) and enhancing nutrition. Improved nutrition is also in synergy with SDG 3, good health, and wellbeing. However, it is also important to understand how nutritional outcomes are distributed within a household. A study that systematically reviewed the nutritional outcomes of urban agriculture found that, despite the contribution of urban agriculture to enhanced household nutrition, it did not make much difference in the children's nutritional outcomes (Masset, 2011). This indicates that the benefits of urban agriculture are not equally distributed among different demographic groups.

As many of the urban agriculture initiatives emphasize the adoption of agroecological production methods, they are directly related to SDG 2.4 (sustainable food production systems) and the reduced use of greenhouse gas-emitting fertilizers and pesticides that also contribute to pollution while being harmful to wildlife make it in synergy with SDG 13 (climate action) and SDG 15, life on land (Nicholls et al., 2020). Urban agriculture contributes to SDG 3 (good health and wellbeing) through enhanced nutrition and physical and mental health. However, it is also important to note that intensive use of agrochemicals as largely practiced in developing countries will negatively impact SDG 3 (good health) and SDG 15 (life on land).

Urban agriculture also contributes to SDG 8 (decent work and economic growth) by creating jobs and employment in the food value chain. The multiple social and environmental benefits of urban agriculture contribute to SDG 11 (sustainable cities and communities), precisely the SDG target 11.7 (safe, inclusive, and accessible, green, and public spaces), which is in synergy with SDG 4. The reduction of food miles facilitated by urban agriculture is in synergy with SDG 13, and the reduced pressure on rural hinterlands is synergistic with SDG 15. The possible trade-off is the potential competition for space for infrastructure and housing in urban areas (SDG 9).

By enabling the sustainable reuse of urban solid wastes and wastewater in food production, urban agriculture contributes to SDG 12 (responsible consumption and production). This is by facilitating the optimum use of resources and reducing the demand for fresh water in food production, which is important in water-scarce areas, thereby being synergistic to SDG 15 again. However, if adequate attention is not given to eliminating the risks from the use of wastewater, it might lead to concerns related to public health and threats from the spread of infectious diseases (trade-off with SDG 3). By improving the urban microclimate, urban agriculture directly contributes to reducing global warming, thereby contributing to SDG 13, climate action. However, quantitative data on the benefits of urban agriculture on climate change are still significantly less.

Urban agriculture also directly contributes to SDG 15 (life on land) by facilitating urban wildlife habitats and reducing the pressure on rural hinterlands to extend intensive food production. Although urban areas have less biodiversity than rural areas, urban agriculture helps improve the ecology, biodiversity, and habitats at the microlevel, especially in those which adopt ecological production methods such as allotment gardens, home gardens, and community gardens. This, in turn, contributes to SDG 2.4 (sustainable food production systems) by facilitating enhanced EES (pollination, natural pest control, etc.).

PATTERN OF SMALL-SCALE URBAN AGRICULTURE SYSTEMS

Urban agriculture includes small-scale, low-budget, nature-based practices and large-scale capital and technology-intensive practices. Pinheiro and Govind (2020) have pointed out that the latest trends in urban agriculture research, especially those from the Global North, focus more on capital-intensive, building-integrated production technologies. They emphasize more on increased production and profits with the least focus on socio-ecological outcomes. Small-scale urban agriculture practices, on the contrary, provide multiple benefits with more focus on socio-ecological benefits and economic benefits to an extent. Unlike capital-intensive production methods, which largely focus on technological solutions, small-scale urban agriculture practices are mostly oriented toward NbS. Also, they are more integrated and adapted to the specific local context of urban areas.

Although urban agriculture includes growing crops and rearing animals, much of the urban agriculture literature largely focuses on crop cultivation. Although urban livestock keeping, beekeeping, and aquaculture are present in many parts of the world, they have received less attention in the academic literature (Pinheiro & Govind, 2020). In addition, certain types of small-scale urban agriculture unique to a specific area are not yet receiving adequate attention in the urban agriculture literature. For instance, sack gardening is a different form of urban agriculture that emerged in the Kibera slums of Nairobi as a livelihood and wellbeing strategy for the deprived sections (Gallaher, 2017; Gallaher et al., 2015). Based on the available literature, various patterns of small-scale urban agriculture practices (Figure 3.2) and their relevance as NbS to sustainability are discussed below.

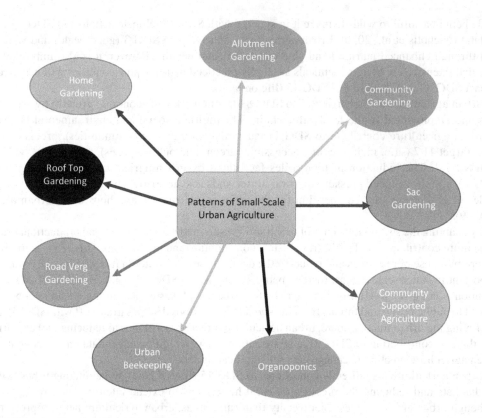

FIGURE 3.2   Patterns of small-scale urban agriculture.

URBAN HOME GARDENING

Urban home gardens are known by many other terms such as 'house-lot gardens', 'house gardens', 'kitchen gardens', 'dooryard gardens', and 'backyard gardens'. In the Global South, urban and rural home gardening is a traditional form of local agriculture systems. Urban home gardening in the North has been overlooked and understudied (Taylor & Lovell, 2015). Despite its critical relevance to the agri-food systems, urban home gardening has received relatively less attention within the urban agriculture literature and policy.

Urban home gardens are managed by a single household on the land of the same property or adjacent vacant lot, which is leased, owned, or borrowed, and also include container gardening on rooftops (Taylor & Lovell, 2014, 2015). Urban home gardening is carried out in every possible private property space, such as back and front yards, patios, and balconies (Kortright & Wakefield, 2011). The establishment of urban home gardens is a vital tool for creating sustainable cities and supporting the marginal sections of the urban population with livelihood, strengthening their cultural identity (Van Veenhuizen, 2006) and social cohesion from sharing food (Thomasson, 1994), nutrition, health, wellbeing, and self-reliance (Marques et al., 2021; Vávra et al., 2018). Urban home gardens are also considered 'small-scale social-ecological systems' (Buchmann, 2009) that contribute to the conservation of local agrobiodiversity (Poot-Pool et al., 2018) and constitute an essential component of UGI (Loram et al., 2007). Urban home gardeners use less agrochemicals in their gardens, yet the crop yields are comparable to or higher than the yield from large-scale conventional farms (Nicholls et al., 2020).

The motivation for setting up an urban home garden may vary. Like most other types of urban agriculture, the kind and purpose of the urban home gardens are diverse and reflect on the household's social milieu and financial capabilities. The major drivers for starting urban home gardening can be categorized as economic motivation (producing nutritious food affordably), environmental motivation (as a response to the concerns of the environment), and social and family health motivation (educational, physical, and mental wellbeing and social cohesion) (Kirkpatrick & Davison, 2018).

The size of a home garden may vary considerably across different countries or within a country (Drescher et al., 2006). Unlike other types of agricultural systems, calculating the area of a home garden may be a difficult task, especially when gardening is practiced in different available spaces of a building and its setback spaces. What is more important is the species abundance and multiple functions that the urban home gardens cater to the households and the environment. However, because of the smaller size, it is possible that the importance of urban home gardens may not get adequate attention from the authorities.

Scientific production methods in urban home gardens help households reduce their dependence on vegetables and make small earnings through surplus sales (Ferdous et al., 2016). Although urban home gardens contribute to lowering the spending on food purchases (Miura et al., 2003), the contribution of food from the garden is not added to the agriculture statistics (Buchmann 2009). In most cases, urban home gardening is practiced by independent initiatives without getting any support from the community or the government.

ALLOTMENT GARDENING

Allotment gardens are managed by individuals or families on the small parcels of land allocated to them on rent by the local authority (Acton, 2011; Drescher et al., 2006; Nicholls et al., 2020). They are prominent in European countries and some parts of North America. The allotments are usually far from the home of the individual who uses it for non-commercial production of fruits, vegetables, and ornamental plants for personal use or recreation (Bell et al., 2016). Usually, people interested in allotment gardening need to apply to the local authority, and owing to the scarcity of sufficient land, there will be a long waiting period until the land parcel is allotted to them (Acton, 2011; Edmondson et al., 2020).

A case study conducted in Leicester, a UK city, shows that productivity in allotment gardens is at par with commercial horticulture production (Edmondson et al., 2020). This study also highlights that allotment gardening carried out in 1.5% of the total land in the city and 2% of the city's total green space can provide fresh produce to 2.6% of the population with a '5-a-day fruit and vegetable diet'. There are many studies that have analyzed the multifunctionality of allotment gardens and their potential to act as NbS to urban sustainability, including EES (Cabral et al., 2017; Hashimoto et al., 2019; Langemeyer et al., 2018), biodiversity conservation (Speak et al., 2015), and its contribution to UGI (Borysiak et al., 2017; Sowińska-Świerkosz et al., 2021). A study conducted by Nicholls et al. (2020) has documented the EES in connection with urban agriculture in Brighton and Hove in the UK. They have documented the pollinators found in the urban allotment and home gardens and calculated the quantity of produce that can be 'owed' to the insect pollinators in these gardens. Although public and private allotment gardens are NbS with positive socio-ecological impacts, depending upon the context, the contribution level may vary among these two categories. For instance, Maćkiewicz and Asuero (2021) point out that in Seville, Spain, public allotment gardens have a more positive environmental impact than private allotment gardens, especially due to their contribution to biodiversity. Despite the potential of allotment gardens (private or public), there could be many barriers, including institutional barriers, that make allotment gardens unsophisticated and hence limit their effectiveness as NbS (Sowińska-Świerkosz et al., 2021).

COMMUNITY GARDENING

Community gardens are gardens in single pieces of open space land collectively managed by local community groups or non-profit organizations (Hess & Winner, 2007). Local authorities do not have any roles in garden management. Gardening may occur on public land, public parks, private land, and school compounds (Hess & Winner, 2007). Some of the benefits of community garden-ing include creating social ties, enhancing social cohesion, and creating therapeutic settings for the elderly population, people with dementia, people with disabilities, people with mental health challenges, refugees, and those who face social isolation (Cumbers et al., 2018; Lovell et al., 2014; Malberg Dyg et al., 2020). Community gardens are prominent in North America, some European countries, and Australia. Several community gardening initiatives in the USA focus on low-income neighborhoods and vulnerable groups.

Much of the functions offered by community gardens as NbS are similar to those of allotment gardens, including the provision of various EES (Cabral et al., 2017; Wu et al., 2022), biodiversity conservation, and stormwater runoff benefits (Gittleman et al., 2017). As community gardens focus on community interactions, one key aspect of community gardens is their ability to provide cultural and provisional EES, which are particularly important for the migrant populations in a country (Camps-Calvet et al., 2016; Clarke & Jenerette, 2015). Camps-Calvet et al. (2016) highlight that the socio-cultural EES enable community gardens to act as NbS to address several urban policy chal-lenges related to socio-environmental development. The authors further emphasize the importance of providing access to vacant lands for community gardening to ensure its effectiveness as an NbS.

SACK GARDENING

Sack gardening is a simpler form of vertical growing of vegetables in sacks filled with soil and other nutrients (Gallaher et al., 2015; Zivkovic et al., 2022). Vertical cultivation in sacks permits the grow-ing of a greater number of plants compared to conventional container cultivation, hence making it suitable for areas with limited availability of space. Unlike other forms of urban agriculture, sack gardening occurs in narrow patches of available space. The sacks are placed along the sides of the houses or in other empty spaces in the informal settlements. Although container gardening (in sacks or other containers) is a common practice across the world, the specific term 'sack gardening' is mainly associated with urban agriculture practiced in densely populated areas in Kenya and Ghana (Gallaher et al., 2015; Peprah et al., 2014). These are mainly interventions of non-profit organizations that aim to enhance the socioeconomic and environmental quality of these informal settlements.

Many studies have discussed sack gardening in informal settlements in African countries, espe-cially in Kibera slums in Nairobi, Kenya (Peprah et al., 2014; Gallaher et al., 2015; Zivkovic et al., 2022). These studies highlight that sack gardening has been crucial for providing socio-ecological and economic benefits such as regreening the slums, an affordable source of increased dietary diversity, food security, creating livelihood opportunities, and reducing food expenses. Considering the space scarcity and population density in informal settlements, sack gardening is the best suitable method for regreening the slums while increasing their food security in an affordable way. Although sack garden-ing constitutes an important form of green infrastructure in informal settlements, it may also cause health hazards from reusing polluted grey water for the irrigation of crops (Gallaher, 2017).

COMMUNITY-SUPPORTED AGRICULTURE

In CSA, a direct economic and social partnership is established between farmers and consumers (Walthall, 2016). In CSA, consumers financially support the farmers with an upfront payment for the 'share' of vegetables they will receive later during the harvest (Mert-Cakal & Miele, 2020; Ostrom, 2007; Walthall, 2016). Unlike allotment gardening and community gardening, the primary intention of CSA is commercial production. However, they are entrusted with the principles of local small-scale production, safe and organic production methods, facilitating short food supply chains, and community building (Walthall, 2016). One study in Brazil shows that CSA can facilitate the adoption of agroecological technologies, including agroforestry among small-scale family farmers (Cechin et al., 2021). These features make CSA a social innovation that supports the social economy in an ecologically sound way (Mert-Cakal & Miele, 2020).

ROOFTOP GARDENS

Rooftop gardening enables the effective utilization of otherwise unused spaces for production in residential and non-residential buildings (Thomaier et al., 2015). This helps to address the scarcity of available space for food production in the space-crunch urban areas. Open-air rooftop cultivation is carried out for commercial and non-commercial purposes (Sanyé-Mengual et al., 2015). Although rooftop gardening is common worldwide (Hamilton et al., 2014), the scale and intensity may vary across regions. Apart from food production, rooftop gardens contribute to urban biodiversity and EES (Grard et al., 2017). In addition, rooftop gardens constitute an important component of UGI (Langemeyer et al., 2020; Lin et al., 2017) for providing crucial environmental and social benefits.

URBAN BEEKEEPING

Compared to other types of urban agriculture, urban beekeeping has received considerably less attention in the literature. Urban beekeeping has crucial ecological significance due to its contribution to biodiversity conservation and other EES, including pollination. Matsuzawa and Kohsaka (2022) cite studies that show that bees may survive more in urban areas due to limited exposure to chemical pesticides. This has been further proven by the study of Patenković et al. (2022), which shows that urban areas serve as vital habitats for free-living wild honey bees while contributing to their significant genetic diversity compared to managed native honey bee colonies. Moreover, urban beekeeping also supports adopting agroecological methods in urban agriculture.

Despite their socio-ecological importance, there has been a decrease in urban bee colonies over the years, which in turn has sparked an interest in the Global North for the expansion of urban agriculture especially in Europe and North America. Some cities, including New York and Los Angeles in the USA and the German state of Bayer, have started to actively adopt urban beekeeping in their environmental policies to become pollinator-friendly cities. However, the increasing level of urban beekeeping also poses challenges, primarily related to safety concerns. Appropriate government rules and the installation of barriers can minimize public safety issues while maximizing the benefits of urban beekeeping (Matsuzawa & Kohsaka, 2022).

ORGANOPONICS

Organoponics or organopónicos is a type of urban agriculture that originated and is mostly practiced in Cuba. Organoponics was popularized as a coping mechanism that uses locally available organic inputs to overcome the scarcity of agrochemicals during the Special Period (an extended period of economic crisis in Cuba due to the collapse of the Soviet Union) in Cuba. Organoponics is a cultivation system in $30 \, \text{m} \times 1 \, \text{m}$ rectangular-walled constructions consisting of raised beds of growth medium that is a mixture of soil and organic material, such as composted bagasse (sugarcane residue) (Koont, 2007, 2008, 2011), animal manure, and household wastes (Sustainable Agriculture in Cuba, 2021). The technology of organoponics was developed by the military division in Cuba before the collapse of the Soviet Union and has been popularized among civilians for urban agriculture during the Special Period (Altieri et al., 1999; Koont, 2007, 2008, 2011). With government interventions to popularize organoponics through extensive extension services (Koont, 2011), it has become one of the primary methods of urban cultivation (Altieri et al., 1999). Organoponics has become a popular method of urban agriculture as a measure to overcome the poor quality of urban soil (Altieri et al., 1999).

The critical feature of organoponics is that it offers an alternative to agrochemical-based food production in urban areas while helping to circularize the waste. Therefore, it can overcome the concerns related to soil contamination from chemicals and can also be carried out in soils with low fertility fields (F. Marshall, 2017; Orsini et al., 2013) while reducing waste.

ROAD VERGE GARDENING

Road verge gardening is gardening by the residents on the road verge, public footpaths, or natural strips in front of their private property (Hsu, 2019; Kingsley et al., 2021). It is an emergent form of non-commercial urban agriculture. Verge gardening has become popular in Australian cities, and it has received policy support from some local governments in Australia for using these spaces for food production (Kingsley et al., 2021). Ulm and Ayllon (2022) argue that verge gardens can be considered small-scale and scalable NbS to address some of the societal challenges, including climate change, food security, and public and environmental health. Other scholars have highlighted the potential of verge gardens to act as NbS to many socio-environmental challenges. For instance, edible verge gardening (verge gardens that are used for food production) can offer socio-ecological benefits for enhancing human and environmental health, climate change adaptation, and mitigation (Mcdougall et al., 2020). The verge gardens add green spaces in cities and also enhance urban biodiversity and EES (Marshall et al., 2020; Säumel et al., 2016). The most important aspect is that verge gardening takes place in otherwise unused spaces for food production and urban greening without competing for urban land (Marshall et al., 2020; Ulm & Ayllon, 2022).

CONCLUSIONS

Thus, we see that multifunctionality is a vital aspect of urban agriculture. Apart from food production, urban agriculture offers multiple functions to urban ecology and social wellbeing. Yet, much of the discussions on urban agriculture are still trapped in the narrative of 'feeding the masses', which further focuses on entrepreneurial-based capita- and technology-intensive production in the urban landscapes. This chapter has shown that various forms of urban agriculture not only help in bridging the green and brown agendas and reducing the food miles by shortening the food supply chains but also synergize with SDGs by extending social, economic, and ecological benefits. The chapter has further demonstrated that, although urban agriculture is a common term, it encompasses varying practices. These varying practices are considered NbS to sustainability. Some of the important forms of small-scale agricultural practices are 'urban home gardening', 'allotment gardening', 'community gardening', 'sac gardening', 'rooftop gardening', 'road verge gardening', 'organoponics', and urban beekeeping. However, the outcomes of these types of gardening vary from context to context, depending upon the way urban agriculture is practiced. Contrary to the dominant narratives, this chapter has explored the potential of small-scale urban agriculture systems that can act as NbS to enhance urban ecology and food systems. Furthermore, the chapter has underlined the urban agriculture practices specific to local or regional contexts that could be considered while addressing the issues related to SDGs.

REFERENCES

Acton, L. (2011). Allotment gardens: A reflection of history, heritage, community and self. Papers from the Institute of Archaeology, 21.

Al-Kodmany, K. (2018). The vertical farm: A review of developments and implications for the vertical city. *Buildings*, 8(2), Article 2. https://doi.org/10.3390/buildings802002

Altieri, M. A., Companioni, N., Cañizares, K., Murphy, C., Rosset, P., Bourque, M., & Nicholls, C. I. (1999). The greening of the "barrios": Urban agriculture for food security in Cuba. *Agriculture and Human Values*, 16(2), 131–140.

Anderson, V., & Gough, W. A. (2022). Nature-based cooling potential: A multi-type green infrastructure evaluation in Toronto, Ontario, Canada. *International Journal of Biometeorology*, 66(2), 397–410. https://doi.org/10.1007/s00484-021-02100-5

Artmann, M., & Sartison, K. (2018). The role of urban agriculture as a nature-based solution: A review for developing a systemic assessment framework. *Sustainability*, 10(6), 1937.

Artmann, M., Sartison, K., & Vávra, J. (2020). The role of edible cities supporting sustainability transformation-A conceptual multi-dimensional framework tested on a case study in Germany. *Journal of Cleaner Production*, 255, 120220.

Badami, M. G., & Ramankutty, N. (2015). Urban agriculture and food security: A critique based on an assessment of urban land constraints. *Global Food Security*, 4, 8–15. https://doi.org/10.1016/j.gfs.2014.10.003

Barthel, S., & Isendahl, C. (2014). Urban gardens, agriculture, and water management: Sources of resilience for long-term food security in cities. *Ecological Economics*, 86, 224–234. https://doi.org/10.1016/j.ecolecon.2012.06.018

Barthel, S., Parker, J., Folke, C., & Colding, J. (2014). Urban gardens: Pockets of social-ecological memory. In K.G. Tidball & M.E. Krasny (Eds.), *Greening in the Red Zone* (pp. 145–158). Springer.

Bell, S., Fox-Kämper, R., Keshavarz, N., Benson, M., Caputo, S., Noori, S., & Voigt, A. (2016). *Urban allotment gardens in Europe*. Routledge.

Benis, K., Turan, I., Reinhart, C., & Ferrão, P. (2018). Putting rooftops to use-A cost-benefit analysis of food production vs. Energy generation under mediterranean climates. *Cities*, 78, 166–179.

Benke, K., & Tomkins, B. (2017). Future food-production systems: Vertical farming and controlled-environment agriculture. *Sustainability: Science, Practice and Policy*, 13(1), 13–26. https://doi.org/10.1080/15487733.2017.1394054

Borysiak, J., Mizgajski, A., & Speak, A. (2017). Floral biodiversity of allotment gardens and its contribution to urban green infrastructure. *Urban Ecosystems*, 20(2), 323–335. https://doi.org/10.1007/s11252-016-0595-4

Buchmann, C. (2009). Cuban home gardens and their role in social-ecological resilience. *Human Ecology*, 37(6), 705–721.

Buehler, D., & Junge, R. (2016). Global trends and current status of commercial Urban rooftop farming. *Sustainability*, 8(11), Article 11. https://doi.org/10.3390/su8111108

Button, R. (2013). Growing communities: Urban agriculture in post-Katrina New Orleans [University of Mississippi]. https://egrove.olemiss.edu/cgi/viewcontent.cgi?article=1862&context=etd

Cabral, I., Keim, J., Engelmann, R., Kraemer, R., Siebert, J., & Bonn, A. (2017). Ecosystem services of allotment and community gardens: A Leipzig, Germany case study. *Urban Forestry & Urban Greening*, 23, 44–53. https://doi.org/10.1016/j.ufug.2017.02.008

Calvet-Mir, L., & March, H. (2019). Crisis and post-crisis urban gardening initiatives from a Southern European perspective: The case of Barcelona. *European Urban and Regional Studies*, 26(1), 97–112. https://doi.org/10.1177/0969776417736098

Camps-Calvet, M., Langemeyer, J., Calvet-Mir, L., & Gómez-Baggethun, E. (2016). Ecosystem services provided by urban gardens in Barcelona, Spain: Insights for policy and planning. *Environmental Science & Policy*, 62, 14–23.

Canet-Martí, A., Pineda-Martos, R., Junge, R., Bohn, K., Paço, T. A., Delgado, C., Alenčikienė, G., Skar, S. L. G., & Baganz, G. F. M. (2021). Nature-based solutions for agriculture in circular cities: Challenges, gaps, and opportunities. *Water*, 13(18), Article 18. https://doi.org/10.3390/w13182565

Cechin, A., da Silva Araújo, V., & Amand, L. (2021). Exploring the synergy between community supported agriculture and agroforestry: Institutional innovation from smallholders in a brazilian rural settlement. *Journal of Rural Studies*, 81, 246–258. https://doi.org/10.1016/j.jrurstud.2020.10.031

Chiffoleau, Y., Millet-Amrani, S., & Canard, A. (2016). From short food supply chains to sustainable agriculture in urban food systems: Food democracy as a vector of transition. *Agriculture*, 6(4), Article 4. https://doi.org/10.3390/agriculture6040057

Clarke, L. W., & Jenerette, G. D. (2015). Biodiversity and direct ecosystem service regulation in the community gardens of Los Angeles, CA. *Landscape Ecology*, 30(4), 637–653. https://doi.org/10.1007/s10980-014-0143-7

Cook, J., Oviatt, K., Main, D. S., Kaur, H., & Brett, J. (2015). Re-conceptualizing urban agriculture: An exploration of farming along the banks of the Yamuna River in Delhi, India. *Agriculture and Human Values*, 32(2), 265–279.

Cumbers, A., Shaw, D., Crossan, J., & McMaster, R. (2018). The work of community gardens: Reclaiming place for community in the city. *Work, Employment and Society*, 32(1), 133–149. https://doi.org/10.1177/0950017017695042

Devi, G., & Buechler, S. (2009). Gender dimensions of urban and peri-urban agriculture in Hyderabad, India. *Women Feeding Cities: Mainstreaming Gender in Urban Agriculture and Food Security*. Practical Action, London, 35–50.

Dewaelheyns, V., Lerouge, F., Rogge, E., & Vranken, L. (2014). *Garden Space: Mapping Trade-Offs and the Adaptive Capacity of Home Food Production*. Belgium: Division of Bioeconomics, University of Leuven.

Diehl, J. A., Sweeney, E., Wong, B., Sia, C. S., Yao, H., & Prabhudesai, M. (2020). Feeding cities: Singapore's approach to land use planning for urban agriculture. *Global Food Security*, 26, 100377. https://doi.org/10.1016/j.gfs.2020.100377

Diekmann, L. O., Gray, L. C., & Thai, C. L. (2020). More than Food: The social benefits of localized Urban food systems. *Frontiers in Sustainable Food Systems*, 4. https://www.frontiersin.org/article/10.3389/fsufs.2020.534219

Dorst, H., van der Jagt, A., Raven, R., & Runhaar, H. (2019). Urban greening through nature-based solutions-Key characteristics of an emerging concept. *Sustainable Cities and Society*, 49, 101620.

Drescher, A. W., Holmer, R. J., & Iaquinta, D. L. (2006). Urban homegardens and allotment gardens for sustainable livelihoods: Management strategies and institutional environments. In B. M. Kumar & P. K. R. Nair (Eds.), *Tropical Homegardens: A Time-Tested Example of Sustainable Agroforestry* (pp. 317–338). Springer Netherlands. https://doi.org/10.1007/978-1-4020-4948-4_18

Edmondson, J. L., Childs, D. Z., Dobson, M. C., Gaston, K. J., Warren, P. H., & Leake, J. R. (2020). Feeding a city-Leicester as a case study of the importance of allotments for horticultural production in the UK. *Science of the Total Environment*, 705, 135930.

Evans, D. L., Falagán, N., Hardman, C. A., Kourmpetli, S., Liu, L., Mead, B. R., & Davies, J. A. C. (2022). Ecosystem service delivery by urban agriculture and green infrastructure - a systematic review. *Ecosystem Services*, 54, 101405. https://doi.org/10.1016/j.ecoser.2022.101405

Faivre, N., Fritz, M., Freitas, T., de Boissezon, B., & Vandewoestijne, S. (2017). Nature-Based Solutions in the EU: Innovating with Nature to address social, economic and environmental challenges. *Environmental Research*, 159, 509–518. https://doi.org/10.1016/j.envres.2017.08.032

Ferdous, Z., Datta, A., Anal, A. K., Anwar, M., & Khan, A. M. R. (2016). Development of home garden model for year round production and consumption for improving resource-poor household food security in Bangladesh. *NJAS-Wageningen Journal of Life Sciences*, 78, 103–110.

Gallaher, C. M. (2017). Regreening Kibera: How urban agriculture changed the physical and social environment of a large slum in Kenya. *Global Urban Agriculture*, 171–183. Wallingford: CABI.

Gallaher, C. M., WinklerPrins, A. M., Njenga, M., & Karanja, N. K. (2015). Creating space: Sack gardening as a livelihood strategy in the Kibera slums of Nairobi, Kenya. *Journal of Agriculture, Food Systems, and Community Development*. https:// doi.org/10.5304/jafscd.2015.052.006

Gauthier, E. (2012). "Green" food processing technologies: Factors affecting consumers' acceptance. In *Green technologies in food production and processing* (pp. 615–641). Springer.

Gittleman, M., Farmer, C. J. Q., Kremer, P., & McPhearson, T. (2017). Estimating stormwater runoff for community gardens in New York City. *Urban Ecosystems*, 20(1), 129–139. https://doi.org/10.1007/s11252-016-0575-8

Glover, T. D. (2021). Healthy garden plots? Harvesting stories of social connectedness from community gardens. *International Journal of Environmental Research and Public Health*, 18(11), 5747.

Goldstein, B., Hauschild, M., Fernández, J., & Birkved, M. (2016). Testing the environmental performance of urban agriculture as a food supply in northern climates. *Journal of Cleaner Production*, 135, 984–994. https://doi.org/10.1016/j.jclepro.2016.07.004

Gómez Martín, E., Giordano, R., Pagano, A., van der Keur, P., & Máñez Costa, M. (2020). Using a system thinking approach to assess the contribution of Nature based solutions to sustainable development goals. *Science of The Total Environment*, 738, 139693. https://doi.org/10.1016/j.scitotenv.2020.139693

Goodman, D. (2003). The quality 'turn' and alternative food practices: Reflections and agenda. *Journal of Rural Studies*, 1(19), 1–7.

Grando, S., Carey, J., Hegger, E., Jahrl, I., & Ortolani, L. (2017). Short Food Supply Chains in Urban areas: Who takes the lead? Evidence from three cities across Europe. *Urban Agriculture & Regional Food Systems*, 2(1), urbanag2016.05.0002. https://doi.org/10.2134/urbanag2016.05.0002

Grard, B. J.-P., Chenu, C., Manouchehri, N., Houot, S., Frascaria-Lacoste, N., & Aubry, C. (2017). Rooftop farming on urban waste provides many ecosystem services. *Agronomy for Sustainable Development*, 38(1), 2. https://doi.org/10.1007/s13593-017-0474-2

Grigorescu, I., Popovici, E.-A., Mocanu, I., Sima, M., Dumitraşcu, M., Damian, N., Mitrică, B., & Dumitrică, C. (2022). Urban and Peri-Urban agriculture as a nature-based solution to support food supply, health and Wellbeing in Bucharest Metropolitan area during the COVID-19 Pandemic. In D. La Rosa & R. Privitera (Eds.), *Innovation in Urban and Regional Planning* (pp. 29–37). Springer International Publishing. https://doi.org/10.1007/978-3-030-96985-1_4

Hamilton, A. J., Burry, K., Mok, H.-F., Barker, S. F., Grove, J. R., & Williamson, V. G. (2014). Give peas a chance? Urban agriculture in developing countries. A review. *Agronomy for Sustainable Development*, 34(1), 45–73.

Hashimoto, S., Sato, Y., & Morimoto, H. (2019). Public-private collaboration in allotment garden operation has the potential to provide ecosystem services to urban dwellers more efficiently. *Paddy and Water Environment*, 17(3), 391–401. https://doi.org/10.1007/s10333-019-00734-1

Hess, D., & Winner, L. (2007). Enhancing justice and sustainability at the local level: Affordable policies for urban governments. *Local Environment*, 12(4), 379–395.

Hoover, B. (2013). White spaces in black and Latino places: Urban agriculture and food sovereignty. *Journal of Agriculture, Food Systems, and Community Development*, 3(4), 109–115.

Horst, M., McClintock, N., & Hoey, L. (2017). The intersection of planning, urban agriculture, and food justice: A review of the literature. *Journal of the American Planning Association*, 83(3), 277–295.

Hovorka, A. J. (2005). The (Re) production of gendered positionality in Botswana's commercial Urban Agriculture Sector. *Annals of the Association of American Geographers*, 95(2), 294–313. https://doi.org/10.1111/j.1467-8306.2005.00461.x

Hsu, J. P. (2019). Public pedagogies of edible verge gardens: Cultivating streetscapes of care. *Policy Futures in Education*, 17(7), 821–843.

Kato, Y. (2020). Gardening in times of urban transitions: Emergence of entrepreneurial cultivation in Post-Katrina New Orleans. *City & Community*, 19(4), 987–1010. https://doi.org/10.1111/cico.12476

Kim, J. E. (2017). Fostering behaviour change to encourage low-carbon food consumption through community gardens. *International Journal of Urban Sciences*, 21(3), 364–384. https://doi.org/10.1080/12265934.2017.1314191

Kingsley, J., Egerer, M., Nuttman, S., Keniger, L., Pettitt, P., Frantzeskaki, N., Gray, T., Ossola, A., Lin, B., Bailey, A., Tracey, D., Barron, S., & Marsh, P. (2021). Urban agriculture as a nature-based solution to address socio-ecological challenges in Australian cities. *Urban Forestry & Urban Greening*, 60, 127059. https://doi.org/10.1016/j.ufug.2021.127059

Kirkpatrick, J. B., & Davison, A. (2018). Home-grown: Gardens, practices and motivations in urban domestic vegetable production. *Landscape and Urban Planning*, 170, 24–33.

Koont, S. (2007). Urban agriculture in Cuba: Of, by, and for the Barrio. *Nature, Society, and Thought*, 20(3/4), 311.

Koont, S. (2008). A Cuban success story: Urban agriculture. *Review of Radical Political Economics*, 40(3), 285–291.

Koont, S. (2011). *Sustainable urban agriculture in Cuba*. University Press, Florida.

Korth, M., Stewart, R., Langer, L., Madinga, N., Da Silva, N. R., Zaranyika, H., van Rooyen, C., & de Wet, T. (2014). What are the impacts of urban agriculture programs on food security in low and middle-income countries: A systematic review. *Environmental Evidence*, 3(1), 1–10.

Kortright, R., & Wakefield, S. (2011). Edible backyards: A qualitative study of household food growing and its contributions to food security. *Agriculture and Human Values*, 28(1), 39–53.

Krishnan, S., Nandwani, D., Smith, G., & Kankarta, V. (2016). Sustainable Urban Agriculture: A Growing Solution to Urban Food Deserts. In D. Nandwani (Ed.), *Organic Farming for Sustainable Agriculture* (pp. 325–340). Springer International Publishing. https://doi.org/10.1007/978-3-319-26803-3_15

Langemeyer, J., Camps-Calvet, M., Calvet-Mir, L., Barthel, S., & Gómez-Baggethun, E. (2018). Stewardship of urban ecosystem services: Understanding the value(s) of urban gardens in Barcelona. *Landscape and Urban Planning*, 170, 79–89. https://doi.org/10.1016/j.landurbplan.2017.09.013

Langemeyer, J., Madrid-Lopez, C., Mendoza Beltran, A., & Villalba Mendez, G. (2021). Urban agriculture-A necessary pathway towards urban resilience and global sustainability? *Landscape and Urban Planning*, 210, 104055. https://doi.org/10.1016/j.landurbplan.2021.104055

Langemeyer, J., Wedgwood, D., McPhearson, T., Baró, F., Madsen, A. L., & Barton, D. N. (2020). Creating urban green infrastructure where it is needed - A spatial ecosystem service-based decision analysis of green roofs in Barcelona. *Science of The Total Environment*, 707, 135487. https://doi.org/10.1016/j.scitotenv.2019.135487

Langergraber, G., Pucher, B., Simperler, L., Kisser, J., Katsou, E., Buehler, D., Mateo, M. C. G., & Atanasova, N. (2020). Implementing nature-based solutions for creating a resourceful circular city. *Blue-Green Systems*, 2(1), 173–185. https://doi.org/10.2166/bgs.2020.933

Lin, B. B., & Egerer, M. H. (2017). Urban agriculture: An opportunity for biodiversity and food provision in urban landscapes. In A. Ossola & J. Niemelä (Eds.), *Urban Biodiversity: From Research to Practice* (71–86). Routledge.

Lin, B. B., Philpott, S. M., & Jha, S. (2015). The future of urban agriculture and biodiversity-ecosystem services: Challenges and next steps. *Basic and Applied Ecology*, 16(3), 189–201.

Lin, B. B., Philpott, S. M., Jha, S., & Liere, H. (2017). Urban agriculture as a productive green infrastructure for environmental and social wellbeing. In P. Yok Tan & C. Yung Jim (Eds.), *Greening Cities: Forms and Functions*. (pp. 155–179). Springer.

London, J. K., Cutts, B. B., Schwarz, K., Schmidt, L., & Cadenasso, M. L. (2021). Unearthing the entangled roots of urban agriculture. *Agriculture and Human Values*, 38(1), 205–220. https://doi.org/10.1007/s10460-020-10158-x

Loram, A., Tratalos, J., Warren, P. H., & Gaston, K. J. (2007). Urban domestic gardens (X): The extent & structure of the resource in five major cities. *Landscape Ecology*, 22(4), 601–615.

Lovell, R., Husk, K., Bethel, A., & Garside, R. (2014). What are the health and wellbeing impacts of community gardening for adults and children: A mixed method systematic review protocol. *Environmental Evidence*, 3(1), 1–13.

Maćkiewicz, B., & Asuero, R. P. (2021). Public versus private: Juxtaposing urban allotment gardens as multifunctional Nature-based Solutions. Insights from Seville. *Urban Forestry & Urban Greening*, 65, 127309.

Malan, N. (2015). Urban farmers and urban agriculture in Johannesburg: Responding to the food resilience strategy. Agrekon, 54(2), 51–75. https://doi.org/10.1080/03031853.2015.1072997

Malberg Dyg, P., Christensen, S., & Peterson, C. J. (2020). Community gardens and wellbeing amongst vulnerable populations: A thematic review. *Health Promotion International*, 35(4), 790–803. https://doi.org/10.1093/heapro/daz067

Mancebo, F. (2018). Gardening the city: Addressing sustainability and adapting to global warming through urban agriculture. *Environments*, 5(3), 38.

Manríquez-Altamirano, A., Sierra-Pérez, J., Muñoz, P., & Gabarrell, X. (2020). Analysis of urban agriculture solid waste in the frame of circular economy: Case study of tomato crop in integrated rooftop greenhouse. *Science of The Total Environment*, 734, 139375. https://doi.org/10.1016/j.scitotenv.2020.139375

Marques, P., Silva, A. S., Quaresma, Y., Manna, L. R., de Magalhães Neto, N., & Mazzoni, R. (2021). Home gardens can be more important than other urban green infrastructure for mental wellbeing during COVID-19 pandemics. *Urban Forestry & Urban Greening*, 64, 127268. https://doi.org/10.1016/j.ufug.2021.127268

Marshall, A. J., Grose, M. J., & Williams, N. S. G. (2020). Of mowers and growers: Perceived social norms strongly influence verge gardening, a distinctive civic greening practice. *Landscape and Urban Planning*, 198, 103795. https://doi.org/10.1016/j.landurbplan.2020.103795

Marshall, F. (2017). Why peri-Urban ecosystem services matter for Urban policy. https://www.researchgate.net/publication/331158518_Why_Peri-urban_Ecosystem_Services_Matter_for_Urban_Policy

Masset, E. (2011). A systematic review of agricultural interventions that aim to improve nutritional status of children. London: EPPI-Centre, Social Science Research Unit, Institute of Education, University of London.

Matsuzawa, T., & Kohsaka, R. (2022). Preliminary Experimental Trial of Effects of Lattice Fence Installation on Honey Bee Flight Height as Implications for Urban Beekeeping Regulations. *Land*, 11(1), Article 1. https://doi.org/10.3390/land11010019

McClintock, N. (2010). Why farm the city? Theorizing urban agriculture through a lens of metabolic rift. *Cambridge Journal of Regions, Economy and Society*, 3(2), 191–207.

Mcdougall, R., Rader, R., & Kristiansen, P. (2020). Urban agriculture could provide 15% of food supply to Sydney, Australia, under expanded land use scenarios. *Land Use Policy*, 94, 104554.

McGranahan, G., & Satterthwaite, D. (2000). Environmental health or ecological sustainability? Reconciling the brown and green agendas in urban development. In Sustainable cities in developing countries (pp. 73–90). Earthscan Publications Ltd. https://scholar.google.com/scholar_lookup?title=Environmental+health+or+ecological+sustainability?+Reconciling+the+Brown+and+Green+agendas+in+urban+development&author=McGranahan,+G.&author=Satterwaithe,+D.&publication_year=2000&pages=73%E2%80%9390

Mert-Cakal, T., & Miele, M. (2020). 'Workable utopias' for social change through inclusion and empowerment? Community supported agriculture (CSA) in Wales as social innovation. *Agriculture and Human Values*, 37(4), 1241–1260. https://doi.org/10.1007/s10460-020-10141-6

Miura, S., Kunii, O., & Wakai, S. (2003). Home gardening in urban poor communities of the Philippines. *International Journal of Food Sciences and Nutrition*, 54(1), 77–88.

Mohan, S. V., Hemalatha, M., Amulya, K., Velvizhi, G., Chiranjeevi, P., Sarkar, O., Kumar, A. N., Krishna, K. V., Modestra, J. A., Dahiya, S., Yeruva, D. K., Butti, S. K., Sravan, J. S., Chatterjee, S., & Kona, R. (2020). Decentralized Urban Farming Through Keyhole Garden: A Case Study with Circular Economy and Regenerative Perspective. *Materials Circular Economy*, 2(1), 12. https://doi.org/10.1007/s42824-020-00011-1

Mok, H.-F., Williamson, V. G., Grove, J. R., Burry, K., Barker, S. F., & Hamilton, A. J. (2014). Strawberry fields forever? Urban agriculture in developed countries: a review. *Agronomy for Sustainable Development*, 34(1), 21–43. https://doi.org/10.1007/s13593-013-0156-7

Mougeot, L. J. (2005). *Agropolis: The social, political, and environmental dimensions of urban agriculture.* IDRC.

Muñoz-Erickson, T. A., Campbell, L. K., Childers, D. L., Grove, J. M., Iwaniec, D. M., Pickett, S. T., Romolini, M., & Svendsen, E. S. (2016). Demystifying governance and its role for transitions in urban social-ecological systems. *Ecosphere*, 7(11), e01564.

Nesshöver, C., Assmuth, T., Irvine, K. N., Rusch, G. M., Waylen, K. A., Delbaere, B., Haase, D., Jones-Walters, L., Keune, H., Kovacs, E., Krauze, K., Külvik, M., Rey, F., van Dijk, J., Vistad, O. I., Wilkinson, M. E., & Wittmer, H. (2017). The science, policy and practice of nature-based solutions: An interdisciplinary perspective. *Science of The Total Environment*, 579, 1215–1227. https://doi.org/10.1016/j.scitotenv.2016.11.106

Newell, J. P., Foster, A., Borgman, M., & Meerow, S. (2022). Ecosystem services of urban agriculture and prospects for scaling up production: A study of Detroit. *Cities*, 125, 103664. https://doi.org/10.1016/j.cities.2022.103664

Newman, P. (2009). Bridging the Green and Brown Agendas. In *Planning Sustainable Cities-Global Report on Human Settlements 2009* (pp. 113–131). Earthscan.

Nicholls, E., Ely, A., Birkin, L., Basu, P., & Goulson, D. (2020). The contribution of small-scale food production in urban areas to the sustainable development goals: A review and case study. *Sustainability Science*, 15(6), 1585–1599. https://doi.org/10.1007/s11625-020-00792-z

Okvat, H. A., & Zautra, A. J. (2014). Sowing seeds of resilience: Community gardening in a post-disaster context. In *Greening in the red zone* (pp. 73–90). Springer.

Orsini, F., Gasperi, D., Marchetti, L., Piovene, C., Draghetti, S., Ramazzotti, S., Bazzocchi, G., & Gianquinto, G. (2014). Exploring the production capacity of rooftop gardens (RTGs) in urban agriculture: The potential impact on food and nutrition security, biodiversity and other ecosystem services in the city of Bologna. *Food Security*, 6(6), 781–792.

Orsini, F., Kahane, R., Nono-Womdim, R., & Gianquinto, G. (2013). Urban agriculture in the developing world: A review. *Agronomy for Sustainable Development*, 33, 695–720.

Ostrom, M. R. (2007). Community supported agriculture as an agent of change. Remaking the North American Food System: *Strategies for Sustainability*, 99–120.

Paiho, S., Wessberg, N., Pippuri-Mäkeläinen, J., Mäki, E., Sokka, L., Parviainen, T., Nikinmaa, M., Siikavirta, H., Paavola, M., & Antikainen, M. (2021). Creating a Circular City-An analysis of potential transportation, energy and food solutions in a case district. *Sustainable Cities and Society*, 64, 102529.

Palassio, C., & Wilcox, A. (2005). *The Edible City*. Coach House Books.

Pascucci, S. (2020). *Building natural resource networks: Urban agriculture and the circular economy*. Burleigh Dodds, Burleigh Dodds Series in Agricultural Science, 1–20.

Passidomo, C. (2014). Whose right to (farm) the city? Race and food justice activism in post-Katrina New Orleans. *Agriculture and Human Values*, 31(3), 385–396. https://doi.org/10.1007/s10460-014-9490-x

Patenković, A., Tanasković, M., Erić, P., Erić, K., Mihajlović, M., Stanisavljević, L., & Davidović, S. (2022). Urban ecosystem drives genetic diversity in feral honey bee. *Scientific Reports*, 12(1), Article 1. https://doi.org/10.1038/s41598-022-21413-y

Pauleit, S., Zölch, T., Hansen, R., Randrup, T. B., & van den Bosch, C. K. (2017). Nature-based solutions and climate change-four shades of green. In Nature-*Based solutions to climate change adaptation in urban areas* (pp. 29–49). Springer, Cham.

Peprah, K., Amoah, S. T., & Akongbangre, J. N. (2014). Sack farming: Innovation for land scarcity farmers in Kenya and Ghana. *Int Ournal Innov Res Stud*, 3(5), 30–44.

Pettygrove, M., & Ghose, R. (2018). From "Rust Belt" to "Fresh Coast": Remaking the city through food justice and Urban agriculture. *Annals of the American Association of Geographers*, 108(2), 591–603. https://doi.org/10.1080/24694452.2017.1402672

Pfeiffer, A., Silva, E., & Colquhoun, J. (2015). Innovation in urban agricultural practices: Responding to diverse production environments. *Renewable Agriculture and Food Systems*, 30(1), 79–91.

Pinheiro, A., & Govind, M. (2020). Emerging global trends in Urban agriculture research: A scientometric analysis of peer-reviewed journals. *Journal of Scientometric Research*, 9(2), 163–173. https://doi.org/10.5530/jscires.9.2.20

Poot-Pool, W. S., Cetzal-Ix, W., Basu, S. K., Noguera-Savelli, E., & Noh-Contreras, D. G. (2018). Urban home gardens: A sustainable conservation model for local plants based on Mexican Urban agri-horticultural practices. In *Urban Horticulture* (pp. 73–88). Springer.

Poulsen, M. N., McNab, P. R., Clayton, M. L., & Neff, R. A. (2015). A systematic review of urban agriculture and food security impacts in low-income countries. *Food Policy*, 55, 131–146.

Puigdueta, I., Aguilera, E., Cruz, J. L., Iglesias, A., & Sanz-Cobena, A. (2021). Urban agriculture may change food consumption towards low carbon diets. *Global Food Security*, 28, 100507.

Reynolds, K. (2015). Disparity despite diversity: Social injustice in New York City's Urban agriculture system. *Antipode*, 47(1), 240–259. https://doi.org/10.1111/anti.12098

Reynolds, K., & Cohen, N. (2016). *Beyond the Kale: Urban Agriculture and Social Justice Activism in New York City*. University of Georgia Press.

Rich, M., Viljoen, A., & Rich, K. M. (2015).The Healing City's social and Therapeutic horticulture as a new Dimension of Urban Agriculture. https://www.researchgate.net/publication/283236532_ THE_'HEALING_CITY'_-_SOCIAL_AND_THERAPEUTIC_HORTICULTURE_AS_A_NEW_ DIMENSION_OF_URBAN_AGRICULTURE.

Rolf, W. (2020). *Peri-urban farmland included in green infrastructure strategies promotes transformation pathways towards sustainable urban development* [PhD Thesis]. Universität Potsdam.

Sanyé-Mengual, E., Orsini, F., Oliver-Solà, J., Rieradevall, J., Montero, J. I., & Gianquinto, G. (2015). Techniques and crops for efficient rooftop gardens in Bologna, Italy. *Agronomy for Sustainable Development*, 35(4), 1477–1488.

Sartison, K., & Artmann, M. (2020). Edible cities - An innovative nature-based solution for urban sustainability transformation? An explorative study of urban food production in German cities. *Urban Forestry & Urban Greening*, 49, 126604. https://doi.org/10.1016/j.ufug.2020.126604

Säumel, I., Reddy, S. E., & Wachtel, T. (2019). Edible City Solutions-One Step Further to Foster Social Resilience through Enhanced Socio-Cultural Ecosystem Services in Cities. *Sustainability*, 11(4). https:// doi.org/10.3390/su11040972

Säumel, I., Weber, F., & Kowarik, I. (2016). Toward livable and healthy urban streets: Roadside vegetation provides ecosystem services where people live and move. *Environmental Science & Policy*, 62, 24–33.

Scharf, N., Wachtel, T., Reddy, S. E., & Säumel, I. (2019). Urban Commons for the Edible City-First Insights for Future Sustainable Urban Food Systems from Berlin, Germany. *Sustainability*, 11(4). https://doi. org/10.3390/su11040966

Schmutz, U., Kneafsey, M., Kay, C. S., Doernberg, A., & Zasada, I. (2018). Sustainability impact assessments of different urban short food supply chains: Examples from London, UK. *Renewable Agriculture and Food Systems*, 33(6), 518–529.

Schoen, V., Blythe, C., Caputo, S., Fox-Kämper, R., Specht, K., Fargue-Lelièvre, A., Cohen, N., Poniży, L., & Fedeńczak, K. (2021). "We Have Been Part of the Response": The Effects of COVID-19 on Community and Allotment Gardens in the Global North. *Frontiers in Sustainable Food Systems*, 5.

Seddon, N., Turner, B., Berry, P., Chausson, A., & Girardin, C. A. J. (2019). Grounding nature-based climate solutions in sound biodiversity science. *Nature Climate Change*, 9(2), 84–87. https://doi.org/10.1038/ s41558-019-0405-0

Simon, D. (Ed.). (2016). *Rethinking sustainable cities*: Accessible, green and fair. Policy Press.

Skar, S. L. G., Pineda-Martos, R., Timpe, A., Pölling, B., Bohn, K., Külvik, M., Delgado, C., Pedras, C. M. G., Paço, T. A., & Ćujić, M. (2020). Urban agriculture as a keystone contribution towards securing sustainable and healthy development for cities in the future. *Blue-Green Systems*, 2(1), 1–27.

Slater, R. J. (2001). Urban agriculture, gender and empowerment: An alternative view. *Development Southern Africa*, 18(5), 635–650. https://doi.org/10.1080/03768350120097478

Sowińska-Świerkosz, B., Michalik-Śnieżek, M., & Bieske-Matejak, A. (2021). Can Allotment Gardens (AGs) Be Considered an Example of Nature-Based Solutions (NBS) Based on the Use of Historical Green Infrastructure? *Sustainability*, 13(2), Article 2. https://doi.org/10.3390/su13020835

Speak, A. F., Mizgajski, A., & Borysiak, J. (2015). Allotment gardens and parks: Provision of ecosystem services with an emphasis on biodiversity. *Urban Forestry & Urban Greening*, 14(4), 772–781. https://doi. org/10.1016/j.ufug.2015.07.007

Stewart, R., Korth, M., Langer, L., Rafferty, S., Da Silva, N. R., & van Rooyen, C. (2013). What are the impacts of urban agriculture programs on food security in low and middle-income countries? *Environmental Evidence*, 2(1), 7. https://doi.org/10.1186/2047-2382-2-7

Sustainable agriculture in Cuba: Urban farming and "Organopónicos". (2021, December 14). Panoramas. https://www.panoramas.pitt.edu/health-and-society/sustainable-agriculture-cuba-urban-farming-and-% E2%80%9Corganop%C3%B3nicos%E2%80%9D

Taylor, J. R., & Lovell, S. T. (2014). Urban home food gardens in the Global North: Research traditions and future directions. *Agriculture and Human Values*, 31(2), 285–305.

Taylor, J. R., & Lovell, S. T. (2015). Urban home gardens in the Global North: A mixed methods study of ethnic and migrant home gardens Renewable *Agriculture and Food Systems*, in Chicago, IL. 30(1), 22–32.

Thomaier, S., Specht, K., Henckel, D., Dierich, A., Siebert, R., Freisinger, U. B., & Sawicka, M. (2015). Farming in and on urban buildings: Present practice and specific novelties of Zero-Acreage Farming (ZFarming). *Renewable Agriculture and Food Systems*, 30(1), 43–54.

Thomasson, D. A. (1994). Montserrat kitchen gardens: Social functions and development potential. *Caribbean Geography*, 5(1), 20.

Ulm, K., & Ayllon, M. J. Z. (2022). Veggies *in verges*: *A policy inventory for footpath food gardens across Greater* Sydney.

UN- Habitat. (2009). *Planning Sustainable Cities: Global Report on Human Settlements.* Earthscan. https://unhabitat.org/planning-sustainable-cities-global-report-on-human-settlements-2009

van der Jagt, A. P. N., Raven, R., Dorst, H., & Runhaar, H. (2020). Nature-based innovation systems. *Environmental Innovation and Societal Transitions*, 35, 202–216. https://doi.org/10.1016/j.eist.2019.09.005

Van Passel, S. (2013). Food miles to assess sustainability: A revision. *Sustainable Development*, 21(1), 1–17. https://doi.org/10.1002/sd.485

Van Veenhuizen, R. (Eds.). (2006). Introduction: Cities farming for the future. In *Cities Farming for Future, Urban Agriculture for Green and Productive Cities*, (pp. 2-17). RUAF Foundation, IDRC and IIRP, ETC-Urban Agriculture, Leusden, The Netherlands.

Vávra, J., Daněk, P., & Jehlička, P. (2018). What is the contribution of food self-provisioning towards environmental sustainability? A case study of active gardeners. Journal *of Cleaner Production*, 185, 1015–1023.

Wakeland, W., Cholette, S., & Venkat, K. (2012). Food transportation issues and reducing carbon footprint. In *Green technologies in food production and processing* (pp. 211–236). Springer.

Walthall, B. (2016). Strengthening city region food systems: Synergies between multifunctional peri-urban agriculture and short food supply chains: a local case study in berlin, Germany. In *Land Use Competition* (pp. 263–277). Springer.

White, M. M. (2011). Sisters of the soil: Urban gardening as resistance in Detroit. Race/Ethnicity: *Multidisciplinary Global* Contexts, 5(1), 13–28.

Wilhelm, J. A., & Smith, R. G. (2018). Ecosystem services and land sparing potential of urban and peri-urban agriculture: A review. *Renewable Agriculture and Food Systems*, 33(5), 481–494. https://doi.org/10.1017/S1742170517000205

World Bank. (2013). *Urban agriculture: Findings from four city case studies.* Washington. https://documents1.worldbank.org/curated/en/434431468331834592/pdf/807590NWP0UDS00Box0379817B00PUBLIC0.pdf

Wu, C., Li, X., Tian, Y., Deng, Z., Yu, X., Wu, S., Shu, D., Peng, Y., Sheng, F., & Gan, D. (2022). Chinese residents' perceived ecosystem services and disservices impacts behavioral intention for Urban community garden: An extension of the theory of planned behavior. *Agronomy*, 12(1), Article 1. https://doi.org/10.3390/agronomy12010193

WWF. (2019). *Nature in All Goals: How nature-based solutions can help us achieve all the sustainable development goals (Nature in All Goals,* p. 20). World Wide Fund. https://sdghelpdesk.unescap.org/sites/default/files/2019-10/nature_in_all_goals_publication__2019__1.pdf

Yacamán Ochoa, C., Matarán Ruiz, A., Mata Olmo, R., Macías Figueroa, Á., & Torres Rodríguez, A. (2020). Peri-Urban organic agriculture and short food supply chains as drivers for strengthening city/region food systems-two case studies in Andalucía, Spain. *Land*, 9(6), Article 6. https://doi.org/10.3390/land9060177

Zivkovic, A., Merchant, E. V., Nyawir, T., Hoffman, D. J., Simon, J. E., & Downs, S. (2022). Strengthening vegetable production and consumption in a Kenyan informal settlement: A feasibility and preliminary impact assessment of a Sack Garden Intervention. *Current Developments in Nutrition*, 6(5), nzac036. https://doi.org/10.1093/cdn/nzac036

4 Multifunctionality of Urban Agriculture in Burkina Faso and Niger

Environmental, Social, Economic, and Political Perspectives

Hamid El Bilali

INTRODUCTION

Burkina Faso and Niger are two landlocked countries in Sahelian West Africa. They have low human development (UNDP, 2019) and are affected by multiple forms of malnutrition (FAO et al., 2021). Agriculture is a leading sector for the economies of both countries, with a significant contribution to the gross domestic product and employment (World Bank, 2021). Nevertheless, agriculture is extensive, poorly mechanized, and vulnerable to climate change (El Bilali, 2021b). Climate change represents a challenge for agriculture (Mainardi, 2011; USAID, 2017) and is also an essential driver of poverty, livelihood vulnerability, and food insecurity (El Bilali, 2021a). Challenges faced by both countries have been exacerbated by political instability, social unrest and conflicts, which increased the number of internally displaced people (IDP). Niger experienced a decrease in the security level due to terrorism induced by non-state armed groups, which led to migratory crises both inside and outside the country (World Food Programme, 2019). The security context is even more critical in Burkina Faso and has progressively deteriorated in recent years, mainly in the eastern and northern regions, which triggered large population movements, including about 1.902 million IDP (OCHA, 2022).

Sub-Saharan Africa is experiencing one of the fastest urbanization worldwide (Parnell & Walawege, 2011; Tiffen, 2006; Weinreb et al., 2020); indeed, the rural population has grown slowly while the urban population has increased dramatically. Urbanization is driven by rural–urban migration. Weinreb et al. (2020) found that rural–urban migration is affected by climate change and variability. Urbanization has several implications, among others, in terms of diets and dietary habits (Casari et al., 2022). Urbanization also means an increased vulnerability to changes, including environmental ones (Parnell and Walawege, 2011). Orsini et al. (2013) argue that *"in many developing countries, the urbanization process goes along with increasing urban poverty and polluted environment, growing food insecurity and malnutrition, especially for children, pregnant and lactating women; and increasing unemployment"* (p. 695). Urbanization also creates additional challenges in terms of the management of food systems and food security (Zhou & Staatz, 2016). Therefore, alternative strategies and solutions are needed to ensure food and nutrition security in a sustainable way in West Africa in general and Burkina Faso and Niger in particular. Urban and peri-urban agriculture (UPA) is widely presented as one of these solutions.

DOI: 10.1201/9781003359425-4

UPA can be defined as

An industry that produces, processes and markets food and fuel, largely in response to the daily demand of consumers within a town, city or metropolis, on land and water dispersed throughout the urban and peri-urban area, applying intensive production methods, using and reusing natural resources and urban wastes, to yield a diversity of crops and livestock.

(UNDP, 1996)

It involves horticulture, animal husbandry, aquaculture, and other activities for producing food or other agricultural goods in urban districts and their surroundings. UPA encompasses all persons, groups, activities, sites, and economies that focus on production in urban and peri-urban areas as well as their positive environmental and societal externalities and co-benefits (Skar et al., 2020).

Urban agriculture has been addressed in many global sustainable development initiatives, such as the Sustainable Development Goals (SDGs), the New Urban Agenda (NUA), and the Milan Urban Food Policy Pact (MUFPP). At the United Nations' Sustainable Development Summit of September 2015, Member States adopted the 2030 Agenda for Sustainable Development, with 17 SDGs at its core (United Nations, 2015). The SDGs aim to end poverty, protect the planet, and ensure prosperity for all developing and developed countries. There is a whole SDG, viz. 11, dedicated to sustainable development in cities and communities (i.e., *Make cities and human settlements inclusive, safe, resilient, and sustainable*). Meanwhile, the NUA was adopted on 20 October 2016 at the United Nations Conference on Housing and Sustainable Urban Development (Habitat III) in Quito (Ecuador). It sets new global standards for planning, managing, and living sustainably in cities (United Nations, 2017). The NUA includes different references to urban agriculture. The SDGs and the NUA are closely connected, with the NUA seen as a vehicle for delivering the SDGs in urban settlements. Moreover, there is a high level of alignment between the individual SDGs and the commitments made in the NUA (Home in Place, 2022). The MUFPP was signed by more than 100 cities in October 2015 in Milan in the framework of EXPO 2015 (MUFPP, 2020). It is a non-binding agreement on urban food policies "designed by cities for cities" and signed so far by about 210 cities from all over the world with more than 450 million inhabitants. The MUFFP Framework for Action identified six areas of work streams: (1) governance, (2) sustainable diets and nutrition, (3) social and economic equity, (4) food production, (5) food supply and distribution, and (6) food waste. The MUFPP has built a monitoring framework process to assess the progress made by cities in achieving more sustainable food systems and, therefore, in the implementation of the Pact (FAO et al., 2019).

Several studies suggested that UPA can address many challenges, such as food insecurity, poverty, malnutrition, and health problems (Orsini et al., 2013; Zezza & Tasciotti, 2010). In their review, Orsini et al. (2013) show that UPA has several environmental, social, and economic benefits. In fact, it contributes to food security, livelihoods and income generation, social inclusion and reduction of gender inequalities, waste reduction, biodiversity and air quality improvement, and reduction of the environmental impacts related to food transport and storage. However, many previous studies show that several constraints hamper the development of urban agriculture in developing countries, such as those of sub-Saharan Africa/West Africa. These relate, inter alia, to insufficient government support, lack of restricted market access, land tenure insecurity, limited access to productive factors, and inequality issues (Houessou et al., 2020).

Despite its numerous benefits, few studies have investigated the importance of UPA in West Africa and the Sahel. In this context, the aim is to analyze the environmental, social, and economic impacts (both positive and negative) of UPA in Burkina Faso and Niger. A further aim is to investigate the governance and policy issues related to UPA and explore their effects on the development of UPA in both countries.

The work is based on a systematic literature review that follows the Preferred Reporting Items for Systematic Reviews and Meta-Analyses guidelines (Moher et al., 2009; Page et al., 2021). It draws upon a search of all documents indexed in the Web of Science, without indicating any range of years or time period, performed in June 2022 using the following search query: *(urban OR city OR* *town OR ville OR urbain) AND (agriculture OR farming OR garden OR horticulture OR growing OR animal OR élevage OR pastoralism) AND (Burkina OR "Niger" OR Sahel OR "West* Africa" OR "Afrique occidentale" OR "Afrique de l'Ouest").* The selected documents (Table 4.1) underwent two analysis steps. First, the dimensions of sustainability (environmental, social, economic, political) addressed were identified. Second, the analysis allowed going further by identifying the impacts (positive or negative) addressed. The classification of the identified impacts under sustainability dimensions and themes was informed by the SAFA (Sustainability Assessment of Food and Agriculture Systems) approach of FAO (2013, 2014), which considers the dimensions of environmental integrity, social well-being, economic resilience, and good governance.

TABLE 4.1
Overview of Documents Included in the Review

Document	Country/Region	City/Urban Area	Dimensions Addressed
Ouédraogo et al. (2022)	Burkina Faso	Bobo-Dioulasso	Economic
Bellwood-Howard et al. (2021)	Mali, Burkina Faso, and Ghana	Bamako (Mali), Ouagadougou (Burkina Faso), and Tamale (Ghana)	Economic
Korbéogo (2021)	Burkina Faso	Ouagadougou	Political
Manka'abusi et al. (2020)	Burkina Faso and Ghana	Ouagadougou and Tamale	Environmental
Ouédraogo et al. (2020)	Burkina Faso	Bobo-Dioulasso	Environmental
Tankari (2020)	Burkina Faso	Countrywide	Social
Dao et al. (2019)	Burkina Faso	Ouagadougou	Environmental
Lompo et al. (2019)	Burkina Faso	Bobo-Dioulasso	Environmental
Manka'abusi et al. (2019)	Burkina Faso	Ouagadougou	Environmental
Schlecht et al. (2019)	Burkina Faso	Ouagadougou	Environmental
Ouédraogo et al. (2019)	Burkina Faso	Bobo-Dioulasso	Environmental
Rossi and Dobigny (2019)	Niger	Niamey	Environmental, political
Roessler (2019)	Burkina Faso	Ouagadougou	Environmental, social
Akoto-Danso et al. (2019)	Ghana and Burkina Faso	Tamale (Ghana) and Ouagadougou (Burkina Faso)	Environmental
Stenchly et al. (2019)	Ghana and Burkina Faso	Tamale (Ghana) and Ouagadougou (Burkina Faso)	Environmental
Roessler et al. (2019)	Burkina Faso	Ouagadougou	Environmental
Bougnom et al. (2019)	Burkina Faso	Ouagadougou	Environmental, social
Dossa et al. (2019)	Burkina Faso	Bobo-Dioulasso	Environmental
Dao et al. (2018)	Burkina Faso	Ouagadougou	Environmental, social
Stenchly et al. (2018)	Burkina Faso	Ouagadougou	Environmental, economic
Korbéogo (2018)	Burkina Faso	Ouagadougou	Environmental, economic, social, political
Bellwood-Howard et al. (2018)	Ghana and Burkina Faso	Tamale (Ghana) and Ouagadougou (Burkina Faso)	Environmental, political
Robineau and Dugué (2018)	Burkina Faso	Bobo-Dioulasso	Environmental, political

(Continued)

TABLE 4.1 (*Continued*)
Overview of Documents Included in the Review

Document	Country/Region	City/Urban Area	Dimensions Addressed
Chagomoka et al. (2018)	Burkina Faso	Ouagadougou	Social
Häring et al. (2017)	Ghana and Burkina Faso	Tamale (Ghana) and Ouagadougou (Burkina Faso)	Environmental
Stenchly et al. (2017)	Burkina Faso	Ouagadougou	Environmental
Chagomoka et al. (2017)	Burkina Faso	Ouagadougou	Social
Stenchly et al. (2017)	Ghana and Burkina Faso	Tamale (Ghana) and Ouagadougou (Burkina Faso)	Environmental
Karg et al. (2016)	Ghana and Burkina Faso	Tamale (Ghana) and Ouagadougou (Burkina Faso)	Social
Roessler et al. (2016)	Ghana and Burkina Faso	Tamale (Ghana) and Ouagadougou (Burkina Faso)	Environmental
Di Leo et al. (2016)	Burkina Faso	Bobo-Dioulasso	Environmental, political
Kaboré et al. (2016)	Burkina Faso		Environmental
Robineau (2015)	Burkina Faso	Bobo-Dioulasso	Political
Dossa et al. (2015a)	Burkina Faso	Bobo-Dioulasso	Political
Dossa et al. (2015b)	Nigeria, Burkina Faso, and Mali	Kano (Nigeria), Bobo-Dioulasso (Burkina Faso), and Sikasso (Mali)	Environmental
Abdulkadir et al. (2015)	Nigeria, Burkina Faso, and Mali	Kano (Nigeria), Bobo-Dioulasso (Burkina Faso), and Sikasso (Mali)	Environmental, economic
Gomgnimbou et al. (2014)	Burkina Faso	Bobo-Dioulasso	Environmental, social
Orsini et al. (2013)	Global		Environmental, social, political
Diogo et al. (2013)	Niger	Niamey	Environmental
Boussini et al. (2012)	Burkina Faso	Ouagadougou	Economic, social
Lompo et al. (2012)	Burkina Faso	Bobo-Dioulasso	Environmental
Amadou et al. (2012)	Nigeria, Burkina Faso, and Mali	Kano (Nigeria), Bobo-Dioulasso (Burkina Faso), and Sikasso (Mali)	Environmental, social, economic
Probst et al. (2012)	Benin, Ghana, and Burkina Faso	Cotonou (Benin), Accra (Ghana), and Ouagadougou (Burkina Faso)	Economic, social
Kiba et al. (2012)	Burkina Faso	Ouagadougou	Environmental
Brinkmann et al. (2012)	Mali, Burkina Faso, Nigeria, and Niger	Sikasso (Mali), Bobo-Dioulasso (Burkina Faso), Kano (Nigeria), and Niamey (Niger)	Environmental, political
Sangare et al. (2012)	Burkina Faso	Bobo-Dioulasso	Environmental
Abdulkadir et al. (2012)	Nigeria, Burkina Faso, and Mali	Kano (Nigeria), Bobo-Dioulasso (Burkina Faso), and Sikasso (Mali)	Environmental, economic
Kiba et al. (2012)	Burkina Faso	Ouagadougou	Environmental, social
Probst et al. (2012)	Benin, Ghana, and Burkina Faso	Cotonou (Benin), Accra (Ghana), and Ouagadougou (Burkina Faso)	Environmental, social
Sanou et al. (2011)	Burkina Faso	Bobo-Dioulasso	Economic
Dossa et al. (2011a)	Nigeria, Burkina Faso, and Mali	Kano (Nigeria), Bobo-Dioulasso (Burkina Faso), and Sikasso (Mali)	Social, economic
Dossa et al. (2011b)	Nigeria, Burkina Faso, and Mali	Kano (Nigeria), Bobo-Dioulasso (Burkina Faso), and Sikasso (Mali)	Social, political
Abdu et al. (2011)	Nigeria, Burkina Faso, and Mali	Kano (Nigeria), Bobo-Dioulasso (Burkina Faso), and Sikasso (Mali)	Environmental, social
Diogo et al. (2011)	Niger	Niamey	Economic

(*Continued*)

TABLE 4.1 (*Continued*)
Overview of Documents Included in the Review

Document	Country/Region	City/Urban Area	Dimensions Addressed
Predotova et al. (2011)	Niger	Niamey	Environmental
Kaboré et al. (2011)	Burkina Faso	Ouagadougou	Environmental, social
Diogo et al. (2010a)	Niger	Niamey	Environmental
Diogo et al. (2010b)	Niger	Niamey	Environmental, social
Predotova et al. (2010a)	Niger	Niamey	Environmental
Predotova et al. (2010b)	Niger	Niamey	Environmental
Hayashi et al. (2010)	Niger	Niamey	Environmental, economic
Bernholt et al. (2009)	Niger	Niamey	Environmental
Graefe et al. (2008)	Niger	Niamey	Environmental, social, political
Millogo et al. (2008)	Burkina Faso	Ouagadougou and Bobo-Dioulasso	Environmental, social
Levasseur et al. (2007)	Ghana, Mali, Senegal, Togo, Burkina Faso, and Cameroon	Accra (Ghana), Bamako (Mali), Dakar (Senegal), Lomé (Togo), Ouagadougou (Burkina Faso), and Yaoundé (Cameroon)	Environmental, economic, social, political
Mattoni et al. (2007)	Burkina Faso	Bobo-Dioulasso	Environmental, economic, social
Rischkowsky et al. (2006)	Cameroon and Burkina Faso	Maroua (Cameroon) and Bobo-Dioulasso (Burkina Faso)	Social, economic
Cissé et al. (2005)	Côte d'Ivoire, Mali, Benin, Senegal, Niger, Mauritania, and Burkina Faso	Abidjan (Côte d'Ivoire), Bamako (Mali), Cotonou (Benin), Dakar (Senegal), Niamey (Niger), Nouakchott (Mauritania), and Ouagadougou (Burkina Faso)	Political
Thys et al. (2005)	Burkina Faso	Ouagadougou	Social, economic
Wang et al. (2005)	Burkina Faso	Ouagadougou	Social
Sidibe et al. (2004)	Burkina Faso	Bobo-Dioulasso	Environmental, economic
Freidberg (2001)	Burkina Faso	Bobo-Dioulasso	Environmental, economic, social, political

Source: Author.

ENVIRONMENTAL DIMENSION

UPA contributes to environmental integrity through its positive effects on the atmosphere, water, land/soil, biodiversity, materials, and energy (Table 4.2). Orsini et al. (2013) suggest that *"urban agriculture has ecological benefits by reducing the city waste, improving urban biodiversity and air quality, and overall reducing the environmental impact related to both food transport and storage"* (p. 695). However, in their analysis of urban of peri-urban agriculture (URA) in Niamey (Niger), Graefe et al. (2008) identified different environmental challenges and weaknesses and put that

> The URA livestock component depended heavily on the supply of live animals and feed from the rural hinterland. The application of organic and inorganic fertilizers in combination with the indiscriminate use of nutrient-loaded wastewater for irrigation purposes led to important nutrient surpluses in URA gardens. First calculations of partial nutrient balances indicated that gaseous emission of nitrogenous compounds as well as leaching of nitrogen, phosphorus and potassium to the groundwater might be large.

(p. 47)

TABLE 4.2

Overview of Selected Documents Addressing Environmental Issues

Environmental Themes	Sub-themes	Positive Impacts and Benefits	Negative Impacts and Risks	Document
Atmosphere	Greenhouse gases (GHG)	Addition of biochar reduces CO_2–C emissions from urban garden soils	Wastewater irrigation and fertilizer application increase GHG emissions from UPA fields	Manka'abusi et al. (2020)
		UPA lowers air temperatures in cities		Di Leo et al. (2016)
		Urban agriculture reduces the impact related to food transport and storage		Orsini et al. (2013)
		Improved animal feeding reduces the risk of N emissions		Diogo et al. (2013)
			Urban gardens generate important GHG emissions	Lompo et al. (2012)
			Emissions of GHG (ammonia, nitrous oxide, and carbon dioxide) from urban gardens	Predotova et al. (2010)
			Important gaseous emissions from urban gardens	Graefe et al. (2008)
Atmosphere	Air quality	UPA mitigates climate change impacts in cities		Di Leo et al. (2016)
		Urban agriculture improves air quality		Orsini et al. (2013)
Water	Water withdrawal		The use of wells and potable water for irrigation raises queries about sustainability	Bellwood-Howard et al. (2018)
Water	Water quality		Excessive use of manure in urban gardens can cause the pollution of water	Gomgnimbou et al. (2014)
			High risk of nutrient leaching and groundwater contamination in urban vegetable gardens	Sangare et al. (2012)
			Nitrogen and phosphorus leaching from urban gardens	Predotova et al. (2011)
			Leaching of nitrogen, phosphorus, and potassium into the groundwater	Graefe et al. (2008)
Land	Soil quality	Urban gardening increases organic matter content		Ouédraogo et al. (2020)
		Gypsum amendment reduces soil pH and improves soil structure in urban gardens irrigated with industrial wastewater		Dao et al. (2019)
		Widespread practice of agro-ecological production methods in urban gardens		Ouédraogo et al. (2019)
			Irrigation with wastewater increases soil salinity risk	Häring et al. (2017)
			Excessive use of manure in urban gardens can cause the pollution of soils	Gomgnimbou et al. (2014)

(Continued)

TABLE 4.2 (*Continued*)

Overview of Selected Documents Addressing Environmental Issues

Environmental Themes	Sub-themes	Positive Impacts and Benefits	Negative Impacts and Risks	Document
			The use of wastewater in urban gardens leads to the accumulation of N, P, and heavy metals in soils	Kiba et al. (2012)
			UPA leads to heavy metal accumulation in the soils of gardens irrigated with untreated wastewater	Abdu et al. (2011)
Biodiversity	Ecosystem diversity		UPA affects regulating ecosystem services provision	Stenchly et al. (2019)
			UPA affects species diversity	Stenchly et al. (2019)
			Usage of industrial wastewater can lead to soil salinization with negative consequences on soil properties and biota	Stenchly et al. (2017)
			UPA can affect the provisioning of ecosystem services	Stenchly et al. (2017)
		Urban agriculture improves urban biodiversity		Orsini et al. (2013)
Biodiversity	Species diversity		UPA can affect the functional diversity of beneficial weed communities	Stenchly et al. (2017)
		Urban agriculture improves urban biodiversity		Orsini et al. (2013)
			Urban agriculture decreases macroinvertebrates diversity and richness	Kaboré et al. (2016)
		UPA contributes to plant species richness and diversity		Bernholt et al. (2009)
	Genetic diversity		Erosion of local genetic resources	Roessler (2019)
Materials and energy	Material use		Inputs exceeded the recommended amounts of nutrients	Lompo et al. (2019)
			Low nutrient use efficiency	
			UPA can lead to nutrient depletion in the feed-supplying rural hinterland	Schlecht et al. (2019)
		Biochar application reduces P and K leaching	Leaching of nitrogen and phosphorus	Akoto-Danso et al. (2019)
			Depletion of potassium	
		Manure collected from urban ruminant production could contribute to urban crop production	Conversion ratios and energy use efficiency are very low for intensive urban ruminant production	Dossa et al. (2019)

(Continued)

TABLE 4.2 (*Continued*)

Overview of Selected Documents Addressing Environmental Issues

Environmental Themes	Sub-themes	Positive Impacts and Benefits	Negative Impacts and Risks	Document
			Poor phosphorus use efficiencies due to excess application	Abdulkadir et al. (2015)
			UPA can lead to long-term potassium depletion	Abdulkadir et al. (2015)
		Improved animal feeding increases nutrient use efficiency	UPA systems are characterized by excessive nutrient inputs	Diogo et al. (2013)
			Poor feed conversion efficiencies and wastage of nutrients in urban livestock farms	Diogo et al. (2010a)
			Surpluses of carbon (C), nitrogen (N), phosphorus (P), and potassium (K) in high-input urban gardens	Diogo et al. (2010b)
			Urban gardens have N and C surpluses due to the excessive application of manure and mineral fertilizers	Lompo et al. (2012)
			Livestock production is heavily dependent on feed supply from the rural hinterland	Graefe et al. (2008)
			Excessive application of fertilizers and use of wastewater for irrigation lead to important nutrient surpluses	Graefe et al. (2008)
Materials and energy	Waste reduction and disposal		UPA systems are characterized by poor handling of manure	Diogo et al. (2013)
		Urban livestock production contributes to the reuse and recycling of by-products such as crop residues, cereal brans, and cottonseed cake		Amadou et al. (2012)
			Nitrogen and carbon losses from dung/manure storage in urban gardens	Predotova et al. (2010)
			Excessive use of manure in urban gardens	Gomgnimbou et al. (2014)
		Urban agriculture reduces the city's waste		Orsini et al. (2013)

ATMOSPHERE AND CLIMATE CHANGE

The effects of UPA mainly relate to greenhouse gas (GHG) emissions (Manka'abusi et al., 2020) and, consequently, climate change. Manka'abusi et al. (2020) found that wastewater irrigation and fertilizer application increase the emissions of CO_2–C, N_2O–N, and NH_3–N from irrigated UPA fields in Ouagadougou (Burkina Faso) and Tamale (Ghana). As for the effects of biochar on GHG emissions, Manka'abusi et al. (2020) suggest that the addition of biochar contributes to reducing emissions from urban garden soils (p. 500), and this can also reduce the negative impacts of waste-water application on underground water quality by reducing nitrate leaching. Manka'abusi et al. (2019) showed that biochar application increased the average total biomass and marketable yield of amaranth, lettuce, and carrot irrigated with wastewater in urban vegetable gardens of Ouagadougou (Burkina Faso). Meanwhile, Di Leo et al. (2016) found that UPA (cf. green infrastructure) lowers air temperatures in urban areas of Bobo-Dioulasso (Burkina Faso) so it can contribute to improving air quality and mitigating the impacts of global warming in cities.

WATER RESOURCES

UPA affects both water withdrawal (Bellwood-Howard et al., 2018) and water quality (Bougnom et al., 2019; Dao et al., 2018; Gomgnimbou et al., 2014; Korbéogo, 2018). While the use of waste-water for irrigation raises concerns regarding food safety and quality, the use of potable water for irrigation in some cities, such as Tamale (Ghana) (Bellwood-Howard et al., 2018), raises queries over long-term sustainability. Similarly, using well water for irrigation in Ouagadougou (Burkina Faso) might cause the depletion of underground water resources (Bellwood-Howard et al., 2018). Gomgnimbou et al. (2014) found that most livestock farms in the urban and peri-urban areas of Bobo-Dioulasso (Burkina Faso) have no manure storage capacity so manure is primarily used for field fertilization; this might lead to an excessive use that causes the pollution of water and soils. Sangare et al. (2012) highlighted that, in vegetable gardens in Bobo-Dioulasso (Burkina Faso), water application exceeded plant requirements, crops tended to be strongly over-fertilized, nutrient supply exceeded crop requirements, and the partial factor productivity of nutrients tended to be low, which combined with significant observed rates of drainage imply a high risk of nutrient leaching and groundwater contamination. Kiba et al. (2012) underlined the increases in nutrient and heavy metals content observed in the 15–30 cm horizon of the soils of urban agriculture fields in Ouagadougou (Burkina Faso) that had been amended with non-sorted solid urban wastes for more than 10 years and pointed out the risks of element transfer to deeper horizons, which might cause water pollution.

BIODIVERSITY

There are linkages between UPA and species diversity (Bernholt et al., 2009; Kaboré et al., 2016; Stenchly et al., 2019), genetic diversity (Mattoni et al., 2007; Millogo et al., 2008; Roessler, 2019; Roessler et al., 2019; Sidibe et al., 2004), and ecosystem and landscape diversity (Bellwood-Howard et al., 2018; Brinkmann et al., 2012; Stenchly et al., 2019), which affects the provision of eco-system services (e.g., pollination, pest regulation/control). The selection decisions and trait prefer-ences of local breeders can affect the contribution of UPA to the genetic diversity of the livestock (Dossa et al., 2015b; Roessler, 2019), as in the case of sheep and goat farmers in Kano (Nigeria), Bobo-Dioulasso (Burkina Faso), and Sikasso (Mali) (Dossa et al., 2015b) or cattle and sheep breeders in the peri-urban area of Ouagadougou (Burkina Faso) (Roessler, 2019); while imported breeds can improve production performances, they can also lead to the erosion of local genetic resources. Kaboré et al. (2016) recorded the lowest macroinvertebrate diversity and richness in urban agriculture sites in Burkina Faso, compared to protected, extensive agriculture and intensive agriculture ones. They argue that the measurements of overall taxonomic richness and diversity efficiently detected the high impoverishment of the urban agriculture sites. Stenchly et al. (2019)

suggest that UPA affects species diversity and ecosystem services provision in Tamale (Ghana) and Ouagadougou (Burkina Faso). In particular, there are trade-offs between pest biocontrol and pollination regulating services. Stenchly et al. (2017) found that wastewater irrigation and related soil pH affected arthropod composition in urban fields in Ouagadougou (Burkina Faso), which may lead to negative effects on important and useful arthropod groups such as predators and pollinators. Furthermore, Stenchly et al. (2017) found that the management of urban okra fields in Ouagadougou (Burkina Faso) and Tamale (Ghana) led to changes in soil properties, which may alter the functional diversity of beneficial weed communities by potentially harming the provisioning of ecosystem services such as pollination and pest control. While UPA contributes to the provision of many eco-system services, it is also dependent on some services, such as pollination (Stenchly et al., 2018) so the changes in land use and landscape can negatively affect the performance of urban farms.

LAND MANAGEMENT AND SOIL FERTILITY

UPA practices can influence the content of soil organic carbon (SOC) (Ouédraogo et al., 2020) and, therefore, soil quality and fertility. Ouédraogo et al. (2020) found that additions of organic amendments in market gardening in Bobo-Dioulasso (Burkina Faso) increased SOC content and promoted SOC storage in the stable fraction, thus contributing to soil quality improvement, with positive effects in terms of climate change mitigation. Dao et al. (2019) found that the amendment of gypsum (rich in calcium) mitigates the effects of sodic alkaline industrial wastewater irrigation in urban agriculture of Ouagadougou (Burkina Faso) by reducing soil pH, reducing sodium absorption rate and improving the stability of soil thus soil structure. Häring et al. (2017) point out that irrigation with wastewater increased exchangeable sodium over time in urban vegetable gardens in Tamale (Ghana) and Ouagadougou (Burkina Faso) so the soil salinity risk. Ouédraogo et al. (2019) show that vegetable farms in the Bobo-Dioulasso region (Burkina Faso) are characterized by the wide-spread practice of agro-ecological production methods such as crop rotations and associations and organic fertilization, but they have weak and imprecise knowledge of the effects of UPA practices on the environment. Kiba et al. (2012) found that using solid waste and wastewater led to the accumulation of N, P, and heavy metals in urban garden soils in Ouagadougou. However, the effects on soil quality depend on fertilization strategies and practices. Abdu et al. (2011) found that there are positive balances of heavy metals, especially cadmium (Cd) and zinc (Zn) in urban vegetable gardens of Kano (Nigeria), Bobo-Dioulasso (Burkina Faso) and Sikasso (Mali) irrigated with untreated wastewater and warned that "*if such balances remain unchanged for another 10–20 years vegetables raised in these garden fields are likely to be unsuitable for human consumption*" (p. 387).

CYCLES OF NUTRIENTS AND MATERIALS

Urban agriculture, both crop production (horticulture) and animal husbandry, plays an important role in the cycles of materials – especially those of nitrogen (Abdulkadir et al., 2015; Akoto-Danso et al., 2019; Lompo et al., 2012, 2019), potassium (Abdulkadir et al., 2015; Akoto-Danso et al., 2019; Lompo et al., 2012, 2019), phosphorus (Abdulkadir et al., 2015; Akoto-Danso et al., 2019; Lompo et al., 2012), and carbon (Lompo et al., 2012, 2019) – as well as the use of energy (Dossa et al., 2019; Schlecht et al., 2019). Lompo et al. (2012, 2019) analyzed the flows of nitrogen, potassium, and carbon in urban vegetable gardens in Bobo-Dioulasso (Burkina Faso) and found that inputs exceeded the recommended amounts leading to horizontal annual surpluses and relatively low nutrient use efficiency. The surplus of nutrients is leached or emitted into the atmosphere thus contributing to GHG emissions (Lompo et al., 2012; Predotova et al., 2010) and climate change. Also, the storage of dung in urban gardens leads to the emissions of GHG such as ammonia, nitrous oxide, carbon dioxide, and methane (Predotova et al., 2010). The analysis of input and output of nutrients in urban and peri-urban livestock holdings of Ouagadougou (Burkina Faso) by Schlecht et al. (2019) suggests that farmers' feed supplied more nitrogen than animals required, and this can lead in the long run to nutrient depletion of the feed-supplying rural hinterland. They also pointed out that dietary imbalances between protein and energy supply contributed to suboptimal feed utilization at the individual animal level, thus reducing economic and biological performance. Likewise, Dossa et al. (2019) found that daily supplies of crude protein and metabolizable energy clearly exceeded the animals' requirements for maintenance plus growth in urban ruminant production farms in Bobo-Dioulasso (Burkina Faso) so that the feed and protein conversion ratios, as well as energy use efficiency, were very low. Diogo et al. (2010a) found that urban sheep/goat and cattle farms in Niamey (Niger) are characterized by poor conversion efficiencies of offered feed and wastage of nutrients. However, urban livestock production contributes to the reuse, thus the valorization, of by-products such as crop residues, cereal brans, and cottonseed cake (Amadou et al., 2012). Roessler et al. (2016) noticed that higher inputs do not necessarily lead to higher outputs in livestock systems whereas specialization leads to higher manure wastages in Ouagadougou (Burkina Faso) and Tamale (Ghana) and recommended linking livestock producers to crop farmers and livestock manure markets to enable recycling of resources and limit the negative externalities of specialized urban livestock production. Akoto-Danso et al. (2019) found that urban farmers' fertilization practices in Tamale (Ghana) and Ouagadougou (Burkina Faso) led to positive annual nutrient balances (surpluses) for nitrogen and phosphorus, which increases the risk of leaching, and negative balance (deficit) for potassium. They also showed that biochar application reduces K leaching.

SOCIAL DIMENSION

UPA affects livelihoods, food security, and nutrition as well as the health of both urban gardeners and urban dwellers (Table 4.3). UPA contributes to the decent livelihoods of tens of thousands of households. For instance, Thys et al. (2005) found that about a quarter of the households in

TABLE 4.3
Overview of Selected Documents Addressing Social Issues

Social Themes	Sub-themes	Positive Impacts and Benefits	Negative Impacts and Risks	Document
Decent livelihood	Quality of life	Livelihoods and food security of urban farms are less vulnerable to rainfall variability		Tankari (2020)
		UPA contributes to nutrition security and dietary diversity		Chagomoka et al. (2018)
			UPA causes several nuisances to neighbors (e.g., smells, flies, noise)	Gomgnimbou et al. (2014)
		Urban agriculture contributes to food security		Orsini et al. (2013)
		UPA contributes to food and nutrition security in West African capitals		Levasseur et al. (2007)
	Fair access to means of production		Land access and tenure Control and use of water resources	Korbéogo (2018)
Equity	Non-discrimination	Urban agriculture favors social inclusion		Orsini et al. (2013)
		UPA favors participation and inclusion		Dossa et al. (2011)
	Gender equality	Urban agriculture reduces gender inequalities		Orsini et al. (2013)

(Continued)

TABLE 4.3 (*Continued*)

Overview of Selected Documents Addressing Social Issues

Social Themes	Sub-themes	Positive Impacts and Benefits	Negative Impacts and Risks	Document
Human safety and health	Public health		Wastewater used represents a high risk for spreading bacteria and antimicrobial resistance	Bougnom et al. (2019)
			Wastewater used in irrigation does not meet safety standards Irrigation water and lettuce contaminated with *Salmonella* spp.	Dao et al. (2018)
			Urban herds can be a source of zoonotic diseases (e.g. tuberculosis, brucellosis)	Boussini et al. (2012)
		Certified organic vegetable production can have a positive impact on food safety in urban West Africa		Probst et al. (2012)
			The use of untreated wastewater for irrigation in urban gardens represents a risk to human health	Abdu et al. (2011)
			Presence of pathogens such as *Salmonella* in produce (lettuce) irrigated with wastewater in urban gardens	Diogo et al. (2010b)
			Urban producers frequently use banned or inappropriate pesticides in excessive quantities	
			Urban producers use polluted irrigation water, wastewater, and untreated sewage sludge	
			Remarkable link between UPA and malaria infections	Wang et al. (2005)

Ouagadougou are currently keeping livestock. Orsini et al. (2013) argue that urban agriculture favors both social inclusion/participation and the reduction of gender inequalities, as a high share of urban farmers are women. Dossa et al. (2011) found that participation in UPA is not affected by household socio-economic status in West Africa (Kano, Nigeria; Bobo-Dioulasso, Burkina Faso; Sikasso, Mali), which implies that UPA is an inclusive sector.

UPA also plays a vital role in many households' food and nutrition security. Urban agriculture contributes to food security by not only boosting the food supply in cities (Boussini et al., 2012; Chagomoka et al., 2018; Karg et al., 2016; Orsini et al., 2013) but also providing urban gardeners with the needed financial means to access food (Orsini et al., 2013). In particular, urban gardens are a valuable source of fresh perishable crops (Karg et al., 2016). Tankari (2020) found that rural farm households are more dependent on rainfall for their livelihoods than urban farm households in Burkina Faso so the food security status of urban agriculture households is less vulnerable to rainfall variability than rural ones. Chagomoka et al. (2018) point out that vegetable production in urban and peri-urban areas of Ouagadougou (Burkina Faso) contributes to household dietary diversity and helps address micronutrient deficiencies.

Nevertheless, UPA also implies some risks in terms of human safety and health (Bougnom et al., 2019a b; Diogo et al., 2010a; Levasseur et al., 2007), which are mainly due to the use of wastewater for irrigation. Referring to vegetable production in West African capital cities, Levasseur et al. (2007) put that *"producers frequently use banned or inappropriate pesticides in excessive quantities, and also polluted irrigation water, wastewater, and untreated sewage sludge"* (p. 245), which raise concerns in terms of food safety and human health. Bougnom et al. (2019) found that wastewater used for urban agriculture in Ouagadougou (Burkina Faso) represents a high risk for spreading bacteria and antimicrobial resistance among humans and animals. The bacteria found in wastewater samples relate to human pathogens that cause diseases such as gastroenteritis and diarrhea. Dao et al. (2018) found that *"in 60% of the cases, irrigation water did not meet the standards of the World Health Organization (WHO) for safe irrigation water, and in 30% of the cases, irrigation water was contaminated with Salmonella spp."*. What is even more alarming is that half of all lettuce samples, irrigated with wastewater in urban gardens of Ouagadougou (Burkina Faso), tested positive for *Salmonella* spp. Likewise, Diogo et al. (2010b) detected the presence of fecal pathogens such as *Salmonella* in produce (lettuce) irrigated with wastewater in urban gardens in Niamey (Niger). Referring to Ouagadougou (Burkina Faso), Wang et al. (2005) suggest that *"a remarkable link was found between urban agriculture activities, seasonal availability of water supply and the occurrence of malaria infections in this semi-arid area"*. As for animal production, Boussini et al. (2012) found that the prevalence of tuberculosis and brucellosis (two zoonotic diseases that can be transmitted from animals to humans) were high in urban and peri-urban dairy cattle farms in Ouagadougou (Burkina Faso) and warned that they could pose a major risk to human health and reduce significantly animal production and productivity.

UPA has become a marker of socio-economic dynamics in many cities, such as Ouagadougou (Korbéogo, 2018). There are interactions between UPA practices and the social context in which it takes place; UPA can be considered a product of the social context and environment, but it also contributes to shaping them. Korbéogo (2018) posits that differences in interests among the stakeholders involved in urban agriculture in Ouagadougou can lead to conflicts regarding, inter alia, land access and tenure, and control and use of water resources. Gomgnimbou et al. (2014) found that livestock farms in the urban and peri-urban areas of Bobo-Dioulasso (Burkina Faso) are located less than 50 m from houses so that neighbors enumerate different nuisances such as bad smells and the presence of flies and the noise. So, they suggest a better structuring for livestock activities, especially by optimizing management to reduce nuisances to neighboring residents.

ECONOMIC DIMENSION

UPA contributes to the generation of income for many urban households and the development of urban and peri-urban areas (Table 4.4). Korbéogo (2018) suggests that UPA represents an important source of revenue for various individuals and groups in Ouagadougou (Burkina Faso). Thys et al. (2005) found that income generation is one of the main motivations for households to keep livestock in Ouagadougou. Abdulkadir et al. (2015) posit that UPA generates important annual gross margins in Kano (Nigeria), Bobo-Dioulasso (Burkina Faso), and Sikasso (Mali) but profitability depends on fertilization practices and nutrient use efficiencies. Bellwood-Howard et al. (2021) found that the profitability of urban vegetable producers (cf. lettuce) in Bamako (Mali), Ouagadougou (Burkina Faso), and Tamale (Ghana) is higher during the hot, dry season. Meanwhile, Diogo et al. (2011) found that the prices, as well as gross margins of vegetables (viz. amaranth, cabbage, lettuce, and tomato) in Niamey (Niger), were generally higher during the rainy season (August–October) with respect to the cool, dry season (November–January). Also, Graefe et al. (2008) found that urban gardening is an economically beneficial occupation in Niamey. Referring again to the context of Niamey (Niger), Hayashi et al. (2010) suggest that the profitability of peri-urban horticulture is much higher than that of traditional, rural crop production. The high profitability of urban animal production might explain why a large share of peri-urban livestock producers in Ouagadougou are willing to pay for improved forages as cash crops

TABLE 4.4

Overview of Selected Documents Addressing Economic Issues

Economic Themes	Sub-themes	Positive Impacts and Benefits	Negative Impacts and Risks	Document
Investment	Profitability	High profitability during the hot, dry season		Bellwood-Howard et al. (2021)
		UPA is an important source of income		Abdulkadir et al. (2015)
		Forage marketing is a profitable activity in urban areas		Sanou et al. (2011)
		UPA profitability is high		Hayashi et al. (2010)
Vulnerability	Risk management		Income of urban gardeners is more vulnerable to the loss of pollination services than rural ones	Stenchly et al. (2018)
Local economy	Value creation	Important source of revenue for various individuals and groups		Korbéogo (2018)
		UPA contributes to income generation and local development		Abdulkadir et al. (2012)
		UPA contributes to job creation and economic development		Graefe et al. (2008)
		UPA is an important source of income in West African capitals		Levasseur et al. (2007)
		Urban livestock production contributes to income generation in cities		Thys et al. (2005)

(Ouédraogo et al., 2022); this, in turn, explains why forage marketing has become a profitable activity in cities such as Bobo-Dioulasso, Burkina Faso (Sanou et al., 2011). Abdulkadir et al. (2012) stress that UPA remains an important economic activity in the livelihoods of urban and peri-urban farmers in Kano (Nigeria), Bobo-Dioulasso (Burkina Faso), and Sikasso (Mali), and, for that, it is important to address constraints hindering its development and sustainability such as high costs of inputs, water shortages and lack of fertilizers in crop production systems, and feeding constraints and animal diseases in livestock production. Considering a rural–urban gradient of Ouagadougou (Burkina Faso), Stenchly et al. (2018) found that the income of urban households is more dependent on pollination services, which makes them more vulnerable to and affected by a decline in insect pollinators than rural households. Robineau and Dugué (2018) suggest that the development of urban agriculture in Bobo-Dioulasso (Burkina Faso) has been affected by many economic factors, including agricultural and regional market dynamics, access to commercial outlets, and availability of agricultural inputs.

POLITICAL DIMENSION

Different political and governance factors enable or hinder the development of UPA in Burkina Faso and Niger. Moreover, UPA represents different challenges in terms of policy and governance. The different patterns of urbanization of cities such as Niamey in Niger (Brinkmann et al., 2012;

Cissé et al., 2005; Rossi & Dobigny, 2019), Ouagadougou (Bellwood-Howard et al., 2018; Cissé et al., 2005; Korbéogo, 2018) and Bobo-Dioulasso (Brinkmann et al., 2012; Di Leo et al., 2016; Robineau, 2015; Robineau & Dugué, 2018) in Burkina Faso, and Tamale in Ghana (Bellwood-Howard et al., 2018) present real challenges for urban decision-makers and planners and might suggest that different governance tools will be needed to ensure sustainable growth and management of cities without sacrificing spaces dedicated to urban agriculture while managing the challenges that agriculture within and in the periphery of cities might imply. Referring to the examples of Sikasso (Mali), Bobo-Dioulasso (Burkina Faso), Kano (Nigeria), and Niamey (Niger), Brinkmann et al. (2012) posit that ex-ante land use planning and monitoring strategies are going to be more and more important in the context of the increasing scarcity of water and land suitable for agriculture. Comparing land use planning and zoning in Tamale (Ghana) and Ouagadougou (Burkina Faso), Bellwood-Howard et al. (2018) noted that

> In Ouagadougou, commercial production was concentrated in open-space farming sites, whereas in Tamale, it was more dispersed, with isolated space farms playing an unexpectedly important market role. This was attributed to Tamale's recent rapid expansion, combined with more relaxed planning implementation and a permissive legislative context.
>
> *(p. 34)*

and called for paying more attention to planning trajectories, which are shaped by historical and geographical contexts, to foster the contribution of UPA to food systems in West African cities.

Korbéogo (2018) suggests that urban agriculture has become a socio-political arena for a wide range of stakeholders (e.g., state representatives, experts, farmers) in Ouagadougou (Burkina Faso). The concerned stakeholders have different stakes, interests, visions, and sources of power and legitimacy, which sometimes leads to conflicts and gives rise to several governance processes negotiated through institutional or informal bargaining. These processes revolve around stakes such as the use of natural resources (e.g., water and land) and residents' nutrition and health. In this respect, Korbéogo (2021), referring to the roles of and land value capture by traditional authorities (cf. moose chieftaincies) in urban spatial planning in Ouagadougou, concluded that *"the dynamic interplay between bureaucratic institutions, traditional authorities and grassroots organisations is contributing to reshaping governance systems, as well as the construction of statehood in Burkina Faso"* (p. 190). Cissé et al. (2005) point out that *"the uncertain legal status of urban agriculture contrasts with the multiplicity of actors who intervene directly or indirectly in its promotion and development"* (p. 143) in West Africa, but the multiplicity and diversity of actors involved demonstrate the added value of UPA in social and economic terms. Meanwhile, Robineau and Dugué (2018) showed that the different forms of UA that emerged in specific locations of Bobo-Dioulasso (Burkina Faso) reflect, among others, the formal or informal land negotiations and transactions between the various categories of actors. Robineau (2015) suggests that changing and dynamic socio-spatial arrangements between the actors (especially market gardeners and pig breeders) are at the core of the persistence of urban agriculture in Bobo-Dioulasso (Burkina Faso).

The multi-functionality of UPA makes it a valuable tool in local policy and planning. In this respect, Di Leo et al. (2016) suggest that UPA is being used by the municipal government in Bobo-Dioulasso (Burkina Faso) for policy formulation in the fields of climate change, food security, and urbanization. This, in turn, makes UPA particularly suitable as an entry point to address holistic, sustainable management in cities. However, Robineau (2015) points out that the lack of knowledge of public authorities about urban agriculture impedes its inclusion in public policies of urban planning. Likewise, Cissé et al. (2005) – who examined the institutional aspects of urban agriculture in seven West African cities [Abidjan (Côte d'Ivoire), Bamako (Mali), Cotonou (Benin), Dakar (Senegal), Niamey (Niger), Nouakchott (Mauritania) and Ouagadougou (Burkina Faso)] – noted that the growing interest by public authorities in UPA failed to lead to its effective consideration in the institutional and legal provisions of most cities in West Africa. Furthermore, decentralization processes have not led to a commitment by city governments to support UPA.

Cissé et al. (2005) point out that coordination problems between various sectors and levels of government, as well as a failure to achieve the necessary functional complementarity between the different sectors (e.g., agriculture, land management, water management, waste management), hamper the development of UPA in West Africa.

CONCLUSIONS

This review explores the environmental, social, and economic impacts (both positive and negative) of UPA in Burkina Faso and Niger. It also analyzes the governance and policy issues related to UPA in both countries and explores their effects on the perspectives of its development. The analysis of the selected literature shows that the lion's share of the documents deals with UPA in Burkina Faso. Moreover, most of the considered studies focus on environmental issues while social issues and, especially, economic and political ones are generally overlooked. However, numerous articles address simultaneously different issues.

UPA contributes to environmental integrity through its positive effects on the atmosphere, water, land/soil, biodiversity, and materials and energy. Regarding the atmosphere, the effects of UPA mainly relate to GHG emissions and, consequently, climate change. Indeed, wastewater irrigation and fertilizer application generate GHG emissions. As for water, UPA affects both water withdrawal and water quality. While the use of wastewater for irrigation raises concerns regarding food safety and quality, the use of potable water for irrigation in some cities raises queries over long-term sustainability. Moreover, the excessive use of manure and inorganic fertilizers might lead to nutrient leaching and underground water pollution. Regarding biodiversity, there are linkages between UPA and species diversity, genetic diversity, and ecosystem and landscape diversity, which affect the provision of ecosystem services. UPA also affects land management and soil fertility. In particular, UPA practices can influence the organic matter content and, therefore, soil quality and fertility. The use of solid waste and wastewater leads to the accumulation of nutrients and heavy metals in urban garden soils. Urban agriculture also plays an essential role in the cycles of materials and nutrients. In general, inputs exceed the recommended amounts, leading to horizontal annual surpluses and relatively low nutrient use efficiency. The surplus of nutrients is leached or emitted into the atmosphere, thus contributing to GHG emissions. However, urban livestock production contributes to the reuse, thus the valorization, of by-products such as crop residues.

UPA affects livelihoods, food security, and nutrition as well as the health of both urban gardeners and urban dwellers. UPA contributes to the decent livelihoods of tens of thousands of households. It also favors both social inclusion/participation and gender equity. UPA also plays an important role in the food and nutrition security of many households not only by boosting the food supply in cities but also by providing urban gardeners with the needed financial means to access food. In particular, urban gardens are a valuable source of fresh, perishable produce. Production in urban and peri-urban areas contributes to household dietary diversity and helps address micronutrient deficiencies. Nevertheless, UPA also implies some risks in terms of human safety and health, which are mainly due to the use of wastewater for irrigation. Wastewater used for urban agriculture represents a high risk for spreading bacteria and antimicrobial resistance among humans and animals. Urban animal production contributes to the spreading of zoonotic diseases that could pose a significant risk to human health. Furthermore, differences in interests among the stakeholders involved in urban agriculture can lead to conflicts, and urban livestock farms can generate nuisances to neighboring residents.

UPA is an important economic activity that contributes to generating income and revenues for many urban households and developing urban and peri-urban areas. Indeed, income generation is one of the main motivations for practicing UPA. Some studies show that the profitability of peri-urban horticulture is much higher than that of traditional, rural crop production. Likewise, the high profitability of urban animal production explains why urban livestock producers are willing to pay for improved forages. However, the economic performance of UPA is affected by many factors, such as the availability and costs of agricultural inputs and access to commercial outlets.

Different political and governance factors affect the development of UPA in Burkina Faso and Niger. The different patterns of urbanization present real challenges for urban decision-makers and planners and suggest that different governance tools will be needed to ensure the sustainable growth of cities without sacrificing spaces dedicated to UPA while managing the challenges it may imply. UPA has become a socio-political arena for a wide range of stakeholders that often have different stakes, interests, and sources of power/legitimacy, which generates several governance processes negotiated through institutional or informal bargaining. The uncertain legal status of UPA contrasts with the multiplicity and diversity of actors intervening in its promotion, which, anyway, demonstrates the added value of UPA in social and economic terms. The multi-functionality of UPA makes it a valuable tool in local policy and planning. Municipal governments are using UPA for policy formulation and is suitable as an entry point to address holistic, sustainable management in cities. However, the lack of knowledge of public authorities about UPA impedes its inclusion in public policies. Moreover, the growing interest by public authorities in UPA failed to lead to its effective consideration in the institutional/legal provisions, while coordination problems between various sectors and government levels, as well as the low level of functional complementarity between the different sectors, hamper its development in West Africa.

All in all, UPA can support poor and marginalized communities, ensure food security and self-sufficiency, and promote sustainable livelihoods in cities. It can also improve waste management, increase energy usage efficiency, and help conserve natural and agricultural biodiversity. However, there may be health, safety, and environmental pollution concerns arising from intensive UPA models. In this context, providing appropriate local governance, institutional environment, and long-term policies, as well as coherent strategic visions at the local level, are necessary to prevent any negative repercussions of UPA while retaining its socio-economic relevance. Finally, while the multifaceted benefits of UPA are widely recognized, different actions are needed to unlock its potential in Burkina Faso, Niger, and West Africa at large. In particular, overcoming socio-cultural, environmental, and technological barriers impeding its persistence/expansion and creating a favorable policy and regulatory environment for agriculture in and around cities are paramount.

REFERENCES

Abdu, N., Abdulkadir, A., Agbenin, J. O., & Buerkert, A., (2011), Vertical distribution of heavy metals in wastewater-irrigated vegetable garden soils of three West African cities. *Nutrient Cycling in Agroecosystems*, *89*(3), 387–397. https://doi.org/10.1007/s10705-010-9403-3

Abdulkadir, A., Dossa, L. H., Lompo, D. J.-P., Abdu, N., & van Keulen, H., (2012), Characterization of urban and peri-urban agroecosystems in three West African cities. *International Journal of Agricultural Sustainability*, *10*(4), 289–314. https://doi.org/10.1080/14735903.2012.663559

Abdulkadir, A., Sangare, S. K., Amadou, H., & Agbenin, J. O., (2015), Nutrient balances and economic performance in urban and peri-urban vegetable production systems of three West African cities. *Experimental Agriculture*, *51*(1), 126–150. https://doi.org/10.1017/S0014479714000180

Akoto-Danso, E. K., Manka'abusi, D., Steiner, C., Werner, S., Haering, V., Lompo, D. J.-P., Nyarko, G., Marschner, B., Drechsel, P., & Buerkert, A., (2019), Nutrient flows and balances in intensively managed vegetable production of two West African cities. *Journal of Plant Nutrition and Soil Science*, *182*(2), 229–243. https://doi.org/10.1002/jpln.201800339

Amadou, H., Dossa, L. H., Lompo, D. J.-P., Abdulkadir, A., & Schlecht, E., (2012), A comparison between urban livestock production strategies in Burkina Faso, Mali and Nigeria in West Africa. *Tropical Animal Health and Production*, *44*(7), 1631–1642. https://doi.org/10.1007/s11250-012-0118-0

Bellwood-Howard, I., Ansah, I. G. K., Donkoh, S. A., & Korbéogo, G., (2021), Managing seasonality in West African informal urban vegetable markets: The role of household relations. *Journal of International Development*, *33*(5), 874–893. https://doi.org/10.1002/jid.3562

Bellwood-Howard, I., Shakya, M., Korbeogo, G., & Schlesinger, J., (2018), The role of backyard farms in two West African urban landscapes. *Landscape and Urban Planning*, *170*, 34–47. https://doi.org/10.1016/j.landurbplan.2017.09.026

Bernholt, H., Kehlenbeck, K., Gebauer, J., & Buerkert, A., (2009), Plant species richness and diversity in urban and peri-urban gardens of Niamey, Niger. *Agroforestry Systems*, 77(3), 159–179. https://doi.org/10.1007/s10457-009-9236-8

Bougnom, B. P., McNally, A., Etoa, F.-X., & Piddock, L. J. V., (2019), Antibiotic resistance genes are abundant and diverse in raw sewage used for urban agriculture in Africa and associated with urban population density. *Environmental Pollution*, 251, 146–154. https://doi.org/10.1016/j.envpol.2019.04.056

Bougnom, B. P., Zongo, C., McNally, A., Ricci, V., Etoa, F. X., Thiele-Bruhn, S., & Piddock, L. J. V., (2019), Wastewater used for urban agriculture in West Africa as a reservoir for antibacterial resistance dissemination. *Environmental Research*, 168, 14–24. https://doi.org/10.1016/j.envres.2018.09.022

Boussini, H., Traore, A., Tamboura, H. H., Bessin, R., Boly, H., & Ouedraogo, A., (2012), Prevalence of tuberculosis and brucellosis in urban and peri-urban dairy cattle farms in Ouagadougou, Burkina Faso. *Revue Scientifique et Technique-Office International Des Epizooties*, 31(3), 943–951.

Brinkmann, K., Schumacher, J., Dittrich, A., Kadaore, I., & Buerkert, A., (2012), Analysis of landscape transformation processes in and around four West African cities over the last 50 years. *Landscape and Urban Planning*, 105(1–2), 94–105. https://doi.org/10.1016/j.landurbplan.2011.12.003

Casari, S., Di Paola, M., Banci, E., Diallo, S., Scarallo, L., Renzo, S., Gori, A., Renzi, S., Paci, M., de Mast, Q., Pecht, T., Derra, K., Kaboré, B., Tinto, H., Cavalieri, D., & Lionetti, P., (2022), Changing dietary habits: The impact of urbanization and rising socio-economic status in families from Burkina Faso in Sub-Saharan Africa. *Nutrients*, 14(9), 1782. https://doi.org/10.3390/nu14091782

Chagomoka, T., Drescher, A., Glaser, R., Marschner, B., Schlesinger, J., & Nyandoro, G., (2017), Contribution of urban and periurban agriculture to household food and nutrition security along the urban-rural continuum in Ouagadougou, Burkina Faso. *Renewable Agriculture and Food Systems*, 32(1), 5–20. https://doi.org/10.1017/S1742170515000484

Chagomoka, T., Drescher, A., Nyandoro, G., Afari-Sefa, V., Schlesinger, J., & Nchanji, E., B., (2018), Contribution of vegetables to household diets along the urban-rural continuum in Ouagadougou, Burkina Faso. *Acta Horticulturae*, 1205, 87–96. https://doi.org/10.17660/ActaHortic.2018.1205.10

Cissé, O., Gueye, N. F. D., & Sy, M., (2005), Institutional and legal aspects of urban agriculture in French-speaking West Africa: from marginalization to legitimization. *Environment and Urbanization*, 17(2), 143–154. https://doi.org/10.1177/095624780501700211

Dao, J., Lompo, D. J.-P., Stenchly, K., Haering, V., Marschner, B., & Buerkert, A., (2019), Gypsum amendment to soil and plants affected by sodic alkaline industrial wastewater irrigation in Urban agriculture of Ouagadougou, Burkina Faso. *Water, Air, & Soil Pollution*, 230(12), 282. https://doi.org/10.1007/s11270-019-4311-x

Dao, J., Stenchly, K., Traoré, O., Amoah, P., & Buerkert, A., (2018), Effects of water quality and post-harvest handling on microbiological contamination of lettuce at Urban and peri-Urban locations of Ouagadougou, Burkina Faso. *Foods*, 7(12), 206. https://doi.org/10.3390/foods7120206

Di Leo, N., Escobedo, F. J., & Dubbeling, M., (2016), The role of urban green infrastructure in mitigating land surface temperature in Bobo-Dioulasso, Burkina Faso. *Environment, Development and Sustainability*, 18(2), 373–392. https://doi.org/10.1007/s10668-015-9653-y

Diogo, R. V. C., Buerkert, A., & Schlecht, E., (2010a), Resource use efficiency in urban and peri-urban sheep, goat and cattle enterprises. *Animal*, 4(10), 1725–1738. https://doi.org/10.1017/S1751731110000790

Diogo, R. V. C., Buerkert, A., & Schlecht, E., (2010b), Horizontal nutrient fluxes and food safety in urban and peri-urban vegetable and millet cultivation of Niamey, Niger. *Nutrient Cycling in Agroecosystems*, 87(1), 81–102. https://doi.org/10.1007/s10705-009-9315-2

Diogo, R. V. C., Buerkert, A., & Schlecht, E., (2011), Economic benefit to gardeners and retailers from cultivating and marketing vegetables in Niamey, Niger. *Outlook on Agriculture*, 40(1), 71–78. https://doi.org/10.5367/oa.2011.0027

Diogo, R. V. C., Schlecht, E., Buerkert, A., Rufino, M. C., & van Wijk, M. T., (2013), Increasing nutrient use efficiency through improved feeding and manure management in urban and peri-urban livestock units of a West African city: A scenario analysis. *Agricultural Systems*, 114, 64–72. https://doi.org/10.1016/j.agsy.2012.09.001

Dossa, L. H., Abdulkadir, A., Amadou, H., Sangare, S., & Schlecht, E., (2011), Exploring the diversity of urban and peri-urban agricultural systems in Sudano-Sahelian West Africa: An attempt towards a regional typology. *Landscape and Urban Planning*, 102(3), 197–206. https://doi.org/10.1016/j.landurbplan.2011.04.005

Dossa, L. H., Buerkert, A., & Schlecht, E., (2011), Cross-location analysis of the impact of household socio-economic status on participation in Urban and peri-Urban agriculture in West Africa. *Human Ecology*, 39(5), 569–581. https://doi.org/10.1007/s10745-011-9421-z

Dossa, L. H., Diogo, R. V. C., Sangare, M., Buerkert, A., & Schlecht, E., (2019), Use of feed resources in intensive Urban ruminant production systems of West Africa: A case study from Burkina Faso. *Animal Nutrition and Feed Technology*, *19*(1), 111. https://doi.org/10.5958/0974-181X.2019.00011.8

Dossa, L. H., Sangaré, M., Buerkert, A., & Schlecht, E., (2015a), Intra-urban and peri-urban differences in cattle farming systems of Burkina Faso. *Land Use Policy*, *48*, 401–411. https://doi.org/10.1016/j.landusepol.2015.06.031

Dossa, L. H., Sangaré, M., Buerkert, A., & Schlecht, E., (2015b), Production objectives and breeding practices of urban goat and sheep keepers in West Africa: regional analysis and implications for the development of supportive breeding programs. *SpringerPlus*, *4*(1), 281. https://doi.org/10.1186/s40064-015-1075-7

El Bilali, H., (2021a), Climate change-food security nexus in Burkina Faso. *CAB Reviews: Perspectives in Agriculture, Veterinary Science, Nutrition and Natural Resources*, *16*(009), https://doi.org/10.1079/PAVSNNR202116009

El Bilali, H., (2021b), Climate change and agriculture in Burkina Faso. *Journal of Aridland Agriculture*, 22–47. https://doi.org/10.25081/jaa.2021.v7.6596

FAO, (2013), *Sustainability Assessment of Food and Agricultural System: indicators.* https://doi.org/10.2144/000113056

FAO, (2014), *SAFA Sustainability Assessment of Food and Agriculture Systems - Guidelines Version 3.0.* https://www.fao.org/3/a-i3957e.pdf

FAO, IFAD, UNICEF, WFP, & WHO, (2021), *The State of Food Security and Nutrition in the World 2021. Transforming food systems for food security, improved nutrition and affordable healthy diets for all.* https://doi.org/https://doi.org/10.4060/cb4474en

FAO, MUFPP Secretariat, & RUAF, (2019), *The Milan Urban Food Policy Pact Monitoring Framework.* https://www.milanurbanfoodpolicypact.org/wp-content/uploads/2019/11/CA6144EN.pdf

Freidberg, S. E., (2001), Gardening on the edge: The social conditions of unsustainability on an African Urban periphery. *Annals of the Association of American Geographers*, *91*(2), 349–369. https://doi.org/10.1111/0004-5608.00248

Gomgnimbou, A. P. K., Nacro, H. B., Sanon, O. H., Sieza, I., Kiendrebeogo, T., Sedogo, M. P., & Martinez, J., (2014), Managing animal manures in the Bobo-Dioulasso peri-urban zone (Burkina Faso): structure of livestock farms, evaluation of their environmental and sanitary impacts, perspectives. *Cahiers Agricultures*, *23*(6), 393–402. https://doi.org/10.1684/agr.2014.0724

Graefe, S., Schlecht, E., & Buerkert, A., (2008), Opportunities and Challenges of Urban and Peri-Urban Agriculture in Niamey, Niger. *Outlook on Agriculture*, *37*(1), 47–56. https://doi.org/10.5367/000000008783883564

Häring, V., Manka'abusi, D., Akoto-Danso, E. K., Werner, S., Atiah, K., Steiner, C., Lompo, D. J. P., Adiku, S., Buerkert, A., & Marschner, B., (2017), Effects of biochar, waste water irrigation and fertilization on soil properties in West African urban agriculture. *Scientific Reports*, *7*(1), 10738. https://doi.org/10.1038/s41598-017-10718-y

Hayashi, K., Abdoulaye, T., & Wakatsuki, T., (2010), Evaluation of the utilization of heated sewage sludge for peri-urban horticulture production in the Sahel of West Africa. *Agricultural Systems*, *103*(1), 36–40. https://doi.org/10.1016/j.agsy.2009.08.004

Home in Place, (2022), *Linking the SDGs with the New Urban Agenda.* https://www.sdgsnewurbanagenda.com

Houessou, M. D., van de Louw, M., & Sonneveld, B. G. J. S., (2020), What Constraints the Expansion of Urban Agriculture in Benin? *Sustainability*, *12*(14), 5774. https://doi.org/10.3390/su12145774

Kaboré, I., Moog, O., Alp, M., Guenda, W., Koblinger, T., Mano, K., Ouéda, A., Ouédraogo, R., Trauner, D., & Melcher, A. H., (2016), Using macroinvertebrates for ecosystem health assessment in semi-arid streams of Burkina Faso. *Hydrobiologia*, *766*(1), 57–74. https://doi.org/10.1007/s10750-015-2443-6

Kaboré, W.-T. T., Hien, E., Zombre, P., Coulibaly, A., Houot, S., & Masse, D., (2011), Organic substrates recycling in the sub-urban agriculture of Ouagadougou (Burkina Faso) for soils fertilization: description of the different actors and their practices. *Biotechnologie Agronomie Societe et Environnement*, *15*(2), 271–286.

Karg, H., Drechsel, P., Akoto-Danso, E., Glaser, R., Nyarko, G., & Buerkert, A., (2016), Foodsheds and City Region Food Systems in Two West African Cities. *Sustainability*, *8*(12), 1175. https://doi.org/10.3390/su8121175

Kiba, D. I., Lompo, F., Compaore, E., Randriamanantsoa, L., Sedogo, P. M., & Frossard, E., (2012), A decade of non-sorted solid urban wastes inputs safely increases sorghum yield in periurban areas of Burkina Faso. *Acta Agriculturae Scandinavica, Section B - Soil & Plant Science*, *62*(1), 59–69. https://doi.org/10.1080/09064710.2011.573802

Kiba, D. I., Zongo, N. A., Lompo, F., Jansa, J., Compaore, E., Sedogo, P. M., & Frossard, E., (2012), The diversity of fertilization practices affects soil and crop quality in urban vegetable sites of Burkina Faso. *European Journal of Agronomy, 38*, 12–21. https://doi.org/10.1016/j.eja.2011.11.012

Korbéogo, G., (2018), Ordering urban agriculture: farmers, experts, the state and the collective management of resources in Ouagadougou, Burkina Faso. *Environment and Urbanization, 30*(1), 283–300. https://doi.org/10.1177/0956247817738201

Korbéogo, G., (2021), Traditional authorities and spatial planning in Urban Burkina Faso: exploring the roles and land value capture by Moose Chieftaincies in Ouagadougou. *African Studies, 80*(2), 190–206. https://doi.org/10.1080/00020184.2021.1920365

Levasseur, V., Pasquini, M. W., Kouamé, C., & Temple, L., (2007), A review of urban and peri-urban vegetable production in West Africa. *Acta Horticulturae, 762*, 245–252. https://doi.org/10.17660/ActaHortic.2007.762.23

Lompo, D. J., Sangaré, S. A. K., Compaoré, E., Papoada Sedogo, M., Predotova, M., Schlecht, E., & Buerkert, A., (2012), Gaseous emissions of nitrogen and carbon from urban vegetable gardens in Bobo-Dioulasso, Burkina Faso. *Journal of Plant Nutrition and Soil Science, 175*(6), 846–853. https://doi.org/10.1002/jpln.201200012

Lompo, D. J.-P., Compaoré, E., Sedogo, M. P., Melapie, M., Bielders, C. L., Schlecht, E., & Buerkert, A., (2019), Horizontal flows of nitrogen, potassium, and carbon in urban vegetables gardens of Bobo Dioulasso, Burkina Faso. *Nutrient Cycling in Agroecosystems, 115*(2), 189–199. https://doi.org/10.1007/s10705-018-9949-z

Mainardi, S., (2011), Cropland use, yields, and droughts: Spatial data modeling for Burkina Faso and Niger. *Agricultural Economics, 42*(1), 17–33. https://doi.org/10.1111/j.1574-0862.2010.00465.x

Manka'abusi, D., Lompo, D. J. P., Steiner, C., Ingold, M., Akoto-Danso, E. K., Werner, S., Häring, V., Nyarko, G., Marschner, B., & Buerkert, A., (2020), Carbon dioxide and gaseous nitrogen emissions from biochar-amended soils under wastewater irrigated urban vegetable production of Burkina Faso and Ghana. *Journal of Plant Nutrition and Soil Science, 183*(4), 500–516. https://doi.org/10.1002/jpln.201900183

Manka'abusi, D., Steiner, C., Akoto-Danso, E. K., Lompo, D. J. P., Haering, V., Werner, S., Marschner, B., & Buerkert, A., (2019), Biochar application and wastewater irrigation in urban vegetable production of Ouagadougou, Burkina Faso. *Nutrient Cycling in Agroecosystems, 115*(2), 263–279. https://doi.org/10.1007/s10705-019-09969-0

Mattoni, M., Bergero, D., & Schiavone, A., (2007), Assessment of structural traits and management related to dairy herds in the peri-urban area of Bobo Dioulasso (South west of Burkina Faso). *Journal of Agriculture and Rural Development in the Tropics and Subtropics, 108*(1), 41–50.

Millogo, V., Ouedraogo, G. A., Agenas, S., & Svennersten-Sjaunja, K., (2008), Survey on dairy cattle milk production and milk quality problems in peri-urban areas in Burkina Faso. *African Journal of Agricultural Research, 3*(3), 215–224.

Moher, D., Liberati, A., Tetzlaff, J., & Altman, D. G., (2009), Preferred reporting items for systematic reviews and meta-analyses: The PRISMA Statement. *PLoS Medicine, 6*(7), e1000097. https://doi.org/10.1371/journal.pmed.1000097

MUFPP, (2020), *The Milan Urban Food Policy Pact (MUFPP) - History.* https://www.milanurbanfoodpolicypact.org/history

OCHA, (2022), *Burkina Faso.* https://www.humanitarianresponse.info/en/operations/burkina-faso

Orsini, F., Kahane, R., Nono-Womdim, R., & Gianquinto, G., (2013), Urban agriculture in the developing world: a review. *Agronomy for Sustainable Development, 33*(4), 695–720. https://doi.org/10.1007/s13593-013-0143-z

Ouédraogo, A., Zampaligré, N., Mulubrhan, B., & Adesogan, A. T., (2022), Assessment of peri-urban livestock producers' willingness to pay for improved forages as cash crops. *Agronomy Journal, 114*(1), 63–74. https://doi.org/10.1002/agj2.20953

Ouédraogo, R. A., Chartin, C., Kambiré, F. C., van Wesemael, B., Delvaux, B., Milogo, H., & Bielders, C. L., (2020), Short and long-term impact of urban gardening on soil organic carbon fractions in Lixisols (Burkina Faso). *Geoderma, 362*, 114110. https://doi.org/10.1016/j.geoderma.2019.114110

Ouédraogo, R. A., Kambiré, F. C., Kestemont, M.-P., & Bielders, C. L., (2019), Caractériser la diversité des exploitations maraîchères de la région de Bobo-Dioulasso au Burkina Faso pour faciliter leur transition agroécologique. *Cahiers Agricultures, 28*, 20. https://doi.org/10.1051/cagri/2019021

Page, M. J., McKenzie, J. E., Bossuyt, P. M., Boutron, I., Hoffmann, T. C., Mulrow, C. D., Shamseer, L., Tetzlaff, J. M., Akl, E. A., Brennan, S. E., Chou, R., Glanville, J., Grimshaw, J. M., Hróbjartsson, A., Lalu, M. M., Li, T., Loder, E. W., Mayo-Wilson, E., McDonald, S., ... Moher, D., (2021), The PRISMA 2020 statement: an updated guideline for reporting systematic reviews. *BMJ*, n71. https://doi.org/10.1136/bmj.n71

Parnell, S., & Walawege, R., (2011), Sub-Saharan African urbanisation and global environmental change. *Global Environmental Change*, *21*, S12–S20. https://doi.org/10.1016/j.gloenvcha.2011.09.014

Predotova, M., Bischoff, W., & Buerkert, A., (2011), Mineral-nitrogen and phosphorus leaching from vegetable gardens in Niamey, Niger. *Journal of Plant Nutrition and Soil Science*, *174*(1), 47–55. https://doi.org/10.1002/jpln.200900255

Predotova, M., Gebauer, J., Diogo, R. V. C., Schlecht, E., & Buerkert, A., (2010), Emissions of ammonia, nitrous oxide and carbon dioxide from urban gardens in Niamey, Niger. *Field Crops Research*, *115*(1), 1–8. https://doi.org/10.1016/j.fcr.2009.09.010

Predotova, M., Schlecht, E., & Buerkert, A., (2010), Nitrogen and carbon losses from dung storage in urban gardens of Niamey, Niger. *Nutrient Cycling in Agroecosystems*, *87*(1), 103–114. https://doi.org/10.1007/s10705-009-9316-1

Probst, L., Adoukonou, A., Amankwah, A., Diarra, A., Vogl, C. R., & Hauser, M., (2012), Understanding change at farm level to facilitate innovation towards sustainable plant protection: a case study at cabbage production sites in urban West Africa. *International Journal of Agricultural Sustainability*, *10*(1), 40–60. https://doi.org/10.1080/14735903.2012.649589

Probst, L., Houedjofonon, E., Ayerakwa, H. M., & Haas, R., (2012), Will they buy it? The potential for marketing organic vegetables in the food vending sector to strengthen vegetable safety: A choice experiment study in three West African cities. *Food Policy*, *37*(3), 296–308. https://doi.org/10.1016/j.foodpol.2012.02.014

Rischkowsky, B., Siegmund-Schultze, M., Bednarz, K., & Killanga, S., (2006), Urban sheep keeping in West Africa: Can socioeconomic household profiles explain management and productivity? *Human Ecology*, *34*(6), 785–807. https://doi.org/10.1007/s10745-006-9011-7

Robineau, O., (2015), Toward a systemic analysis of city-agriculture interactions in West Africa: A geography of arrangements between actors. *Land Use Policy*, *49*, 322–331. https://doi.org/10.1016/j.landusepol.2015.08.025

Robineau, O., & Dugué, P., (2018), A socio-geographical approach to the diversity of urban agriculture in a West African city. *Landscape and Urban Planning*, *170*, 48–58. https://doi.org/10.1016/j.landurbplan.2017.09.010

Roessler, R., (2019), Selection decisions and trait preferences for local and imported cattle and sheep breeds in Peri-/Urban livestock production systems in Ouagadougou, Burkina Faso. *Animals*, *9*(5), 207. https://doi.org/10.3390/ani9050207

Roessler, R., Mpouam, S. E., & Schlecht, E., (2019), Genetic and nongenetic factors affecting on-farm performance of peri-urban dairy cattle in West Africa. *Journal of Dairy Science*, *102*(3), 2353–2364. https://doi.org/10.3168/jds.2018-15348

Roessler, R., Mpouam, S., Muchemwa, T., & Schlecht, E., (2016), Emerging Development Pathways of Urban Livestock Production in Rapidly Growing West Africa Cities. *Sustainability*, *8*(11), 1199. https://doi.org/10.3390/su8111199

Rossi, J.-P., & Dobigny, G., (2019), Urban landscape structure of a fast-growing African City: The Case of Niamey (Niger), *Urban Science*, *3*(2), 63. https://doi.org/10.3390/urbansci3020063

Sangare, S. K., Compaore, E., Buerkert, A., Vanclooster, M., Sedogo, M. P., & Bielders, C. L., (2012), Field-scale analysis of water and nutrient use efficiency for vegetable production in a West African urban agricultural system. *Nutrient Cycling in Agroecosystems*, *92*(2), 207–224. https://doi.org/10.1007/s10705-012-9484-2

Sanou, K. F., Nacro, S., Ouédraogo, M., Ouédraogo, S., & Kaboré-Zoungrana, C., (2011), Marketing of forage in the urban zone of Bobo-Dioulasso (Burkina Faso): Market activities and economic returns. *Cahiers Agricultures*, *20*(6), 487–493. https://doi.org/10.1684/agr.2011.0530

Schlecht, E., Plagemann, J., Mpouam, S. E., Sanon, H. O., Sangaré, M., & Roessler, R., (2019), Input and output of nutrients and energy in urban and peri-urban livestock holdings of Ouagadougou, Burkina Faso. *Nutrient Cycling in Agroecosystems*, *115*(2), 201–230. https://doi.org/10.1007/s10705-019-09996-x

Sidibe, M., Boly, H., Lakouetené, T., Leroy, P., & Bosma, R. H., (2004), Characteristics of Peri-urban Dairy Herds of Bobo-Dioulasso (Burkina Faso), *Tropical Animal Health and Production*, *36*(1), 95–100. https://doi.org/10.1023/B:TROP.0000009525.23669.c2

Skar, S. L. G., Pineda-Martos, R., Timpe, A., Pölling, B., Bohn, K., Külvik, M., Delgado, C., Pedras, C. M. G., Paço, T. A., Ćujić, M., Tzortzakis, N., Chrysargyris, A., Peticila, A., Alencikiene, G., Monsees, H., & Junge, R., (2020), Urban agriculture as a keystone contribution towards securing sustainable and healthy development for cities in the future. *Blue-Green Systems*, *2*(1), 1–27. https://doi.org/10.2166/bgs.2019.931

Stenchly, K., Dao, J., Lompo, D. J.-P., & Buerkert, A., (2017), Effects of waste water irrigation on soil properties and soil fauna of spinach fields in a West African urban vegetable production system. *Environmental Pollution*, *222*, 58–63. https://doi.org/10.1016/j.envpol.2017.01.006

Stenchly, K., Hansen, M., Stein, K., Buerkert, A., & Loewenstein, W., (2018), Income vulnerability of West African farming households to losses in pollination services: A case study from Ouagadougou, Burkina Faso. *Sustainability*, *10*(11), 4253. https://doi.org/10.3390/su10114253

Stenchly, K., Lippmann, S., Waongo, A., Nyarko, G., & Buerkert, A., (2017), Weed species structural and functional composition of okra fields and field periphery under different management intensities along the rural-urban gradient of two West African cities. *Agriculture, Ecosystems & Environment*, *237*, 213–223. https://doi.org/10.1016/j.agee.2016.12.028

Stenchly, K., Waongo, A., Schaeper, W., Nyarko, G., & Buerkert, A., (2019), Structural landscape changes in urban and peri-urban agricultural systems of two West African cities and their relations to ecosystem services provided by woody plant communities. *Urban Ecosystems*, *22*(2), 397–408. https://doi.org/10.1007/s11252-018-0811-5

Tankari, M. R., (2020), Rainfall variability and farm households' food insecurity in Burkina Faso: nonfarm activities as a coping strategy. *Food Security*, *12*(3), 567–578. https://doi.org/10.1007/s12571-019-01002-0

Thys, E., Oueadraogo, M., Speybroeck, N., & Geerts, S., (2005), Socio-economic determinants of urban household livestock keeping in semi-arid Western Africa. *Journal of Arid Environments*, *63*(2), 475–496. https://doi.org/10.1016/j.jaridenv.2005.03.019

Tiffen, M., (2006), Urbanization: Impacts on the evolution of "mixed farming" systems in sub-Saharan Africa. *Experimental Agriculture*, *42*(3), 259–287. https://doi.org/10.1017/S0014479706003589

UNDP, (1996), *Urban agriculture: Food, jobs and sustainable cities.* New York: UNDP.

UNDP, (2019), *Human Development Report 2019.* New York: United Nations Development Programme (UNDP).

United Nations, (2015), *Transforming Our World: The 2030 Agenda for Sustainable Development. Resolution adopted by the General Assembly on* 25 September 2015. New York. https://sustainabledevelopment.un.org/content/documents/7891Transforming Our World.pdf

United Nations, (2017), *New Urban Agenda.* https://habitat3.org/wp-content/uploads/NUA-English.pdf

USAID, (2017), *Climate Change Risk in West Africa Sahel: Regional Fact Sheet.* https://www.climatelinks.org/sites/default/files/asset/document/2017 April_USAID ATLAS_Climate Change Risk Profile - Sahel.pdf

Wang, S.-J., Lengeler, C., Smith, T. A., Vounatsou, P., Diadie, D. A., Pritroipa, X., Convelbo, N., Kientga, M., & Tanner, M., (2005), Rapid urban malaria appraisal (RUMA) I: Epidemiology of urban malaria in Ouagadougou. *Malaria Journal*, *4*(1), 43. https://doi.org/10.1186/1475-2875-4-43

Weinreb, A., Stecklov, G., & Arslan, A., (2020), Effects of changes in rainfall and temperature on age- and sex-specific patterns of rural-urban migration in sub-Saharan Africa. *Population and Environment*, *42*(2), 219–254. https://doi.org/10.1007/s11111-020-00359-1

World Bank, (2021), *World Bank Open Data.* https://data.worldbank.org

World Food Programme, (2019), *Niger Country Strategic Plan (2020 - 2024).* https://docs.wfp.org/api/documents/WFP-0000108569/download/?_ga=2.172546470.427657304.1657202174-2029699135.1657202174

Zezza, A., & Tasciotti, L., (2010), Urban agriculture, poverty, and food security: Empirical evidence from a sample of developing countries. *Food Policy*, *35*(4), 265–273. https://doi.org/10.1016/j.foodpol.2010.04.007

Zhou, Y., & Staatz, J., (2016), Projected demand and supply for various foods in West Africa: Implications for investments and food policy. *Food Policy*, *61*, 198–212. https://doi.org/10.1016/j.foodpol.2016.04.002

5 Urban Home Gardening and Agri-Food System Transitions Toward Sustainability in Kerala

Anita Pinheiro and Madhav Govind

INTRODUCTION

Sustainability transitions include physical technology-based transitions (El Bilali & Allahyari, 2018; Geels, 2002; Sutherland et al., 2015b) and 'soft' transitions that involve transitions in the practices and small-scale technologies (Sutherland et al., 2015a). The research field of sustainability transitions recognizes that most of the interventions on sustainability focus more on incremental changes, therefore adding to the lock-ins, and fundamental changes are necessary (Geels & Schot, 2007; Köhler et al., 2019; Loorbach et al., 2017). Sustainability in urban and agri-food systems is increasingly attracting attention from researchers.

The dominant notion of agriculture as a rural activity is rapidly changing within the discourse of sustainable urban and agri-food systems. Urban agriculture is increasingly recognized for its multifunctionality in creating edible urban green infrastructures while addressing multiple sustainability challenges (Evans et al., 2022; Govind & Pinheiro, Forthcoming; Maćkiewicz & Asuero, 2021; Sartison & Artmann, 2020; Walthall, 2016). It can bridge the brown (human environment) and green agendas (the natural environment) of urban sustainability and contribute to various Sustainable Development Goals (SDGs) (Nicholls et al., 2020). Historically, urban agriculture has also shown its potential to contribute to food production and resilience in times of crisis (Altieri et al., 1999; Barthel et al., 2015; Maltz, 2015). Urban agriculture ranges from simple nature-based solutions to complex technological systems (Al-Chalabi, 2015; Al-Kodmany, 2018; Artmann & Sartison, 2018; Orsini et al., 2014; Sartison & Artmann, 2020), and the outcomes vary depending upon the specific practice and context.

Much of the idea of urban agriculture from academic literature primarily reflects Global North countries' practices (Pinheiro & Govind, 2020). Hence, urban agriculture's varying practices and intricacies in the Global South have not received adequate attention in urban agriculture literature. This warrants special efforts to explore locally specific forms of urban agriculture in the Global South and its overall contribution to sustainability within the specific context. Urban agriculture in India includes many practices ranging from commercial urban agriculture to small home gardens and agrochemical-based conventional agriculture in the farmlands to technology-oriented soil-less cultivation such as hydroponics (Cook et al., 2015; Pinheiro, 2023, 2022b; Sharma et al., 2023; Zasada et al., 2020). There are growing private initiatives to build community gardening and allocate small parcels of land for individuals to grow vegetables. Broadly, urban agriculture in India rests on the opposite sides of the spectrum. It is either a livelihood strategy for the economically weaker sections or a means to self-supply fresh organic vegetables for the higher-income group (Pinheiro et al., 2022). Government interventions are crucial to overcome this divide and explore urban agriculture's benefits to urban and agri-food systems.

DOI: 10.1201/9781003359425-5

The term 'home gardening' is used to indicate food production in and on the residential building and its immediate premises. The home garden food production may include the cultivation of food crops, livestock, and other food items. Within the growing state-level initiatives in India for promoting urban agriculture (in private residential premises) that focus on large cities, Kerala is arguably the first state where the government promotes urban agriculture across the state irrespective of the size and population of urban areas (Pinheiro, 2022a, Forthcoming). In other states, most of the government interventions for urban agriculture focus primarily on large cities while small towns and cities are left out (Pinheiro, 2022a, Forthcoming). In this context, it is essential to understand how government interventions and people's movements shape urban home gardening as a nature-based solution and how urban home gardening contributes to sustainability transitions in agri-food systems. Building upon the niche–regime interactions of sustainability transition theory, the following issues are explored: (a) urban home gardening in the context of agri-food system transitions in Kerala, (b) the government interventions and people's movement that promote urban home gardening niche in Kerala, and (c) enablers of the development of urban home gardening niche and regime's resistance. The analytical framework of the multi-level perspective (MLP) of sustainability transitions theory is discussed in the next section followed by methodology. In the fourth section, the context of agri-food system transitions is discussed. The next sections analyse the empirical observations followed by the conclusion.

SUSTAINABILITY TRANSITIONS

Transitions towards sustainability entail large-scale societal shifts towards more sustainable practices and production that take place as a response to societal challenges (Avelino et al., 2016; Geels, 2002; Köhler et al., 2019; Loorbach et al., 2017). In some cases, the same type of transition may have different impacts in different contexts. The non-linearity of the sustainability transition process entails that transitions cannot be considered as forward in a particular direction; it may also involve forward-moving steps while some others may be backward moving or in a different direction if the belief systems and objectives of actors are changed during the process (Geels, 2020; Geels et al., 2016).

MULTI-LEVEL PERSPECTIVE

MLPs are the prominent approach used in sustainability transitions research, and they help to understand the processes of coevolution of niche innovation and policy change (Derwort et al., 2021). MLP distinguishes transitions at three levels: landscapes, regimes, and niches (Geels, 2011, 2020). Niche is the breeding ground for radical innovations that comprise radical novelties in technologies, rules and social practices, concepts, and ideas (Loorbach, 2007). A regime maintains a stable position and is characterized by lock-ins and a wide distribution of technology (e.g., industrialized agriculture). The landscape exhibits exogenous socio-cultural characteristics at the macro-level for the long term (Geels & Schot, 2007), and the new technologies and practices gradually improve in this phase. Innovations at niche ignite internal momentum, landscape-level changes elicit pressure on the regime, and windows of opportunities are created for new niche-level innovations as a result of the destabilization of regimes (Geels & Schot, 2007). Alignment of processes at niche, regime, and landscape levels makes a transition to sustainability successful.

Anchoring is considered a stepping stone for linking niche and regime, and it can take place in a niche, regime, or the overlapping hybrid forum (Karanikolas et al., 2015). Although anchoring is a continuous process, it does not ensure stability and permanent embedding of a niche into the regime. Elzen et al. (2012) have proposed three types of anchoring as analytical concepts: technological, institutional, and network anchoring. The breakdown of technological, network, and institutional anchoring helps to understand how anchoring of novelties occurs in each of these components. The alignment between these three types of anchoring is important for transitions.

Depending upon the institutional-level support mechanisms, the niche innovations having radical potential may take either fit-and-conform patterns of transitions or stretch-and-transform pathways of transition (Smith & Raven, 2012). The categorization of fit-and-conform and stretch-and-transform patterns of transitions entails that to produce sustainable results, a niche has to be both radical and reforming (Smith, 2006). However, in most of the transitions, the fit-and-conform pattern is the dominant narrative as it is easier for the niche advocates to convince the powerful actors to avail resources (Raven et al., 2016). This highlights the crucial role of local actors (Belmin et al., 2018) and strategic niche management (SNM) that focuses on nurturing niche innovations (Raven et al., 2016) and preventing niche innovations from reverting to the rules of the regime. It is also possible that while adopting a 'stretch and transform' pattern, radical innovations can occasionally 'conform' to the incumbent regime (Audet et al., 2017) or take up 'hybrid fit and stretch' pathway that possesses the characteristics of both pathways (Mylan et al., 2019).

GOVERNMENT ACTORS IN SUSTAINABILITY TRANSITIONS

Transforming the dominant regime demands major policy changes and societal mobilization (Levidow et al., 2014). Among the different actors, public policies have a key role and centrality in shaping the speed and direction of sustainability transitions (Köhler et al., 2019). Moreover, government actors play diverse roles that might vary at different stages of the niche innovation (Wright et al., 2018). However, in transition, the interests of actors are shaped by various meanings, interpretations, and beliefs (Geels, 2020). Different actors have different interpretations and belief systems regarding a particular niche innovation. Therefore, the legitimacy of policy efforts for a particular niche innovation and its social acceptance (Geels, 2020) is also influenced by these meanings, interpretations, and beliefs (Geels, 2020).

URBAN AGRICULTURE IN TRANSITIONS RESEARCH

Among the various theoretical strands within sustainability transitions in agri-food systems, MLP is the most prominently used one (El Bilali, 2018). Hisschemöller (2016) argues that small-scale agriculture systems like urban agriculture have the potential to accelerate the transition towards a sustainable agri-food system worldwide. To facilitate urban agri-food system transitions, it is important to incorporate the rural–urban linkage in the processes (Sarabia et al., 2021), strengthening learning processes (Davids & De Olde, 2014), and self-organization from local conditions (Bell & Cerulli, 2012) for further development of the urban agriculture niches.

METHODOLOGY

Kerala is located on the southwest coast of the Indian Peninsula. It has a high population density (860 people/kilometre) and has an urban population comprising 47.70% of the total population (Census 2011, n.d.-a). The settlement pattern of Kerala represents a rural–urban continuum with a blurred divide between rural and urban areas (Rijesh et al., 2022). Kerala has the highest literacy rate in India (Census 2011, n.d.-b) and has a strong history of social movements (Dreze & Sen, 2002).

This chapter contributes to the growing body of urban agriculture literature within transition studies, especially the niche–regime interactions. The chapter thrusts primarily on qualitative methods for data collection and analysis as there is very little previous knowledge available on the topic of urban agriculture in the study area. The policy documents and other secondary literature were analysed to understand the themes to understand the meanings of the data. The data used in this chapter was collected over a period of time, using multiple methods, including household and online surveys. A household survey conducted in Thiruvananthapuram Corporation, the capital city of Kerala, focused on 50 well-functioning home gardens (carried out in 2015) to gather insights into the potential of well-functioning home gardens. The online survey of 311 urban and peri-urban home gardens across

Kerala (carried out in 2022) provides a broad picture of home gardens of multiple types and scales. Additionally, experts' insights on the topic were gathered using interviews, telephone calls, and other methods. MLP of sustainability transitions theory was used to infer these data.

THE AGRI-FOOD REGIME OF KERALA AND LANDSCAPE-LEVEL CHANGES

The agri-food regime of Kerala is dominated by intensive cultivation of export-oriented commercial crops, with a limited focus on food crops. The research and policy emphasis on intensive commercial cultivation and monoculture has impacted the scale and structure of traditional, small-scale, mixed agriculture systems such as agroforestry systems and home gardens in Kerala (Fox et al., 2017; Guillerme et al., 2011). Vegetables have received comparatively little policy attention until a few years ago.

The negative externalities associated with the conventional agri-food system and food dependency in Kerala act as landscape-level pressures for transitions in the agri-food systems. Because of the food dependency, Kerala has become vulnerable to many challenges including poor quality of the food products, price hikes, and unexpected supply cuts. The main reason is the health concerns raised by the periodic government reports on the detection of higher than the permitted levels of pesticides or the presence of banned pesticides in commercial vegetables in Kerala, even in vegetables sold under the 'organic' category (Kerala Agricultural University, 2014; Pinheiro, 2022a). The summary of agri-food system transition in Kerala is portrayed in Figure 5.1.

In Kerala, the pressure to enhance local production of safe-to-eat vegetables has opened up windows of opportunities for two broad types of niche innovations: a capital-intensive technological solution that primarily focuses on enhanced productivity and nature-based solutions adopting organic and agroecological methods of cultivation that focuses on multiple aspects related to sustainability. The capital- and technology-intensive solutions mainly include hi-tech cultivation such as greenhouse cultivation in polyhouses or precision agriculture in open farms. The promotion of

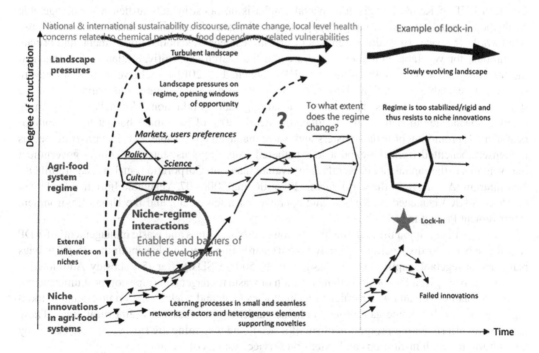

FIGURE 5.1 Multi-level perspectives of agri-food system transitions in Kerala. (Source: Authors' adaptation of Geels (2002) based on empirical data collected by the authors Anderson et al. 2019, Ollivier et al. 2018, and Schiller et al. 2020.)

these technologies in the policy and research agenda is emphasized in the 12th and 13th Five Year Plans (FYPs) of Kerala (Kerala State Planning Board, 2013, 2018) and the approach paper to the 14th FYP (Kerala State Planning Board, 2022). These technological solutions can be considered as incremental approaches to bring sustainability to agri-food systems.

The nature-based pathway of policy interventions focused on bringing transformative changes in sustainable agri-food systems in Kerala and focuses on organic agriculture and food system localization. Instead of productivity, the focus of this line of government intervention was on enhanced production by bringing additional areas including barren farmlands and otherwise unused, unconventional spaces such as residential spaces and premises of schools, government, or private institutions under vegetable cultivation. Although the export market (mainly cash crops) is one of the main aims of the certified organic farmers, the organic agriculture policy and the related government interventions also promote non-certified organic farmers who focus on biodiversity conservation and food crop cultivation for domestic markets.

URBAN HOME GARDENING IN KERALA

This chapter places urban home gardening as one of the niches that emerged in the transition towards organic agriculture and food system localization in the state. In a transition, multiple niches may emerge in response to the landscape-level changes (Geels, 2002). Among the two major niches of nature-based and technological-based solutions, urban home gardening can be considered a sub-niche of the umbrella niche of organic agriculture, which also includes commercial-scale organic cultivation practices in land parcels of varying sizes.

GOVERNMENT INTERVENTIONS FOR PROMOTION OF URBAN HOME GARDENING

The 12th FYP of Kerala has given a special emphasis on boosting self-sufficient local vegetable production through enhanced budget support and adoption of innovative measures to make use of all available spaces (including premises of private residences, schools, government, and private institutions) for vegetable cultivation to overcome the scarcity of cultivable land. Consequently, the Vegetable Development Programme (VDP), launched in 2012 to achieve self-sufficiency in safe-to-eat vegetable production, focuses on urban home garden vegetable cultivation across the state, among other measures, to accelerate local vegetable production. Urban home gardening receives special attention in VDP by allocating nearly 10% of its annual budget to the component named 'promotion of urban clusters and waste management' (Table 5.1). The programme has immensely benefited from the experiences gathered from previous state and central government interventions in the capital city of Kerala, Thiruvananthapuram Corporation, such as Nagarathil Oru Nattinpuram (A Village in the City) during 2005–06 and 2006–07, Vegetable Initiative for Urban Clusters (VIUC) launched in 2011–12 and operated for a few years, and the Urban Environment Improvement Project in 2013–14.

The major intervention through the 'promotion of urban clusters and waste management' of VDP was the subsidized distribution of ready-to-start gardening units each containing 25 grow-bags planted with vegetable saplings at a cost as low as Rs. 500 (~USD 6) after 75% subsidy. Additionally, support has been given for subsidized distribution of waste management (pipe compost units, portable biogas plants, etc.) and water-efficient technologies (drip and wick irrigation units) and marketing support for urban home gardening and home gardening in space-constraint rural areas. These government efforts for the promotion of small-space home gardening are further complemented by agricultural research institutions and extension service systems of Kerala.

There are many policy documents that endorse the incorporation of rural and urban home gardening as an agri-food system sustainability intervention in Kerala. These include Kerala State Environment Policy 2009, Kerala State Organic Farming Policy, Strategy and Action Plan 2010, Kerala Agricultural Development Policy 2015, 13th FYP 2017–22, 13th FYP approach paper, 13th

TABLE 5.1

Details of Fund Allocation and Number of Beneficiaries under the Component 'Promotion of Urban Clusters and Waste Management' of VDP

Year	Number of Beneficiary Households for Grow-Bag Distribution	Total Financial Allocation for 'Promotion of Urban Clusters and Waste Management' (Rs. lakhs)	Total Allocation for VDP (Rs. lakhs)
2012–13	26,500	400	4,400
2013–14	28,000	470	3,846
2014–15	28,334	525	4,111
2015–16	28,334	525	5,695
2016–17	40,000	750	7,430
2017–18	37,000	850	7,900
2018–19	42,000	813.53	6,177.34
2019–20	39,000	850	6,390
2020–21	22,000	850	6,947
2021–22	56,333	850	6,348.25
Total	**347,501**	**6,833.3**	**59,244.59**

Sources: Authors' compilation of data from various sources (Directorate of AD & FW, 2021; Directorate of Agriculture, 2012, 2015, 2017, 2018).

FYP working group report on biodiversity, 13th FYP working group report on environment, 13th FYP working group report on agriculture, 13th FYP working group report on agriculture research and Information and Communication Technology (ICT) in agriculture, and 13th FYP working group report on cooperation and agriculture finance. Overall, these policy documents consider home gardening more as nature-based solutions supported with small-scale technologies (such as mini-polyhouse) for building-integrated production.

Urban home gardening was promoted as an extension of food production into the urban areas and not as an integrated part of the urban fabric. Therefore, urban home gardening or any form of urban agriculture has not received any place in policy documents on urban development and climate change (13th FYP, working group on urban issues, and 13th FYP, working group on climate change and disaster management). In these documents, urban greening has been mentioned only as the traditional practices of parks, gardens, and urban forestry.

From 2022 to 2023 onwards, the component 'promotion of urban clusters and waste management' was discontinued from VDP, while other activities for the promotion of home garden vegetable cultivation across the state were continued with a few additions that mostly focused on technological solutions. Figure 5.2 summarizes the differences in the focus of government interventions of VDP before and after 2022–23.

URBAN HOME GARDENING BECOMING A PEOPLE'S MOVEMENT

Social mobilization is the crux of the success of any government intervention. When home gardening across Kerala started to increase due to government interventions and concerns about pesticide residues in commercial vegetables, interested people started to form networks on the social media platform Facebook. Among the active Facebook-based (FB) agriculture groups in the local language Malayalam, more than ten focus primarily on home garden cultivation, especially those with space and resource constraints. The visions of the FB groups are grounded on building a safe and healthy eating practice through agroecology-based home garden cultivation of vegetables and fruits. These collectives quickly became a platform for learning from each other's knowledge and experience. Because of the availability of a multitude of experts, both trained agricultural personnel including scientists and expert home gardeners to respond to the queries, the newcomers in home

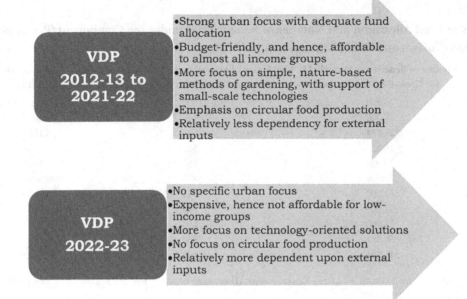

VDP 2012-13 to 2021-22
- Strong urban focus with adequate fund allocation
- Budget-friendly, and hence, affordable to almost all income groups
- More focus on simple, nature-based methods of gardening, with support of small-scale technologies
- Emphasis on circular food production
- Relatively less dependency for external inputs

VDP 2022-23
- No specific urban focus
- Expensive, hence not affordable for low-income groups
- More focus on technology-oriented solutions
- No focus on circular food production
- Relatively more dependent upon external inputs

FIGURE 5.2   Shift in the focus of the Vegetable Development Programme in terms of urban home garden vegetable cultivation.

gardening immensely benefited from these collectives. These collectives were helpful in gathering the basics of gardening, less dependence on external inputs, the adoption of less expensive gardening, the development of innovative gardening technologies and practices, resource-efficient gardening, and the effective utilization of available space. Figure 5.2d shows an example of a low-cost passive hydroponics system developed by a respondent of an in-person survey.

Before COVID-19, many of the FB collectives organized competitions and challenges to further encourage minimal-space home gardening, especially in urban areas and households with minimal space. These initiatives were held in connection with the traditional festivals of Kerala, especially during *Onam*, the harvest festival of Kerala, so that people can have a safe-to-eat feast. The FB collectives also foster a culture of sharing and exchange of gardening inputs, especially seeds and seedlings of traditional and local vegetable varieties through postal services or during physical meet-ups.

What if the home gardeners want to sell their surplus harvest? In the early years of the home gardening movement, especially when everyone was more interested in gifting the surplus, this question seemed to be a surprise to many, if not strange. The admins of the FB group named '*Karshika Vipani* Online Organic Agricultural Market' pursued this idea of linking home garden surplus with the urban short food supply chain. In 2014, they established *nattuchantha* – a weekly market in Thrissur where even small quantities of home-grown surplus can be sold directly to customers. The '*nattuchantha*' entrusts upon selling completely organically grown produce from urban and peri-urban home gardens and 'knowing the address of the farmer'; i.e., all the products are reliable and can be tracked. This niche market aims to bring the independent initiatives of urban and peri-urban home garden production into the larger fabric of urban food systems.

TECHNOLOGIES AND PRACTICES OF URBAN HOME GARDENING

There are many households with home garden vegetable cultivation for more than 20 years, and a few of them were beneficiaries of the government initiative 'A Village in the City' during 2005–07. The study (both in-person and online survey) underlines that pesticide-related health concerns of commercially available vegetables are the major reason for starting home garden vegetable cultivation. Home gardens enable them to provide healthy vegetables for family members, especially children, pregnant women, or cancer patients.

Respondents have used all possible spaces of their dwelling to grow vegetables, including rooftops, balconies, boundary walls, and modified parapet walls (Figure 5.3a, b, c, and h). They grow a variety of vegetables, fruits, and other crops irrespective of their growing season, and, in many instances, fruit trees (mango tree, plantain, etc.) or paddy are grown in containers placed on the roof (Figure 5.3c). Due to health concerns of chemical pesticides, the majority of the respondents have adopted agroecological measures (Figure 5.3e and f) to ensure safe-to-eat food for the families. The pest management systems and fertilizers are developed by the respondents using locally available resources or recycling household wastes, purchased from the government or private outlets or received as gifts from other enthusiastic gardeners.

FIGURE 5.3 Select technologies and practices of urban home gardening in Thiruvananthapuram Corporation. (a) Rooftop vegetable garden. (b) Gardening in modified parapet walls on a rooftop.

(Continued)

FIGURE 5.3 (*Continued*) (c) Paddy cultivation for the festive season in containers placed on a rooftop. (d) Passive hydroponics developed by a respondent.

(*Continued*)

FIGURE 5.3 (*Continued*)   (e) Home-made fertilizer prepared by fermentation of dry leaves at a respondent's home garden. (f) Insects stuck on a sticky trap at a respondent's home garden.

(*Continued*)

FIGURE 5.3 (*Continued*) (g) Author visiting a rooftop rain shelter that re-use containers for growing vegetables for the family. (h) Growing eggplants in hanging plastic cans on a rooftop.

Although most of the respondents use simple methods of growing on the ground or on the building, there are a few respondents who have installed greenhouse technologies such as polyhouse, their miniature form called mini-polyhouse, or rain shelters (Figure 5.3g). Respondents also use small-scale water-saving technologies such as drip or wick irrigation units in their home gardens. Soil-less cultivation technologies such as aquaponics, hydroponics, and passive hydroponics are also used by some of the respondents in both in-person and online surveys. Although these technologies are expensive at the commercial scale, respondents apply the same idea to develop low-budget small-scale systems suitable for small spaces (Figure 5.3d).

SUSTAINABILITY CONTRIBUTIONS OF URBAN HOME GARDENS IN KERALA

Urban home gardening in Kerala contributes significantly to social and ecological sustainability, whereas the economic benefits are limited. Some of the major social contributions of these home gardens include contributions to household vegetable supply, food sovereignty, enhanced livability, strengthening social relations, and enhanced mental health. Major ecological contributions include the circular use of kitchen wastes for gardening, conservation of local agrobiodiversity, enhanced ecosystem services, and reduced or near-zero food miles. Most of the respondents of the online survey (94%, sample across Kerala) make use of their kitchen wastes for vegetable gardening in various ways. By integrating food production with kitchen waste management, these urban home gardens facilitate waste reduction at the source and recycling of resources within the urban system. It also helps to reduce the external dependence on inputs and reduce the cost of vegetable cultivation.

The in-person survey of well-functioning home gardens highlights that home gardens, irrespective of size, can contribute to more than 90% of the vegetable consumption of the family. This indicates that home gardens have the potential to contribute significantly to the vegetable consumption of families if adequately planned. However, we get a different picture of this from the online survey. It shows that only a small proportion of home gardens (19%) are able to contribute to more than 50% of vegetable consumption of the families (Figure 5.4). The difference in the home garden contributions from in-person and online surveys indicates that despite the potential of home gardens (as evident from the in-person survey), the potential of urban home gardening is yet to explore its optimum. This points out towards the requirement for further efforts to strengthen urban home gardening to explore its optimum potential. Irrespective of the garden size and scale, most of the respondents gift the surplus to their neighbours, friends, and relatives to strengthen their social relations. Nearly 93% of respondents across Kerala grow local and traditional vegetable varieties in their home gardens, contributing to the conservation of local vegetable varieties and food sovereignty. The overall contributions of urban home gardening in Kerala to SDGs are summarized in Table 5.2.

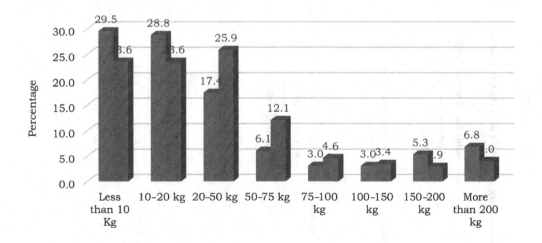

Quantity of vegetables

■ URBAN ■ PERI-URBAN

FIGURE 5.4 Annual production of vegetables harvested from home gardens across Kerala.

TABLE 5.2
Urban Home Gardening in Kerala and Its Contributions to Sustainable Development Goals

Category	Sustainable Development Goal	Relevant Targets of SDGs	Direct Contribution of Urban Home Gardening to the Target	Synergies with Other SDGs
Urban sustainability	1. End poverty in all its forms everywhere	Provides income and/or reduces the need to purchase food	Reduced food purchase, self-sufficiency in vegetable production, small additional income from sales of surplus produce	Increased availability of safe, nutritious, and diverse food (SDG 2)
Agri-food system sustainability	1. End poverty in all its forms everywhere	1.5 By 2030, build the resilience of the poor and those in vulnerable situations and reduce their exposure and vulnerability to climate-related extreme events and other economic, social, and environmental shocks and disasters	Enhanced food resilience in times of crisis such as COVID-19 lockdown periods, unexpected food supply cuts, and food price hikes	Reduced food miles (SDG 11, SDG 13)
Agri-food system sustainability	2. End hunger, achieve food security and improved nutrition, and promote sustainable agriculture	2.1 By 2030, end hunger and ensure access by all people, in particular, the poor and people in vulnerable situations, including infants, to safe, nutritious, and sufficient food all year round	Increased access to and availability of fresh, safe, diverse, nutritious, and tasty food for urban and peri-urban residents including children, pregnant women, and elderly	Good health and well-being (SDG 3)
Agri-food system sustainability		2.2 By 2030, end all forms of malnutrition, including achieving, by 2025, the internationally agreed targets on stunting and wasting in children under 5 years of age and address the nutritional needs of adolescent girls, pregnant and lactating women, and older persons	Increased access to and availability of fresh, safe, diverse, nutritious, and tasty food for urban and peri-urban residents including children, pregnant women, and elderly	
Agri-food system sustainability		2.3 By 2030, double the agricultural productivity and incomes of small-scale food producers, in particular, women, indigenous peoples, family farmers, pastoralists, and fishers, including through secure and equal access to land, other productive resources, and inputs, knowledge, financial services, markets, and opportunities for value addition and non-farm employment	Urban home gardening can be highly productive through making use of otherwise unused spaces, building-integrated production, and being open to diverse sectors of society. Also offers the opportunity to earn a small additional income	

(Continued)

TABLE 5.2 (Continued)
Urban Home Gardening in Kerala and Its Contributions to Sustainable Development Goals

Category	Sustainable Development Goal	Relevant Targets of SDGs	Direct Contribution of Urban Home Gardening to the Target	Synergies with Other SDGs
Agri-food system and urban sustainability	2. End hunger, achieve food security and improved nutrition, and promote sustainable agriculture	2.4 By 2030, ensure sustainable food production systems and implement resilient agricultural practices that increase productivity and production, that help maintain ecosystems, strengthen capacity for adaptation to climate change, extreme weather, drought, flooding, and other disasters, and that progressively improve land and soil quality	Urban home gardening in most cases adopts low-external input ecological production methods, reduces environmental damages by recycling waste at source, helps maintain urban ecosystems, and helps enhance urban microclimate	Increased pollination and pest services (SDG 2), improved green and clean surroundings (SDG 11), improved climate resilience (SDG 13), life on land (SDG 15), fewer emissions, e.g., from synthetic fertilizers and zero distance between the site of food production and consumption (SDG 13), circular urban resource flow and effective waste management (SDG 12)
Urban sustainability	3. Ensure healthy lives and promote well-being for all at all ages	3.4 By 2030, reduce by one-third premature mortality from non-communicable diseases through prevention and treatment and promote mental health and well-being	Improved health from access to safe, fresh, and nutritious food, enhanced vegetable intake, improved physical activity and mental health, strong social relations	Quality education on nutrition, children learning skills on sustainable food production (SDG 4)
Urban sustainability	8. Promote sustained, inclusive, and sustainable economic growth, full and productive employment, and decent work for all	8.4 Improve progressively, through 2030, global resource efficiency in consumption and production and endeavour to decouple economic growth from environmental degradation, in accordance with the 10-year framework of programmes on sustainable consumption and production	Urban home gardening can be both productive and low-input dependent, facilitating sustainable consumption and production. Unlike conventional agriculture, urban home gardening is not linked with environmental degradation.	Sustainable production and consumption (SDG 12), sustainable food production (SDG 2)

(Continued)

TABLE 5.2 (*Continued*)
Urban Home Gardening in Kerala and Its Contributions to Sustainable Development Goals

Category	Sustainable Development Goal	Relevant Targets of SDGs	Direct Contribution of Urban Home Gardening to the Target	Synergies with Other SDGs
Urban sustainability	11. Make cities and human settlements inclusive, safe, resilient, and sustainable	11.3 By 2030, enhance inclusive and sustainable urbanization and capacity for participatory, integrated, and sustainable human settlement planning and management in all countries	Enhanced livability through improved green and clean surroundings and effective waste management	Reduced pressure on rural land (SDG 15) Reduced food miles (SDG 13, SDG 11), sustainable production and consumption (SDG 12)
Agri-food system and urban sustainability	12. Ensure sustainable consumption and production patterns	12.2 By 2030, achieve the sustainable management and efficient use of natural resources	Provides a sustainable form of food production; effective waste management for circular resource flow within the urban systems, reducing waste export to urban peripheries	More sustainable cities (SDG 11)
		12.3 By 2030, halve per capita global food waste at the retail and consumer levels and reduce food losses along production and supply chains, including post-harvest losses	Short urban food supply chain; reduced wastage through gifting and exchange of surplus produce, direct interaction between home gardeners and consumers through local weekly urban markets	Increased access resulting from greater supply (and lower cost) of safe, fresh, and nutritious food (SDG 2), strong social relations (SDG 3)

Source: Authors' work inspired by Nicholls et al. (2020).

URBAN HOME GARDENING AS AN EMERGING NICHE

The urban home gardens explored in this study represent socio-ecological–technological systems and have different characteristics than the traditional home gardens. Urban home gardening per se is not a niche novelty emerged in recent times. It already existed. As Markard and Truffer (2008) point out, a niche comprises old technologies that have already existed for a while and have adopted new technological characteristics. Niches can also comprise a historic method or technology that is revived or revitalized (Sutherland et al., 2015b) or retro-innovations comprising retro-technologies and retro-practices (Slee & Pinto-Correia, 2015). The present urban home gardening in Kerala is revived and revitalized forms of traditional home gardening practices, with the addition of small technologies to suit the space-congested urban residential premises. The nature and scale of the technological component may, however, vary across the home gardens. The addition of technological aspects can be primarily attributed to space scarcity in the residential premises, the availability of new technologies and practices, and policy interventions. Like any other upcoming niches, urban home gardens in Kerala also received considerable subsidies and other support from mainstream actors such as the government and research institutions.

Urban home gardening was already practiced in Kerala by interested individuals, without much support received from the mainstream actors. Considering the high level of urbanization in Kerala, it was necessary to focus on urban food production in the policy interventions for boosting local vegetable production. The health concerns related to pesticide-contaminated commercial vegetables have acted as major pressure for people to take up home garden vegetable production and to make it a movement. Such turn to 'quality' on one side acts as landscape-level pressures and at the same can be considered as a characteristic of the niche (Darrot et al., 2015).

ANCHORING OF NICHE INNOVATIONS

In the case of urban home gardening in Kerala, we can see varying elements of institutional, network, and technological anchoring. New rules, values, and belief systems were developed in relation to urban home gardening, leading to institutional anchoring. Cognitive anchoring can be seen with the changes towards a 'quality turn' that further endorsed the notion that 'every little bit helps' for enhanced food production in contrast to the dominant question of 'available space' for enhanced food production. The belief systems also changed, favouring food production in the urban areas in contradiction to the common disconnection between urban areas and food production. The normative anchoring is primarily characterized by the changes in the policies and the research support that incorporated private residential spaces as sites of interventions for a collective goal of food system self-sufficiency. Economic anchoring is yet to become visible other than a few exceptions.

Transitions to sustainable agri-food systems may not necessarily be (physical) technology-driven (Peneva et al., 2015). With technological anchoring, traditional home gardening practices have become nature-based innovations based on retro-innovations that are supported with adequate physical technologies. The technologies and practices of urban home gardening have become more defined (circular, low-budget, nature-based, less technology-intensive, etc.). The deep-rooted cultural festivals and traditions have also played a major part in facilitating technological anchoring.

Network anchoring is characterized by people's movements and networking with the support of information technology, especially social media platforms. Normative anchoring was instrumental in further extending the network anchoring and therefore expanding and strengthening the niche innovations. The network anchoring has played a crucial role in making the independent initiatives a collective movement and thereby strengthening social relations that wouldn't have been otherwise possible.

The anchoring of niche innovations is not a linear process. It is a continuous process and does not necessarily ensure the stability of a niche or its embedding into the regime. As Elzen et al. (2012) demonstrate, these links can be broken down due to various reasons. Both top-down and bottom-up approaches have played significant roles in the anchoring of urban home gardening in Kerala. However, in agri-food systems that involve 'soft' forms of transitions, the boundaries of niche and anchoring processes cannot be identified clearly (Sutherland et al., 2015b). Therefore, unlike physical technology-based transition processes, we can see some overlapping in the niche development and different anchoring processes.

ENABLERS OF NICHE DEVELOPMENT

Different kinds of niches will require different types of support (Sutherland et al., 2015b). Various actors from inside and outside the agri-food regime act as enablers of the development of urban home gardening niches. These include state and non-state actors and the specific socio-cultural context of Kerala.

POLICY ACTORS

MLP considers policy actors as part of the prevailing regime (Sutherland et al., 2015b). However, governments at national and regional levels also have crucial roles in establishing a niche and its development (Sutherland et al., 2015b). In Kerala, policy actors have played a major role in the development and wide dissemination of niche innovations and making it a movement by facilitating a favourable support system. Their choice of interventions was instrumental in reaching its benefits to a wide section of society irrespective of their income status or availability of space in their residential premises. The overall interventions of the policymakers that focused on circular food production, less cost to the beneficiaries after subsidy, distribution of ready-to-start gardening units, facilitating research and extension systems, and agro-service centres for input availability in urban areas have played significant roles in the development of the niche. Along with the state interventions, the central government initiative VIUC also endorsed the urban agriculture niche in Kerala.

The policy actors act as hybrid actors in the agri-food system transitions of Kerala. They belong to the regime yet endorse niche innovations. Staying within the agri-food regime, the policy actors tried to bring substantial changes to challenge the basic notion of the regime itself, i.e., centralized and intensive commercial production using chemical inputs. The policy actors have shown that private residential spaces, including those in urban areas, can also play a huge role in contributing to the overall development of agriculture.

SCIENTISTS

Scientists are also hybrid actors in the agri-food system transitions in Kerala. Being part of the regime, a section of scientists engages in research and development activities that foster the development of urban home garden niches. The mainstream researchers from Kerala Agricultural University (KAU) (Resp. ID #19, personal communication, 2022; Resp. ID #22, personal communication, 2022) underline the increase in research on urban agriculture-related topics in KAU. Though there are variations in the way scientists support the development of technologies and practices of urban home gardening, they are convinced by the fact that urban areas need to be the sites of food production. Although scientists belong to the incumbent regime, they have played a major role in fostering the urban home gardening niche and bringing it to the mainstream agenda.

EXPECTATIONS AND VISIONS, PEOPLE'S NETWORKS, AND LEARNING

Expectations and vision, development of networks, and learning are the three aspects that are crucial for the SNM of niches in providing direction for the transition and enabling protection of the niche (El Bilali, 2018). In Kerala, the development of expectations and common visions are grounded on overcoming the vulnerabilities associated with food dependency, which has resulted in the formation of networks using Facebook as a medium. These networks have been instrumental in expanding the niche innovations to larger geographic areas, which would have been difficult otherwise, resulting in the home gardening movement.

Learning processes can take place at multiple dimensions such as technical aspects, societal and environmental effects, cultural meanings, and government policy (Schot & Geels, 2008). In Kerala, learning primarily involves the social and environmental constraints of the prevailing conventional agri-food regime and the everyday knowledge of the practice of home gardening especially in urban areas and other areas with space constraints. Both strong niche actors (people's movement, bottom-up learning) and hybrid actors (policy actors and scientists, top-down learning) have contributed to the learning process. The expectations and vision, building networks, and learning have further strengthened the niche by linking home garden vegetable cultivation with the urban short food supply chains.

SOCIO-CULTURAL FACTORS

The socio-cultural factors such as a high literacy rate, strong history of social movements, strong presence of print and television media, wide circulation of agricultural weeklies, and deep-routed tradition of home gardening have contributed to the niche development and its strengthening. The strong tradition of home gardening helped to provide a starting point instead of developing a completely new niche. The rich cultural festivals provide a common goal to both the state and people to aim for the collective harvesting of safe vegetables from the home gardens. All the above-mentioned factors have further contributed to making independent urban home gardening practices carried out in private residences a state-level movement.

TRAPPED IN THE LOCK-INS: THE REGIME'S RESISTANCE

Lock-ins and path dependency are the two major barriers to sustainability transitions. Regime actors can simultaneously act as niche creators (Rajagopalan & Breetz, 2022) and barriers to its development. Like in many other places, in Kerala as well, the sustainability transitions in agri-food systems face regime resistance that is trapped in the notions of productivity and technological paradigm. Although the focus of this study is specifically on the urban home gardening niche, the larger resistance elicited on the umbrella niche of organic agriculture has implications for the further development of urban home gardening and its incorporation into sustainability interventions in urban and agri-food systems in Kerala. In addition, the lock-ins of urban planning systems that do not consider multifunctional urban agriculture impose regime resistance.

The strong regime supporters are part of the policy-making system in the state. The two key actors that presently oppose the organic agriculture niche were once part of facilitating a support system for the niche. These are the Kerala State Planning Board and the Kerala State Biodiversity Board. Over the period of time, the belief systems of the then-niche supporters have changed primarily due to the changes in individuals who form these bodies. Strong resistance is also coming from a section of agricultural scientists who endorse the regime in its status quo. They consider all types of chemical-free agriculture as irrational practices, do not consider the environmental impact of chemical pesticides, and, hence, do not endorse reducing the use of chemical pesticides in Kerala (Kerala State Planning Board, 2021, 2022). These arguments are in complete rejection of one of the major reasons that triggered food system localization including home garden food production in Kerala. Instead, they endorse intensification in agriculture and embrace the adoption of technological silver bullets (biotechnology, gene editing, and nanobiotechnology) for reduced use of chemical pesticides and for enhancing 'productivity, profitability, and sustainability' (Kerala State Planning Board, 2022) in agriculture. The path dependencies and lock-ins of agricultural research systems result in favouring of these technological regimes while hindering the development of agroecological alternatives (Vanloqueren & Baret, 2009).

The resistance of regime actors on the umbrella niche of organic agriculture has implications for the government's promotion of urban home gardening and what trajectory it may take. Accordingly, we can see that recently the government has discontinued the exclusive focus given to urban home gardening and waste management as part of VDP. With such exclusionary practices, government interventions are acting as barriers to further popularization and adoption of urban home gardening practices to a wide mass. One major reason for backing out from focusing on urban areas could be the dominant regime perspectives that do not consider food production as an integral activity of urban areas.

In Kerala, another regime resistance is the lack of policy consideration of urban home gardening as one of the interventions to enhance urban sustainability. Urban planning in Kerala is still trapped in the lock-ins of infrastructural development without much consideration for enhancing the urban ecology. This makes urban agriculture a concern of only agriculture department, which is further influenced by the values and belief systems of agriculture scientists. Agriculture scientists, irrespective of their support of regime or niche, have different belief systems when it comes to the question of linking urban home garden cultivation with the urban short food supply chains. This is another barrier to mainstreaming the urban home gardening niche into an integral part of urban areas. The continued regime resistance will pose serious challenges to the further development of urban home gardening as a niche.

URBAN HOME GARDENING AND AGRI-FOOD SYSTEM TRANSITIONS IN KERALA

The mutual interactions and reinforcements between different social groups and the government provide the internal momentum for the sustainability transition (Belz, 2004). The bottom-up and top-down actors in Kerala reinforced the urban home gardening niche and impacted each other in the transition. Instead of taking a parallel pathway, the FB initiatives tried to fill the loopholes of mainstream intervention and also strengthen it. The role of government was important for complementing the grassroots initiatives with policy and extension support. However, the main challenge for niche development comes from the prevailing regime only.

This chapter does not suggest that urban home gardening will replace the conventional agri-food regime. Emerging urban home gardens simultaneously challenge the conventional notion of resource-dependent urban systems and the prevailing agri-food regime that relies upon intensive centralized production. Private residences can also be the sites of intervention for enhancing the decentralized local food production. Urban home gardening specifically challenges the notion of an incumbent regime that excludes urban areas as sites of food production, which also offers multiple sustainability benefits. Unlike the sustainability transitions that are based on physical technologies (such as sensors, automation, drones, and artificial intelligence), the urban home gardening niche enables 'soft' forms of urban and agri-food system transitions that emphasize nature-based solutions while also embedding some elements of physical technologies.

Given the rural–urban continuum, small-scale collective production through urban home gardening has significant roles to play in reducing the dependency of Kerala on vegetable supplies and enhancing sustainability. We have examples from history (World Wars and Special Period in Cuba) that urban agriculture can significantly contribute to overall vegetable production (Koont, 2008; Lawson, 2014). Supporting decentralized small-scale production with key centralized services is crucial to overcoming the barriers faced in small-scale production (Swaminathan, 2013). The key is to provide adequate institutional-level support that can facilitate the mainstreaming of urban home gardening as part of the food system and urban sustainability.

The urban home gardening niche is still in its pre-development stage. The take-off will require strong policy-level changes and social mobilization. However, the government support system for urban home gardening does not tend to deviate from the prevailing regime rules and institutional structure. Regardless of the potential to make transformative changes, urban home gardening in Kerala has taken neither the 'fit-and-conform' pattern nor the 'stretch-and-transform' pattern as the institutional support system has suddenly been discontinued. The dichotomous classification of the 'fit-and-conform' pattern and 'stretch-and-transform' patterns of transition may not necessarily be relevant in this case. A similar observation has been made by Dutt (2022) from a Global South perspective. The latest policy changes will have a huge impact on the niche development and its take-off. However, with the strong presence of social mobilization, it is highly unlikely that the niche will fail.

CONCLUSION

Using the MLPs of sustainability transitions theory, this study analysed the urban home gardening niche in the context of agri-food system transitions in Kerala. The chapter shows that the state actors and non-state actors have played significant roles in the niche development. Despite the sustainability contributions of urban home gardening, the recent shifts in beliefs and value systems of some of the state actors act as barriers for niche development, which further resulted in taking up neither 'fit and conform' patterns nor 'stretch and transform' patterns of transitions. However, with the strong and continued involvement of social actors who try to fill the gaps in government interventions, it is highly unlikely that niche innovation will fail. The study highlights the problematic application of the dichotomous classification of neither 'fit and conform' patterns nor 'stretch and transform' patterns of transitions in the Global South. The chapter contributes to the growing literature on agri-food system transitions and urban agriculture in the Global South. This chapter focuses on 'soft' transitions in agri-food system practices and policies that are built upon traditional practices and have local-specific relevance. The chapter also highlights the relevance of urban home gardening in the Global South that has been hugely overlooked in the urban agriculture literature and sustainability transitions research.

REFERENCES

Al-Chalabi, M. (2015). Vertical farming: Skyscraper sustainability? *Sustainable Cities and Society, 18,* 74–77. https://doi.org/10.1016/j.scs.2015.06.003

Al-Kodmany, K. (2018). The vertical farm: A review of developments and implications for the vertical city. *Buildings, 8*(2), Article 2. https://doi.org/10.3390/buildings8020024

Altieri, M. A., Companioni, N., Cañizares, K., Murphy, C., Rosset, P., Bourque, M., & Nicholls, C. I. (1999). The greening of the "barrios": Urban agriculture for food security in Cuba. *Agriculture and Human Values, 16*(2), 131–140.

Anderson, C. R., Bruil, J., Chappell, M. J., Kiss, C., & Pimbert, M. P. (2019). From transition to domains of transformation: Getting to sustainable and just food systems through agroecology. *Sustainability, 11*(19), Article 19. https://doi.org/10.3390/su11195272

Artmann, M., & Sartison, K. (2018). The role of urban agriculture as a nature-based solution: A review for developing a systemic assessment framework. *Sustainability, 10*(6), 1937.

Audet, R., Lefèvre, S., Brisebois, É., & El-Jed, M. (2017). Structuring tensions and key relations of montreal seasonal food markets in the sustainability transition of the agri-food sector. *Sustainability, 9*(3), 320.

Avelino, F., Grin, J., Pel, B., & Jhagroe, S. (2016). The politics of sustainability transitions. *Journal of Environmental Policy & Planning, 18*(5), 557–567. https://doi.org/10.1080/1523908X.2016.1216782

Barthel, S., Parker, J., & Ernstson, H. (2015). Food and green space in cities: A Resilience lens on gardens and Urban environmental movements. *Urban Studies, 52*(7), 1321–1338. https://doi.org/10.1177/0042098012472744

Bell, S., & Cerulli, C. (2012). Emerging community food production and pathways for Urban landscape transitions. *Emergence: Complexity & Organization, 14*(1). 31–44.

Belmin, R., Meynard, J.-M., Julhia, L., & Casabianca, F. (2018). Sociotechnical controversies as warning signs for niche governance. *Agronomy for Sustainable Development, 38*(5), 1–12.

Belz, F.-M. (2004). A transition towards sustainability in the Swiss agri-food chain (1970–2000): Using and improving the multi-level perspective. *System Innovation and the Transition to Sustainability,* 97–114.

Census 2011. (n.d.-a). *Kerala Population Sex Ratio in Kerala Literacy rate data* 2011–2022. Retrieved 13 September 2022, from https://www.census2011.co.in/census/state/kerala.html

Census 2011. (n.d.-b). *Literacy Rate of India-Population Census* 2011. Retrieved 17 September 2022, from https://www.census2011.co.in/literacy.php

Cook, J., Oviatt, K., Main, D. S., Kaur, H., & Brett, J. (2015). Re-conceptualizing urban agriculture: An exploration of farming along the banks of the Yamuna River in Delhi, India. *Agriculture and Human Values, 32*(2), 265–279.

Darrot, C., Diaz, M., Tsakalou, E., & Zagata, L. (2015). 'The missing actor': Alternative agri-food networks and the resistance of key regime actors. In *Transition pathways towards sustainability in agriculture: Case studies from Europe* (pp. 143–155). CABI Wallingford UK.

Davids, P., & De Olde, E. (2014). Urban landscape transitions: Analyzing Urban agriculture initiatives in four European cities. Council of Educators in Landscape Architecture. Baltimore, MD.

Derwort, P., Jager, N., & Newig, J. (2021). How to explain major policy change towards sustainability? Bringing together the multiple streams framework and the multilevel perspective on socio-technical transitions to explore the German "Energiewende". *Policy Studies Journal, n/a*(n/a). https://doi.org/10.1111/psj.12428

Directorate of AD & FW, K. (2021). *Scheme for Vegetable Development Mission - Vegetable Development Programme 2021–22. Administrative Sanction accorded - Working Instructions issued - Reg.* https://keralaagriculture.gov.in/wp-content/uploads/2021/06/Final-VDP-2021-22-Working-Instructions.pdf

Directorate of Agriculture. (2012). *VC 20622(3)/12.* Circular. *Agriculture Department-Scheme for Development of Vegetables 2012-13.* Government of Kerala. https://www.keralaagriculture.gov.in/pdf/wi_39.pdf

Directorate of Agriculture. (2015). *VC 10533/15, Circular. Agriculture Department-Scheme for Development of Vegetables 2015-16.* Government of Kerala. https://www.keralaagriculture.gov.in/APS_2015-2016/wi_2015_pdf/2015_VC10533-15.pdf

Directorate of Agriculture. (2017). *VC 50176/16. Scheme for Development of Vegetables 2017-18 - Green Book Components-Administrative Sanctions and Working Instructions.* Government of Kerala. https://keralaagriculture.gov.in/wp-content/uploads/2019/05/50176_2017.pdf

Directorate of Agriculture. (2018). *Scheme for Development of Vegetables 2018-19 -Amber Book Components - Continuos Administrative Sanction and Working Instructions.* Government of Kerala. https://keralaagriculture.gov.in/wp-content/uploads/2019/01/circular_ps_2018-19_8484_02.pdf

Dreze, J., & Sen, A. (2002). *India: Development and Participation.* Oxford University Press on Demand.

Dutt, D. (2022). How power and politics shape niche-regime interactions: A view from the Global South. *Environmental Innovation and Societal Transitions, 43,* 320–330.

El Bilali, H. (2018). Transition heuristic frameworks in research on agro-food sustainability transitions. *Environment, Development and Sustainability, 22*(3), 1693–1728.

El Bilali, H., & Allahyari, M. S. (2018). Transition towards sustainability in agriculture and food systems: Role of information and communication technologies. *Information Processing in Agriculture, 5*(4), 456–464.

Elzen, B., Van Mierlo, B., & Leeuwis, C. (2012). Anchoring of innovations: Assessing Dutch efforts to harvest energy from glasshouses. *Environmental Innovation and Societal Transitions, 5,* 1–18.

Evans, D. L., Falagán, N., Hardman, C. A., Kourmpetli, S., Liu, L., Mead, B. R., & Davies, J. A. C. (2022). Ecosystem service delivery by urban agriculture and green infrastructure - a systematic review. *Ecosystem Services, 54,* 101405. https://doi.org/10.1016/j.ecoser.2022.101405

Fox, T. A., Rhemtulla, J. M., Ramankutty, N., Lesk, C., Coyle, T., & Kunhamu, T. K. (2017). Agricultural land-use change in Kerala, India: Perspectives from above and below the canopy. *Agriculture, Ecosystems & Environment, 245,* 1–10.

Geels, F. W. (2002). Technological transitions as evolutionary reconfiguration processes: A multi-level perspective and a case-study. *Research Policy, 31*(8-9), 1257–1274.

Geels, F. W. (2011). The multi-level perspective on sustainability transitions: Responses to seven criticisms. *Environmental Innovation and Societal Transitions, 1*(1), 24–40.

Geels, F. W. (2020). Micro-foundations of the multi-level perspective on socio-technical transitions: Developing a multi-dimensional model of agency through crossovers between social constructivism, evolutionary economics and neo-institutional theory. *Technological Forecasting and Social Change, 152,* 119894.

Geels, F. W., Kern, F., Fuchs, G., Hinderer, N., Kungl, G., Mylan, J., Neukirch, M., & Wassermann, S. (2016). The enactment of socio-technical transition pathways: A reformulated typology and a comparative multi-level analysis of the German and UK low-carbon electricity transitions (1990-2014). *Research Policy, 45*(4), 896–913.

Geels, F. W., & Schot, J. (2007). Typology of sociotechnical transition pathways. *Research Policy, 36*(3), 399–417.

Govind, M., & Pinheiro, A. (Forthcoming). Multifunctional urban agriculture and sustainability: A global overview. In *Sustainable Urban Agriculture: New Frontiers.* CRC Press, Taylor and Francis Group.

Guillerme, S., Kumar, B. M., Menon, A., Hinnewinkel, C., Maire, E., & Santhoshkumar, A. V. (2011). Impacts of public policies and farmer preferences on agroforestry practices in Kerala, India. *Environmental Management, 48*(2), 351–364.

Hisschemoller, M. (2016). Cultivating the glocal garden. *Challenges in Sustainability, 4*(1), 28–38.

Karanikolas, P., Vlahos, G., & Sutherland, L. A. (2015). Utilizing the multi-level perspective in empirical field research: Methodological considerations. In *Transition Pathways towards Sustainability in Agriculture: Case Studies from Europe;* Sutherland, L.-A., Darnhofer, I., Wilson, GA, Zagata, L., (Eds) (pp. 51–66). CABI Wallingford UK.

Kerala Agricultural University. (2014). *Report No: PAMSTEV 7/2014, April-June 2014.* https://kau.in/document/1052

Kerala State Planning Board. (2013). *Twelfth Five Year Plan 2012–17*. https://www.spb.kerala.gov.in/images/pdf/five_year_plan/five_y_plan_12_17.pdf

Kerala State Planning Board. (2018). *Thirteenth Five Year Plan 2017–22*. Government of Kerala. https://spb.kerala.gov.in/sites/default/files/2021-09/13PlanEng.pdf

Kerala State Planning Board. (2021). *Kerala Development Report: Initiatives, Achievements, Challenges* (p. 338). Government of Kerala. https://spb.kerala.gov.in/sites/default/files/inline-files/Kerala-Development-Report-2021.pdf

Kerala State Planning Board. (2022). *Fourteenth Five Year Plan (2022–27) Approach Paper Draft*. Government of Kerala. https://spb.kerala.gov.in/sites/default/files/2022-06/approach%20paper_vc_final_09062022%20english_website.pdf

Köhler, J., Geels, F. W., Kern, F., Markard, J., Onsongo, E., Wieczorek, A., Alkemade, F., Avelino, F., Bergek, A., & Boons, F. (2019). An agenda for sustainability transitions research: State of the art and future directions. *Environmental Innovation and Societal Transitions, 31*, 1–32.

Koont, S. (2008). A Cuban success story: Urban agriculture. *Review of Radical Political Economics, 40*(3), 285–291.

Lawson, L. J. (2014). Garden for victory! The American victory garden campaign of World War II. In *Greening in the Red Zone* (pp. 181–195). Springer.

Levidow, L., Pimbert, M., & Vanloqueren, G. (2014). Agroecological research: Conforming-or transforming the dominant agro-food regime? *Agroecology and Sustainable Food Systems, 38*(10), 1127–1155.

Loorbach, D. (2007). Transition Management. *New Mode of Governance for Sustainable Development*. Utrecht: International Books.

Loorbach, D., Frantzeskaki, N., & Avelino, F. (2017). Sustainability transitions research: Transforming science and practice for societal change. *Annual Review of Environment and Resources, 42*(1), 599–626. https://doi.org/10.1146/annurev-environ-102014-021340

Maćkiewicz, B., & Asuero, R. P. (2021). Public versus private: Juxtaposing urban allotment gardens as multifunctional Nature-based Solutions. Insights from Seville. *Urban Forestry & Urban Greening, 65*, 127309. https://doi.org/10.1016/j.ufug.2021.127309

Maltz, A. (2015). "Plant a victory garden: Our food is fighting:" Lessons of food resilience from World War. *Journal of Environmental Studies and Sciences, 5*(3), 392–403.

Markard, J., & Truffer, B. (2008). Technological innovation systems and the multi-level perspective: Towards an integrated framework. *Research Policy, 37*(4), 596–615.

Mylan, J., Morris, C., Beech, E., & Geels, F. W. (2019). Rage against the regime: Niche-regime interactions in the societal embedding of plant-based milk. *Environmental Innovation and Societal Transitions, 31*, 233–247.

Nicholls, E., Ely, A., Birkin, L., Basu, P., & Goulson, D. (2020). The contribution of small-scale food production in urban areas to the sustainable development goals: A review and case study. *Sustainability Science, 15*(6), 1585–1599. https://doi.org/10.1007/s11625-020-00792-z

Ollivier, G., Magda, D., Mazé, A., Plumecocq, G., & Lamine, C. (2018). Agroecological transitions: What can sustainability transition frameworks teach us? An ontological and empirical analysis. *Ecology and Society, 23*(2). https://doi.org/10.5751/ES-09952-230205

Orsini, F., Gasperi, D., Marchetti, L., Piovene, C., Draghetti, S., Ramazzotti, S., Bazzocchi, G., & Gianquinto, G. (2014). Exploring the production capacity of rooftop gardens (RTGs) in urban agriculture: The potential impact on food and nutrition security, biodiversity and other ecosystem services in the city of Bologna. *Food Security, 6*(6), 781–792.

Peneva, M., Draganova, M., Gonzalez, C., Diaz, M., & Mishev, P. (2015). High nature value farming: Environmental practices for rural sustainability. In *Transition pathways towards sustainability in agriculture: Case studies from Europe* (pp. 97–111). CABI Wallingford UK.

Pinheiro, A. (2022a). *Technology and Policy Landscpe for Urban Agriculture in Kerala: Exploring the Sustainability Implications*. Jawaharlal Nehru University.

Pinheiro, A. (2023). Policy Interventions for Strengthening Urban Home Gardens in Kerala: Exploring the Case of Thiruvananthapuram Corporation. In *Sowing Sustainable Cities: Lessons for urban agriculture practices in India* (pp. 136–141). Indian Institute for Human Settlements. https://iihs.co.in/knowledge-gateway/wp-content/uploads/2023/04/UPAGrI_Final-_Digital-PDF.pdf

Pinheiro, A. (2022b). *Urban home gardening movement in Kerala-Role of social media collectives*. LEISA-India. https://leisaindia.org/urban-home-gardening-movement-in-kerala-role-of-social-media-collectives/

Pinheiro, A. (2023). Scaling urban agriculture initiatives: Need for supportive policy ecosystems. In *Cultivating Hope: Exploring food growing possibilities in Indian cities* (edited by Dutta, D. and Hazra, A.).

Pinheiro, A., & Govind, M. (2020). Emerging global trends in Urban agriculture research: A scientometric analysis of peer-reviewed journals. *Journal of Scientometric Research*, *9*(2), 163–173. https://doi.org/10.5530/jscires.9.2.20

Pinheiro, A., Saxena, A., & Fatima, M. (2022, December 7). *Urban agriculture: Towards an inclusive approach to make Delhi an 'edible city'*. Down To Earth. https://www.downtoearth.org.in/news/agriculture/urban-agriculture-towards-an-inclusive-approach-to-make-delhi-an-edible-city--83670

Rajagopalan, S., & Breetz, H. L. (2022). Niches, narratives, and national policy: How India developed off-grid solar for rural electrification. *Environmental Innovation and Societal Transitions*, *43*, 41–54. https://doi.org/10.1016/j.eist.2022.02.004

Raven, R., Kern, F., Verhees, B., & Smith, A. (2016). Niche construction and empowerment through socio-political work. A meta-analysis of six low-carbon technology cases. *Environmental Innovation and Societal Transitions*, *18*, 164–180.

Resp. ID #19. (2022). *Response from an agricultural scientist ID #19* [Survey conducted through Google forms].

Resp. ID #22. (2022). *Response from an agricultural scientist ID #22* [Survey conducted through Google forms].

Rijesh, N. M., Mohammed, F. C., Susan, C., & Sruthi, K. V. (2022). Sustainable Spatial Planning for the Rural-Urban Continuum Settlements of Kerala, India. In *Sustainable Urbanism in Developing Countries*. CRC Press.

Sarabia, N., Peris, J., & Segura, S. (2021). Transition to agri-food sustainability, assessing accelerators and triggers for transformation: Case study in Valencia, Spain. *Journal of Cleaner Production*, *325*, 129228.

Sartison, K., & Artmann, M. (2020). Edible cities - An innovative nature-based solution for urban sustainability transformation? An explorative study of urban food production in German cities. *Urban Forestry & Urban Greening*, *49*, 126604. https://doi.org/10.1016/j.ufug.2020.126604

Schiller, K., Godek, W., Klerkx, L., & Poortvliet, P. M. (2020). Nicaragua's agroecological transition: Transformation or reconfiguration of the agri-food regime? *Agroecology and Sustainable Food Systems*, *44*(5), 611–628. https://doi.org/10.1080/21683565.2019.1667939

Schot, J., & Geels, F. W. (2008). Strategic niche management and sustainable innovation journeys: Theory, findings, research agenda, and policy. *Technology Analysis & Strategic Management*, *20*(5), 537–554. https://doi.org/10.1080/09537320802292651

Sharma, S., Dhanda, N., & Verma, R. (2023). Urban vertical farming: A Review. *2023 13th International Conference on Cloud Computing, Data Science & Engineering (Confluence)*, Noida, India, 432–437.

Slee, B., & Pinto-Correia, T. (2015). Understanding the diversity of European rural areas. In *Transition pathways towards sustainability in agriculture: Case studies from Europe*. Sutherland, L.-A., Darnhofer, I., Wilson, GA, Zagata, L., (Eds.) (pp. 33–50). CABI Wallingford UK.

Smith, A. (2006). Green niches in sustainable development: The case of organic food in the United Kingdom. *Environment and Planning C: Government and Policy*, *24*(3), 439–458.

Smith, A., & Raven, R. (2012). What is protective space? Reconsidering niches in transitions to sustainability. *Research Policy*, *41*(6), 1025–1036.

Sutherland, L. A., Peter, S., & Zagata, L. (2015). On-farm renewable energy: A 'classic case' of technological transition. *Transition Pathways towards Sustainability in Agriculture. Case Studies from Europe*, Sutherland, L.-A., Darnhofer, I., Wilson, GA, Zagata, L., (Eds.) (pp. 113–126). CABI Wallingford UK.

Sutherland, L., Darnhofer, I., Wilson, G. A., & Zagata, L. (2015a). *Transition Pathways Towards Sustainability in Agriculture: Case Studies from Europe*. CABI.

Sutherland, L.-A., Zagata, L., & Wilson, G. A. (2015b). Conclusions. In *Transition pathways towards sustainability in agriculture: Case studies from Europe*, Sutherland, L.-A., Darnhofer, I., Wilson, GA, Zagata, L (pp. 205–220). CABI Wallingford UK.

Swaminathan, M. S. (2013). The cooperative pathway of enhancing rural livelihood and nutrition security. *International Journal of Rural Management*, *9*(1), 1–15.

Vanloqueren, G., & Baret, P. V. (2009). How agricultural research systems shape a technological regime that develops genetic engineering but locks out agroecological innovations 1. In *Food sovereignty, agroecology and biocultural diversity* (pp. 57–92). Routledge.

Walthall, B. (2016). Strengthening city region food systems: Synergies between multifunctional peri-urban agriculture and short food supply chains: a local case study in berlin, Germany. In *Land Use Competition* (pp. 263–277). Springer.

Wright, S., Sharpe, S., & Giurco, D. (2018). Greening regional cities: The role of government in sustainability transitions. In *Sustainable development research in the Asia-Pacific region* (pp. 327–343). Springer.

Zasada, I., Weltin, M., Zoll, F., & Benninger, S. L. (2020). Home gardening practice in Pune (India), the role of communities, urban environment and the contribution to urban sustainability. *Urban Ecosystems*, *23*(2), 403–417.

6 In Pursuit of a Traditional Livelihood

Insights from Singapore's Farmers

Jessica Ann Diehl

INTRODUCTION

Growing urban populations' increased demand for food, coupled with the inherent risks of relying on the global food system, has spurred city governments to implement strategies for integrating urban agriculture at different scales. As a city-state that imports 90% of its food, Singapore is vulnerable to food insecurity, particularly during the COVID-19 pandemic with global shipping disruptions. Development is shrinking arable land in Singapore, yet there is increasing interest in domestic food production. The Singapore Green Plan 2030 '30 by 30' Grow Local target is a government initiative to increase domestic food production to 30% by 2030 (Singapore Government Agency, 2023). To bolster the initiative, the government is phasing out traditional farming and offering grants and incentives to new high-tech agricultural startups. As a result, food production is shifting toward high-tech operations (rooftop hydroponics, vertical systems, and indoor LED farms). Can it achieve '30 by 30' without traditional farming? And, with a generation of new entrepreneurs joining the industry, is it sustainable?

Agriculture is a practice directly driven by farmers as a livelihood pursuit and indirectly a government-led effort in sustainable food production. The sustainability of the local food system is directly impacted by decisions made by farmers embedded within multiple, cross-scale urban systems. Therefore, we must first identify the incentives and barriers that farmers face as primary drivers for behavior and decision-making. This chapter explores agriculture in Singapore as an embedded practice. The aim is to bridge the conceptual and practical divide between localized human experience and abstracted structural conditions that determine the sustainability of the urban food system. The lens of political ecology is applied to better understand Singapore's food security from a traditional farmers' perspective.

Political ecology serves as the frame. What is political ecology? It posits that environmental changes do not affect people homogeneously; an unequal distribution of costs and benefits reinforces, increases, or reduces social and economic inequalities (Robins, 2012). More importantly, adopting a political ecology approach is a

> theoretical commitment to critical social theory and a post-positivist understanding of nature and the production of knowledge about it, which views these as inseparable from social relations of power … [it is] a methodological commitment to in-depth, direct observation involving qualitative research of some sort, often in combination with quantitative methods and/or document analysis … [and, it is] a normative political commitment to social justice and structural political change … [highlighting] the struggles, interests, and plight of marginalized populations.

> *(Perreault et al., 2015, pp. 7–8)*

Aligned with this, I drew from sociology, anthropology, geography, urban planning, and environmental sciences for relevant theory and research in participatory planning, vulnerability and resiliency, and social network research.

DOI: 10.1201/9781003359425-6

SINGAPORE: A SMALL ISLAND NATION

Located in Southeast Asia off the southern tip of the Malaysian peninsula (Figure 6.1), Singapore is a 279-square-mile sovereign city-state island. It is one of the densest cities in the world, with 21,647 persons per square mile, and is categorized as 100% urban (Worldometer, 2023). With less than 1% of land allocated for agriculture, Singapore depends on overseas hinterlands in Malaysia, China, Australia, and other countries. It ranks high on the Global Food Security Index for availability and affordability of food (2/113 and 3/113, respectively) yet ranks low for sustainability and adaptation (92/113) (The Economist Group, 2019). The high rank is because of purchasing power due to high per capita income of $58,770 (US Dollar; 2018) (The World Bank, 2019b), the low rank is due to low domestic food production. Historically, the reverse was true.

Only 200 years ago, Singapore was covered in dense tropical forest surrounded by coastal mangroves. The island supported a tiny population of 1,000 people dependent on fishing livelihoods (Swee-Hock, 2012). With British arrival in 1819, the landscape changed rapidly as primary forests were cleared for large rubber and gambier plantations and settlements to accommodate the rapidly growing population, which reached 100,000 by 1871. What became evident with

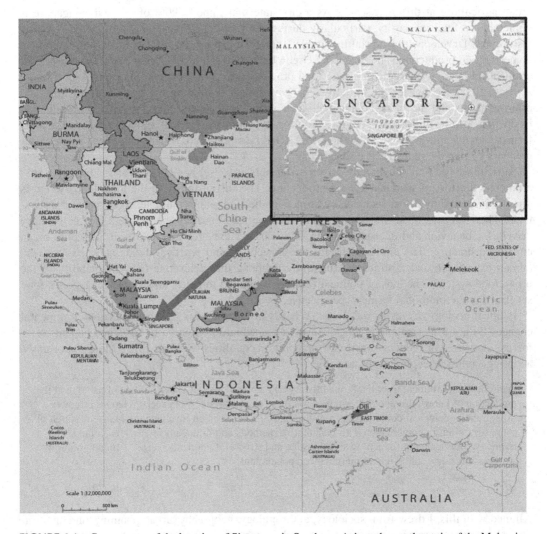

FIGURE 6.1   Context map of the location of Singapore in Southeast Asia at the southern tip of the Malaysian peninsula. (Source: http://images.google.com.)

large-scale land clearing, peaking in 1889 with 90% cleared, was the island's poor soil quality (O'Dempsey et al., 2014). Singapore tropical forest soils are highly weathered, acidic, and nutrient poor. Unlike other terrestrial ecosystems, the environmental conditions of tropical rainforests promote rapid break-down and decomposition of organic matter (Osborne, 2012). Rapid turnover of nutrients and high precipitation lead to high rates of soil leaching. Under natural conditions, leached nutrients and minerals are taken up by trees and other plants' extensive root networks. However, when forests in Singapore were cleared for plantation preparation, the thin topsoil layer could not hold nutrients. This led to the collapse of the plantation system and the end of colonial economic interest.

Singapore became an independent sovereign nation in 1965, established as a unitary parliamentary republic. At that time, 19% of the land area remained in agriculture. As an island with few natural resources and facing severe unemployment and a housing crisis, the new government had a vision of modernization predicated on establishing a manufacturing industry, developing large public housing estates, and significant investment in public education. Human capital was touted as Singapore's most important resource, but the rapid economic growth can also be attributed to natural capital. Located two degrees north of the equator, the island is protected from seasonal monsoons and experiences low climate change variability. It sits at the apex of East-West global shipping routes with a naturally deep channel, and 'friendly' trade policies have allowed it to become a hub of global shipping.

Fast forward to today: Singapore is classified as 100% urban and the remaining farms are located in the northwest, occupying less than 1% of land still allocated for agriculture (Figure 6.2). Commercial farming (at the time of this chapter) produces 10% of domestic consumption and is

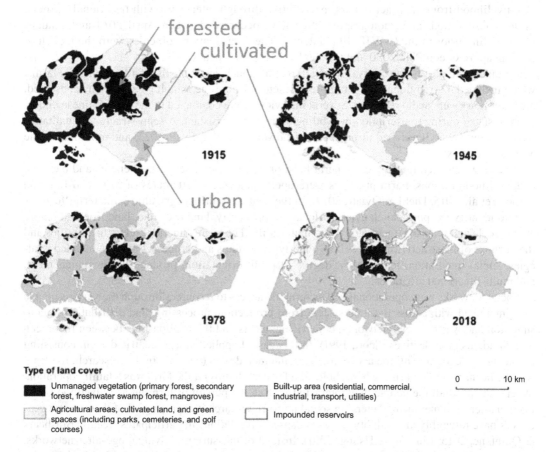

Type of land cover

- ■ Unmanaged vegetation (primary forest, secondary forest, freshwater swamp forest, mangroves)
- ▨ Agricultural areas, cultivated land, and green spaces (including parks, cemeteries, and golf courses)
- ▨ Built-up area (residential, commercial, industrial, transport, utilities)
- □ Impounded reservoirs

0   10 km

FIGURE 6.2   Change in forested, cultivated, and urban land cover in Singapore, 1915–2018 (based on Fong, Leng, & Taylor, 2020).

restricted to poultry, eggs, fish, and vegetables (Agri-Food & Veterinary Authority of Singapore, 2017). Realizing the risk to national food security due to low domestic production, the Singapore Green Plan 2030 lays out a '30 by 30' Grow Local target to build up agri-food industry capability and capacity to produce 30% of nutritional needs locally and sustainably and ultimately help build a more resilient food future.

It is important to note that three-quarters of land in Singapore is government owned and managed by the Singapore Land Authority (SLA). Arable land continues to be developed, and the interest in domestic food production is focused on high-tech, high-producing, land-limited or landless farms (hydroponics, vertical systems). SLA collaborates with other government agencies to offer entrepreneurs of non-traditional, multi-functional urban farms flexibility in land use restrictions—such as co-locating with commercial and residential uses (Diehl et al., 2020). As food production shifts toward high-tech operations, the question remains as to whether Singapore is likely to achieve '30 by 30' while phasing out traditional farming.

DOCUMENTING TRADITIONAL FARMERS' PERSPECTIVES

Based on a mixed-methods research project, this chapter summarizes an investigation of urban agriculture using the lens of political ecology to better understand food security in Singapore from the perspective of traditional commercial farmers. Traditional farming in the Singapore context is defined as open-space production of market-oriented, high-value products on undeveloped land that is predominantly government owned and leased, managed by individual farmers, often as a family operation. Specifically, the author sought to understand farming as a sustainable livelihood from the practitioners' perspective through interviews with traditional farmers, in-field observation, and photography. Interviews occurred between April 2019 and January 2020, ending prior to the COVID-19 pandemic. The investigation coincided with the launch of the Singapore Green Plan 2030 and the '30 by 30' Grow Local target.

A random sample of 16 farms was selected from a publicly available list of 48 commercial farms, which included 42 traditional and 6 high-tech (excluded from the sample). Participants consented, and interviews were audio recorded. Several interviews were conducted in the local Chinese dialect and translated during transcription by a bilingual research assistant. A semi-structured qualitative interview guide was developed through prior fieldwork (Diehl & Bose, 2023) and modified for the Singapore context.

Interviews covered four topics: general farm practices, social networks, benefits and barriers, and livelihood options. Farm practices were operationalized as attributes of form and function (Crooks et al., 2015; Diehl & Oviatt, 2019). In the context of urban agriculture, the term 'form' is used to identify the physical elements of location, proximity, land use zone, land title and tenure status, scale and intensity, technology, and materiality. The term 'function' refers to activities and the purpose of those activities, as well as underlying systems and processes. These included strategies, social integration, temporality, degree of profit orientation, product, water management, and multi-functional attributes.

Social networks were operationalized as farmers' access to resources through the various people they interacted with across the food system from production, processing, and distribution to consumption and waste. Social networks, comprising agents and the relationships between them, can be individuals or collectives (Scott, 1991). In this case, I applied an ego-centric design, consisting of one agent ('ego') and all the agents the 'ego' interacted with ('alters'). In this research, the 'ego' was the farmer—or the farming household (aka farmer's family) if the farm was a family operation. 'Alters' included all the people they interacted with in the pursuit of farming as a livelihood. The main challenge in measuring 'alters' in social network research is capturing all of them; name generators have reliability and validity issues because of limitations in participants' recall (Brashears & Quintane, 2015; Lin, Fu, & Hsung, 2001). Instead of measuring individual ego-alter networks, I assumed that 'alters' could be generalized into types based on their role in the food system.

For example, a farmer might hire multiple laborers to help cultivate the land, but the laborers would generally provide access to a similar range of resources; I did not need to ask about the farmers' relationships with each laborer. Using this logic, I developed a list of social network 'alter' types based on the food system: other farmers, hired laborers, vendors, etc. (similar to Shakya, Christakis, & Fowler, 2017). Refer to Box 6.1 for details. Figure 6.3 is a diagram of social network variables used to develop the interview guide.

BOX 6.1   MEASURING SOCIAL NETWORKS QUALITATIVELY

This study applied a theoretical framework for measuring the mobilization of resources through social networks that combined the Sustainable Livelihoods Framework (Scoones, 2009) with cultural anthropologist Eric Wolf's theories of relational power (Wolf, 2001) and social network theory (Knoke & Yang, 2008) (represented in Figure 6.3). Social networks were operationalized using two key social network theory assumptions according to Knoke and Yang (2008):

- Assumption #1: Structural relations exist at specific time-place locales. Structural relations are defined as the interaction between two agents. Variables included where, when, and frequency of farmer interactions.
- Assumption #2: The type of information exchanged (transaction) and degree of influence depends on whether the relation is direct or indirect. Variables include type of transaction, degree of influence, and direct/indirect relation.

The type of transaction was defined in terms of livelihood assets comprising five capitals from the Sustainable Livelihoods Framework (Scoones, 2009):

- Human capital, defined as knowledge and labor potential;
- Natural capital, defined as ecosystem assets;
- Financial capital, defined as money;
- Physical capital, defined as tangible 'things';
- Social capital, defined as positive social connections (a type of social network, which can be comprised of positive or negative social connections).

The degree of influence was defined as the ways in which social structures facilitate or constrain access to resources (i.e., capitals) between agents. The conceptualization in this research is based on Wolf's (1990) theory of relational power. The degree of influence was measured as:

- Investing, defined as a transaction that provides resources to an agent with the expectation of a future return;
- Withdrawing, defined as gaining resources;
- Exchanging, defined as a transaction in which resources were both provided and gained;
- Blocking, defined as an agent preventing access to resources.

Direct relations were defined as bonding ties, a strong tie between immediate family members (micro), neighbors, friends, and peers (meso), whereas indirect relations were defined as bridging ties, a connection with a person or people of different socioeconomic and/or cultural backgrounds (macro) (Lin, 2001).

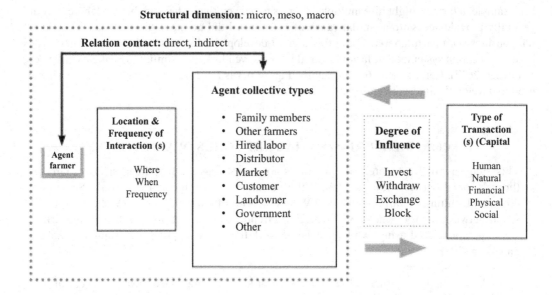

FIGURE 6.3   Diagram of social network variables (Diehl, 2020).

To understand the benefits and barriers of traditional farmers, I asked participants to describe the benefits of farming in their current location, any organizations they interacted with, pressing challenges, and what would make it easier to farm. Finally, I asked about livelihood alternatives including motivation to become a farmer, if it impacted their own dietary or other habits, any other income sources, and what they would do if they could no longer farm. Refer to Box 6.2 for the full interview guide.

BOX 6.2 INTERVIEW GUIDE FOR SINGAPORE FARMERS

1. **To summarize baseline characteristics (micro-scale form and function) of urban agriculture**
   1a. **Can you tell me a little about your farm—basic characteristics including size, types of crops, and techniques? (Take photos)**
      How do you plant/till the soil? Irrigate? Fertilize? Manage insects and pests?
      What type of land use is your farm occupying? How long is the lease on your farm?
      Do you use any high-tech systems? If yes, why? If no, have you considered it?
2. **To create a cognitive map of household-level social networks and access to resources**
   2a. **[Individual] How long have you been involved in farming? Can you tell us about your experience when you first came here?**
   2b. **[Household] Is there any history of farming in your family? Do any family members work here or help out in any way?**
   2c. **[Other farmers] Do farmers generally know each other? What kinds of things do you talk about?**
      Do you ask other farmers or tell other farmers what to grow, where to sell, and how to get any government funding?
      Do farmers generally help each other? How? Do you share workers, equipment, or any financial support?
      How do you network with other farmers? How do you meet new farmers?

Do you ever talk to other farmers about leases or land development? What is discussed?

If not, do you want to be involved in discussions? Why aren't you? Or why do you not want to be involved?

2e. **[Workers] Who else works on this farm? Are they paid or volunteers? How many?**

How do you find out who to hire? Where/how do you advertise?

Do you teach any skills or provide resources to workers?

Do workers ever suggest improvements to your farm?

Do they connect you with other people who can help improve your farm?

How do you decide what to pay them? Do you ever negotiate? Would you share the wage/rate or range of pay?

Do you talk about anything else besides work tasks?

- How many farmers do you have working on your farm?
- What is the percentage of local vs foreign?
- Do you mind sharing with me what is the pay scale like for farmers on your farm?

2f. **[Vendor and/or Consumer] How does your crop go from field to consumer?**

Does a vendor come or do you take it to a market? Where?

What is your relationship with the vendor?

Do they give you a good price? How do you know what a good price is?

- Do they ever refuse to buy your produce? What do you do if that happens?

Does (s)he give you advice regarding your farm? For example, suggest what to grow.

Does (s)he connect you with any people that might improve your farm?

When do you receive payment for the harvest—in advance, on collection, after distribution? Is there any issue with this system? Like, what happens if a crop fails or if you have surplus?

Do you talk about anything else with your vendor(s) or customers?

How do you promote/market? Why this method?

What is your relationship with the consumer?

- Do you ever participate in farmers' markets or other events? Who do you talk to, and what do you talk about there?

Do customers connect you with any people who might offer resources and/or support to you/your farm

2g. **[Government] What is your interaction with authorities including the leasing agency?**

What authorities, boards, or associations do you interact with regularly?

How did you acquire the lease? What was the process like?

Is the rent set or can it change unexpectedly? What happens if you are late or cannot pay?

Have you ever applied for any support to make improvements on your farm?

What authorities do you contact if you have any issues or need something?

Do you ever talk to authorities about leases or land development? What is discussed?

Do you participate in any educational or skill-building programs offered by the government?

Do they ask you for or do you offer any feedback or opinion on the state of farming in Singapore?

- Have you initiated any conversations with the authorities before to give feedback to them?

- Do you have to meet a certain level of productivity?
- Do you get the freedom to decide what to grow on the farm?
- Are you able to determine how much area you want to utilize for farming or ancillary purposes?
- Do you carry out R&D on the farm?
- How often does AVA carry out inspections?
- To carry out changes on the farm, do the authorities have to be informed?

2h. [Others] Does anyone else visit this farm?
Where do they come from?
What do you talk about?
Do they tell you new things that could improve your farm?
Do they connect you with other people who can help improve your farm?

3. To describe the benefits and barriers to urban agriculture practice

3a. What are the benefits to you to farm here?

3b. What organizations, associations, or other supportive networks are available to you as a farmer?

3c. What are the most pressing challenges to sustaining farming here?
Consumer demand, government policy, land development, climate change, etc.

3d. What would make it easier, more productive, or more profitable to farm?
Policy, funding, laws, skills, labor, etc.
How do policies about labor (and where you can hire from) affect your ability to farm?
How do you think government education programs to encourage more young people to study urban farming (i.e., Republic Polytech Earn and Learn Urban Agriculture Technology program) might impact your ability to farm, and farming in Singapore in general?

(4) To identify alternative livelihood options

4a. Why did you decide to become a farmer?
Beyond income, what motivates you to grow your food?
- What kind of impact do you think your act of farming has on society?
- Does this impact your diet or the way you purchase or look at food?

4b. Are there any other income sources?
If yes, what proportion of your income is directly from farming?
Do you rely on government funding?

4c. What would or will you do if you can't farm anymore?

Interviews, documented field observations, memos, and photographs were analyzed qualitatively. Audio recordings were transcribed and cleaned. Interviews, notes, memos, and photographs were imported into Atlas.ti version 8.4.4 for Mac (qualitative coding software). A research assistant conducted the first level of coding to identify words and phrases related to demographic characteristics, form and function variables, and social network 'alter' types. Farm traits were output as an Excel spreadsheet and summarized categorically. The second level of coding qualified social networks related to location and frequency of interaction, type of transaction, degree of influence, whether the relationship was direct or indirect, and micro- to macro-structural dimension. Social networks were grouped by 'alter' types and exported as a Word document for a third level of coding to summarize by structural dimension (micro, meso, macro). Farmers' social networks were summarized quantitatively and qualitatively to identify how they provided access or constraints to livelihood assets and mapped using ArcGIS Pro 2.5. Benefits and barriers and alternative livelihood options were summarized qualitatively.

TRADITIONAL FARMS: CHARACTERISTICS AND PRACTICES

More than half of farms combined traditional and high-tech systems ($n=9$; 56%) including raised-bed hydroponics, automated milking machines, and fish feeding systems. Ten grew vegetables (63%), two raised fish, two produced eggs (chicken and quail), one raised bullfrogs, and one was a goat dairy (refer to Table 6.1). All farms occupied leased land. Notably, every traditional farm that participated in this study and several that had adopted some high-tech practices had a lease expiring within the next three years without renewal (i.e., end of 2021); two farms were in the process of closing at the time of the interview. Six farms (38%) were organic or applied agroecological principles; 10 (62%) used conventional practices. Interviewees were diverse in age; however, most were university educated, Singaporean born, had a family history of farming, and grew vegetables

---

TABLE 6.1
Summary of Farm Practices Based on Form and Function

		Category	n (%)
Form	Location	Lim Chu Kang	14 (88%)
	Proximity	Off-plot	15 (94%)
		On-plot	1 (6%)
	Land use zone	Agriculture	15 (94%)[a]
	Land title and tenure status	Lease[b]	16 (100%)
	Scale and intensity	<5 hectares	11 (69%)
		>5 hectares	4 (25%)[a]
	Technology	Traditional	7 (44%)
		Traditional plus high-tech	9 (56%)[c]
	Materiality	In-soil	9 (56%)
		Raised beds	3 (19%)
		Hen-house	2 (13%)
		Tanks	2 (13%)
Function	Strategy	Organic	6 (38%)
		Conventional	10 (63%)
	Social integration	Low	15 (94%)
		High	1 (6%)
	Degree of profit orientation	Farmer's market	9 (56%)
		Restaurants	8 (50%)
		Grocery store	12 (75%)
		Wet market	2 (13%)
		Online	3 (19%)
		Home delivery	1 (6%)
		Hospitals/hotels	2 (13%)
		Wholesale	4 (25%)
		Export	1 (6%)[c]
	Product	Vegetables[d]	10 (63%)
		Frog	1 (6%)
		Dairy (goat)	1 (6%)
		Fish[d]	2 (13%)
		Eggs (quail, chicken)[d]	2 (13%)

(Continued)

TABLE 6.1 (*Continued*)
Summary of Farm Practices Based on Form and Function

	Category	n (%)
Water management	City water	13 (81%)
	Pond	2 (13%)
	Sea	1 (6%)
Multi-functional	R&D	2 (13%)
	Educational/farm tours	11 (69%)
	Farm to table	1 (6%)
	Therapy garden	1 (6%)
	employ autistic persons	1 (6%)[c]

[a] Missing data.
[b] All traditional farms that participated in this study had a lease to expire with no further renewal within the next 3 years (end of 2021), and two farms were in the process of closing at the time of the interview.
[c] Farms could have more than one category.
[d] Included under 'core' food production.

on two- to three-hectare farms. In addition to commercial orientation, some farms provided other services—specifically, public and educational farm tours.

In discussions about farm practices, several themes emerged. The first related to product: type grown and organic practices. As part of Singapore's Food Security Roadmap published in 2012, the Agri-Food and Veterinary Association (AVA; replaced by the Singapore Food Agency (SFA) in 2019) reinstated local food production as a 'core strategy' for food security and identified three focus areas—eggs, leafy greens, fish—based on challenges with diversifying import sources for those three foods (Diehl et al., 2020). Farm products fall under 'core' food production or 'other.' Farms with 'other' products cannot transition to 'core' food products or vice versa without approval. Regarding 'organic' practices, a vegetable farmer reported, 'We are not organic, but we are under this Good Agricultural Practice (GAP), which a lot of people don't know. GAP means Good Agriculture Practice, which is supported by SFA.' The GAP program was developed in 2021 by SFA as a guide on holistic farm management for local farms in the areas of food safety, produce quality, environmental management, and workers' health, safety, and welfare. Farms that meet the standards are awarded a certification logo to be used on market packaging. Thus, farms not certified 'organic' can still meet GAP requirements.

Several farmers also elaborated on the transition to high-tech. One farmer acknowledged the limitation of what can be grown in high-tech operations—particularly indoor conditions, which was a second theme:

> I saw in the papers, they're going to put high-tech for growing vegetables [on the land bordering the Straits of Singapore] … the limit is only indoor growing of leafy vegetables, it's very hard to grow other types, fruit … the fruit berries ones, leafy vegetables you can go high-tech, things like brinjal, lady's finger, these are what you call the fruit vegetables, these need insects to pollinate, maybe you can hand pollinate, but do you know how much labor is going to do that? When nature provides all the insects and everything, and we spray and kill them, these are free labor, now then how? We're going to get robots to go and pollinate all of them—at the moment, I have not heard of robot pollinating but okay so it's a reductionist way of growing, so in the end, you reduce growing, only leafy vegetables, then I say you got to use what? Air-con, temperature control, LED lights, instead of sunlight energy now is LED lights, and all that equipment … got a life to it, so in the end you are creating more thing, more trash, a lot more trash. (Interview 105, vegetable farm, food forest)

Conversations about high-tech crops pointed to a second theme: labor and lifecycle costs. Although high-tech reduces the labor required to water and maintain plants, it can create a need for new kinds of labor for tasks traditionally 'provided by nature' including maintenance of LED lights (replacing sunlight) and pollination (replacing bees, etc.). The lifecycle costs of farming increase with high-tech operations. Traditional farming relies on 'free' ecosystem services for inputs of sun, rain, and nutrients. As an aside, even Singapore traditional farmers who rely on city water for irrigation benefit from natural hydrological cycles that ensure replenishment of soil nutrients (despite leeching, nutrients still flow through soils and are temporarily available to plants). High-tech systems internalize previously 'free' natural processes using materials with additional lifecycle costs.

On the topic of cost, another farmer pointed out the economics of converting to and maintaining a high-tech operation:

> [T]oday is about LED lighting and all that. But the high cost? ... [S]o talk about growing vegetables, how much will our vegetables cost? That's the biggest debate we have and our internal struggle also. So, if I have so much money to grow vegetables, knowing that the high operations cannot sustain, so I will not be wise to want to go inside there [convert to high-tech]. A lot of these [other traditional farmers] are not going in because we know that it cannot last. As a researcher myself, I always ask, where are the first few technological farms in Singapore? Aero-Green? [W]here is the mushroom farm [Mycofarm] that you see in Jalan Kayu? They're gone. But are we real enough to identify that it's not feasible because you can't produce those kinds of vegetables to sell to make you continue to maintain the high operation cost, high investment capital. Where are the returns? So, we have to be very real in it. (Interview 106, vegetable farm, converting to hydroponics)

This farmer alludes to the limitation of local markets to tolerate price increases, which is discussed later in this chapter in the section on benefits and barriers. Note that Aero-Green Technology was the first in Asia to adopt aeroponics technology; they were permanently closed at the time of this interview. Mycofarm was Singapore's only specialist mushroom grower. They grew a variety of mushrooms free of chemicals and offered guided tours and unique value-added products that included mushroom soap, mushroom beer, and cordyceps green tea. They were also permanently closed at the time of this interview.

SOCIAL NETWORKS AND ACCESS TO RESOURCES

Social networks were operationalized as farmers' access to resources through the various people they interacted with across the food system. See Figure 6.4 for examples of social network diagrams.

Micro Social Network: Individual, Family

The micro dimension of farmers' social networks comprised the family itself including socio-demographic attributes and farmers/family access to resources enabling them to sustain their agrarian livelihoods including personal achievements/education, family composition, and history of farming. I dichotomized the micro social network into strong ($n=7$; 44%) and weak ($n=5$; 31%) (Figure 6.5). Two incomplete interviews were excluded.

In general, farmers were highly educated representing a diverse skillset—many from non-agriculture disciplines. One was an accountant before joining the family farming operation, and another was previously a researcher at Nanyang Technological University. One farmer had a residence in England and had completed a Diploma in Education, whereas another was formerly a marine engineer and currently a pastor. A professional photographer had been drawn into the family business after photo-documenting it (she also had a Real Estate degree from the National University of Singapore (NUS)); another university graduate wanted to pursue medicine but completed a degree in engineering to support the family farm operation. A second farmer with an engineering degree had just completed a Masters in Aquaculture at Wageningen, Netherlands. Two other farmers came from Information Technology and Business Management, respectively, and, finally, one was from Sabah, Malaysia, and had studied agriculture at the local university specializing in pest control.

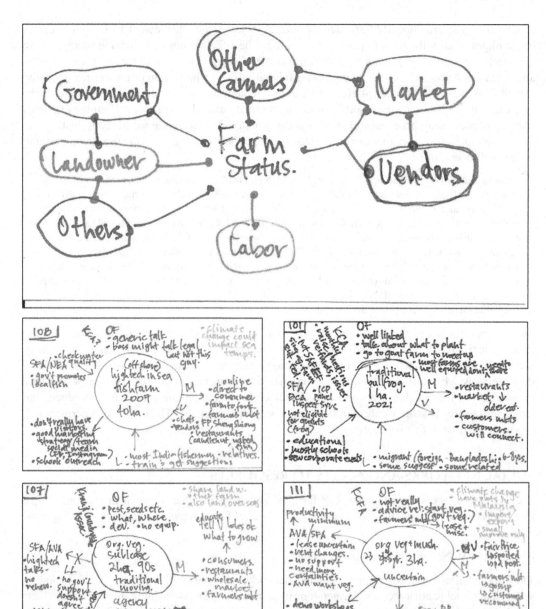

FIGURE 6.4 (Top) Template used to organize data by 'agent collective types' around the interviewed farmer's details including farm status. (Bottom) Examples of individual farmer social network diagrams.

Roughly half of farm operations were family businesses (*n*=9; 56%). While many farms in Singapore now grow vegetables, family histories revealed that few started that way:

> My dad and my uncles, they started this [farm]. Before that we had a pig farm and mixed farm in Yio Chu Kang. Yeah [we switched when the government got rid of the pig farms]. [In] 1991 [we started as a vegetable farm]. In between I think there was like a year of experimenting and learning. (Interview 109, vegetable farm, raised-bed hydroponics)

The transition away from poultry and livestock farms occurred between 1980 and 1989, with farmland reduced from 12% to 3% of the total land area of Singapore (Lepoer, 1991; The World Bank, 2019a). A shift toward intensive farms maximized production; grain, pulses, roots, fodder, and oil

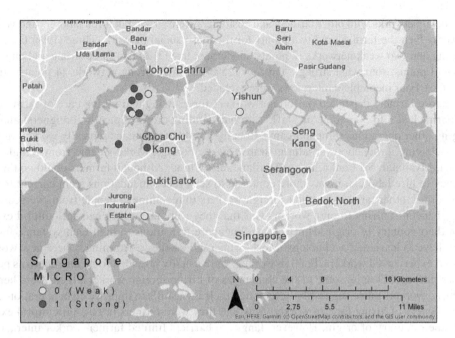

FIGURE 6.5 Farmer's micro social networks.

crops became impractical (Swee-Hock, 2012). In 1984, the government decided it was more economical to import frozen and live pigs than to treat the waste of locally raised pigs—which produce
five times more waste than humans; pig farms were phased out (Shim, 1987). Farms had a choice:
cease operations or switch to a government-mandated option:

> [Or family history is] very simple, we started off as poultry farmers, so we had pigs, ducks, chicken,
> crocodiles—you name it we had it. We are one of the top five largest farms in Singapore in the 60s and
> 70s. When farming was phased out due to swine waste, the government gave us three options and one of
> the options was dairy or goat farming. We chose goat because at the point of time Singapore had already
> five dairy cattle farms. (Interview 102, goat dairy farm)

During the transition away from poultry and livestock, some saw a new opportunity to start farming:

> Actually, my dad started the business, I mean he was an oil rig engineer, then in the 1970s … I suppose
> that's when he saved up some money also—when he was an engineer—to pay off the house with the
> middle flat that he got with my mum, and then I think he probably just received a window of opportu-
> nity to pursue frog farming industry, because then the government was reducing some land as well and
> also were eradicating pig farming some time in the late 1970s. (Interview 101, bullfrog farm)

In summary of the micro social network dimension, farmers were highly educated, about half of the
farms were family operations, and (while not described in this section) farmers had good quality
permanent housing separate from the farm. Notably, over 80% of Singapore residents live in public
housing built and managed by the Housing Development Board, which was established in 1960 to
deal with the high number of Singapore's population living in unhygienic slums and crowded squatter settlements. The government considers quality housing a human right and homelessness is rare.
The main differentiation between strong and weak micro networks related to family integration,
which enabled the transfer of long-term business relationships and knowledge and skills.

MESO SOCIAL NETWORK: HIRED LABORERS, OTHER FARMERS, VENDORS/DISTRIBUTORS, AND CUSTOMERS

The meso social network dimension comprised the community that the farmer/family interacted
with as part of farm operations—in this case, other farmers, hired laborers, vendors, and customers. The meso dimension was calculated based on strong or weak networks across the four groups;

e.g., if social networks with other farmers and customers were strong but hired laborers and vendors were weak, then the farm scored two points for the meso social network. Three (19%) farms had two strong and two weak networks (2 points; moderate), five (31%) had three strong and one weak network (3 points; somewhat strong), and four (25%) had four strong networks (4 points; strong) (Figure 6.6). Four incomplete interviews were excluded.

Singapore farmers depended on foreign workers for growing and taking care of farm operations. Farms had 3 to 30 people on staff, with half tending crops (typically foreigners) and half managing other operations (Singaporeans). Locally hired Singaporeans typically worked indoors doing packaging, production, or administrative work. Foreign workers came from Bangladesh, Malaysia, Myanmar, Sri Lanka, India, Thailand, and China. Although farmers could go through an agency to hire laborers, it was more common to seek recommendations from current foreign workers or ask other farmers for their foreign workers to recommend people from overseas. The primary benefit of hiring foreign workers was that they were farmers or had agricultural experience in their country of origin. However, as Singapore transitions to high-tech systems, foreign workers have to learn new technology. Most farms used on-the-job training, where new workers shadowed experienced workers. Foreign workers were usually provided room and board as part of the work package. Despite some of the challenges of hiring foreign workers, there were benefits when hired workers made suggestions or applied their own skills to improve farm operations. The primary benefit of hiring foreign workers was that they were farmers or had agricultural experience in their country of origin. However, language barriers limited farmer–worker interactions, creating a distinct social hierarchy.

The younger generation of Singaporeans was reported as not interested in farming—the long hours, the hot tropical outdoor climate, and the weekend tasks did not fit the career expectations of the highly educated society. A vegetable farmer converting to hydroponics explained, 'There are many [young Singaporean] interns coming here, but after they intern you ask them if anyone wants to be a farmer, there is none.' Compounded barriers to hiring Singaporean workers were the isolation of farms from transportation and other urban amenities, the high cost of hiring them, and the lack of agricultural knowledge and skills. Some farmers looked for alternative local options: one farmer hired ex-offenders and had an autistic employee.

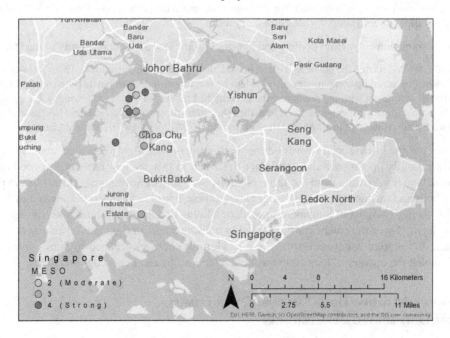

FIGURE 6.6 Farmer's meso social networks.

In comparison to their network of hired laborers in which farmers withdrew human capital and few other resources, the networks among farmers themselves provided access to a variety of resources and bolstered bonding and bridging networks:

> We are such a small community here, out in Kranji [in northwest Singapore]; the farmers definitely know each other and even I think in the cities, the urban farmers know each other. The urban farmers and us out here, we also know each other. So at some point if you go and attend the events or the talks that AVA holds or the networking events or even just the sort of external private events, the farmers all meet each other, and also we've got an association as you know, the Kranji Countryside Association, and recently there's also a new association called the SAFEF, so many of the farmers are members of one or two or one or both, and they would then follow each other. (Interview 103, vegetable farm, agroecology)

The Singapore Agro-Food Enterprises Federation Limited (SAFEF) is made up of farms from the food fish, livestock, and vegetable sectors of the industry, as well as industry associates. SAFEF is a not-for-profit organization, but it is not a grassroots organization like the Kranji Countryside Association (KCA). Most farmers reported membership in the KCA—only one farmer said they were not a member. As described by two farmers:

> [The KCA] was initially formed [by a group of farmers] with the aim to promote local farming, to gather resources to share information, then slowly it became a place where we ... organize dialogues with relevant state boards, the agencies, to discuss what to do with the land lease. (Interview 102, goat dairy)
>
> [Furthermore], one thing beautiful about Kranji Countryside [Association] members is that it brings back the kampung spirit, I do feel the kampung spirit there even though it could not be my problem, or this problem is not mine, but because together this association, it becomes our problem. (Interview 110, quail egg farm)

Note that kampungs are traditional villages and part of the cultural heritage of Singapore. 'Kampong spirit' refers to a sense of community and solidarity. The KCA was a forum to discuss environmental (weather, pests) and market challenges. It was a place to seek advice and brainstorm; where farmers could say:

> 'I have an idea,' you know, but alone [we] are quite powerless, but once [we] bring it up to the association and the [KCA] sees some merit in it ... and other farmers come onboard the idea, or other farmers say 'Oh yes I have the challenge too' ... it legitimises because it's not just one person saying that one thing. (Interview 103, vegetable farm, agroecology)

Outside the KCA, farmers sought advice from each other. They were a critical support system for adopting new products or practices. For example, a farmer switching to vegetable crops to meet government requirements said:

> The difficult thing is that we were not [vegetable] farmers originally, it was only in the recent two years that we are required to grow vegetables. So, we try to match this and learn about it, we also asked other farms how they do it. Yes, we will ask other farms how to grow vegetables and we will try it out. Initially some of those that we grew are small, some were a little weird, the shapes, so [other farmers] told us we have to do this and that, make changes, and we changed slowly. (Interview 111, vegetable farm, agroecology)

Farmers also reported sharing products. For example, one farmer said he went to the goat dairy for fresh milk for his kids. At the same time, he talked to the farmer about current happenings including leases and changes to the Master Plan for the area. Farmers also filled gaps in orders if a crop or product was not ready in time to meet a market contract—they would step in for each other as an informal loan. While farming was a business, one farmer explained how the conflict of competition was negotiated through sharing and support:

> There is always that friendly competition especially when the market is so small and there are only, say, for example, three egg farms and four serious vegetable farmers and one goat farm and one frog farm. There's always that little bit of friendly competition, but I think, by and large because they are all

friends, they do share a lot of knowledge and I think equipment-wise there is a lot of borrowing going around. Probably, but not for the critical things that you need for your business operations everyday obviously, but things like transport. We don't have a van, for example, so when we need something delivered, another farmer is just like a phone call away. Of course, the other day when we had 390 people on our farm, we didn't have enough chairs, so we had to take chairs from the resort, you know they gave us 60 chairs, Kranji Resort. We didn't have chafing dishes, so the restaurant down the road, Garden Asia, lent us chafing dishes. You know, so yes, there is a lot of sharing. (Interview 103, vegetable farm, agroecology)

Farmers only shared minor equipment, and they did not share hired laborers—mainly due to legal employment constraints. Anything needed for major productive operations had to be owned or formally contracted. At the intersection of farmers' networks as both supportive and competitive, one farmer said he sourced seedlings from a local farm, thereby avoiding overseas inputs with less traceable quality and at the same time providing a reliable market for the other farm. On the contrary, due to the small pool of local farmers, another farmer sought overseas advice and knowledge:

Now I use Google and YouTube, such a good way of connecting to other people who are farmers, who have gone through that process, and I learn. In a way when I was doing farming and I was so much into farming, that intuitively I walked the path that other people who are going into sustainable farming are walking, but I am here in Singapore quite isolated because we don't have a big farming community, whereas in other countries they've got many more farmers so they can network, here I don't. That's why this I can say, technology works for us. (Interview 105, vegetable farm, food forest)

An often-overlooked part of the food system is the relationship farmers establish with vendors—the people who collect and transport the products to the market. Being a small island nation, Singapore has a short domestic supply chain. Interviewed farmers primarily sold directly to a market/supermarket and/or directly to customers/consumers. A vendor was required to transport products to destinations. Farmers described three situations. First, some farms had a delivery team on staff. One vegetable farm described, 'We deliver it ourselves. We control the whole chain all the way, from farming to harvesting to cold room, packing, and then back to cold room, and then into cold truck, and then to the particular places.' The second situation was to hire a third-party distributor. And third, markets and supermarkets arrange collection themselves.

Generally, the food system places producers and consumers several degrees away from each other—connected indirectly through vendors/distributors and markets. For large-scale farms, it is impractical to sell directly to consumers, at the same time small-holder farmers are more typically located in rural contexts geographically far away from consumers. Yet, in Singapore, farms are small scale and close to consumers. While feasible, it is not always economical to sell directly to consumers. One farm started as a direct-to-customer but switched the business model to contract with major supermarkets. On the contrary, another farm diversified through direct on-farm sales to customers (including an on-site restaurant), institutional customers including schools, government departments, several corporate companies, and travel agents. A third farm offered a subscription service to customers similar to a community-supported agriculture (CSA where customers pay a monthly or regular fee in exchange for a portion of the weekly or seasonal harvest).

To reach customer social networks directly, farmers attended the KCA farmers' market. It was considered a place to educate customers more than a place to earn a profit—despite being well-attended by the public: it was reported to have drawn 4,000 people the first time it was convened and with a current attendance at 12,000 to 13,000. One farmer explained:

[At the KCA farmer's market], customers ask what we do, how we do [it], that kind of thing, so we educate. So far, we take it as education trip, educate our visitors, our customers about what we do. (Interview 102, goat dairy)

For unique products, education made customers aware of a product not known to be available domestically. Farmers also took the opportunity to distinguish their product in a saturated market, which was the case with a chicken egg farmer—the average Singaporean consumes an estimated 390 eggs per year! According to the farmer:

> The main reason why I took part in farmers' markets was to let people understand, experience, and cherish the egg, because many people took for granted the benefits of the egg, they think that it's a very, very cheap product, so they just don't show any appreciation. Hence I want them to understand the difference between a good and a bad egg. So it's more education-centric, to let them understand how to handle the egg, how to eat, how to preserve et cetera. (Interview 113, chicken egg farm)

Singapore is well-established as a food lovers paradise, boasting 52 Michelin-starred restaurants in 2022, yet there is an under-appreciation for locally grown products. One farmer lamented:

> [I]t's so sad to ask kids where their food comes from and they will tell you that it comes from the supermarket, and 'Do you know about farms in Singapore?' They will be like, 'No, Singapore there's no farms.' It's so sad to hear about such things. (Interview 115, vegetable farm)

Farmers repeatedly mentioned education during interviews. Beyond the visibility of local farms, it was also about acceptance of products not always uniform:

> There are customers who have been with us for long long time, there's one who has been with us more than 15 years. Some of them really understand what is really organic farming and what to expect from the vegetables that we sell to them. But there are some who don't, who expect it to be untarnished, no blemish at all. We deliver to them, sometimes they don't quite understand, sometimes they do, refuse to pay. We've got to explain to them and then sometimes some of them are convinced and they pay, otherwise sometimes they just drag on, but it's usually after one or two ... we just don't deliver to them anymore. (Interview 105, vegetable farm, food forest)

In fact, one farm—now a well-known tourist destination—took on the challenge of education head-on:

> I think our role here at Bollywood Veggies is very much as the mothership, so you know we open the doors to the world of farming, to the layman, the layman comes here because of our restaurant, the layman comes here because of our school programs ... but then it is their first experience of a farm in Singapore. Then they will get to know that there are so many other farms in the area because we promote the other farms, because on the weekends a few of them are here also, and through that, just a bit of that knowledge, can lead to a lot of other things, they will then start to look for local produce in the supermarkets, better still they will get involved in farms in people ways rather than just consumption and supporting of local produce. So, our role cannot be understated because connecting people that are inherently disconnected to farming, to the land, and to Singapore's very indigenous parts of Singapore's history and heritage is what we do, and no one else does that. (Interview 103, vegetable farm, agroecology)

In summary of the meso social network dimension, farmers employed and often provided room and board for foreign workers who were already skilled in traditional farming practices, they knew many of the other traditional and new high-tech farmers through events convened through formal organizations, had fixed contracts with vendors and markets, and made regular contact with customers through on-farm educational tours and farmers markets.

Macro Social Network: Government, Landlord, Market, and Others

The macro social network dimension extended beyond the operational farming community. In the context of this project, it comprised the landowner, i.e., SLA, other government agencies, food markets, and two emergent 'other' agents: tourists and overseas consultants. The macro social network was calculated based on strong or weak networks across the three groups; e.g., if social networks with government agencies were strong but weak with markets and there were no 'other' agents, then the farm scored one point for the macro social network. Two (13%) farms had one strong and two weak networks (1 point; moderate), four (25%) had two strong and one weak network (2 points; somewhat strong), and six (38%) had three strong networks (3 points; strong) (Figure 6.7). Four incomplete interviews were excluded.

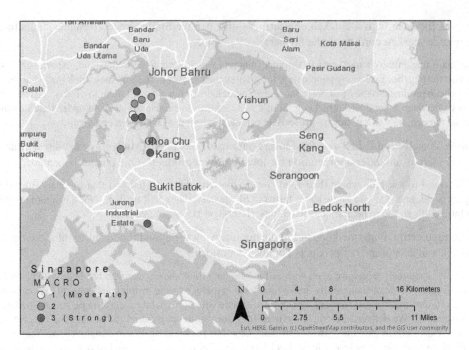

FIGURE 6.7   Farmer's macro social networks.

Traditional farmers reported regularly interacting with SFA at the government level (through an intermediary like SAFEF) because of expiring land leases and tenure:

> If we have problems like with the land … SAFEF helped us a lot with [the land lease problem]. So SFA actually wanted to give out the land lease at only like 5 years or 10 years land lease but all the farmers refused to continue farming because it's very short and for farming, the payback and all is at least 5 years—at least, so 10 years we cannot generate any profit at all, so nobody wants to do, so if 20 years we still can do, we still can have some profit. All the farmers told SAFEF, problems and why they are not going to bid for the land, everything, so SAFEF considered everything, and they told the Ministers, and the Ministers discussed with SFA or something like that. (Interview 115, vegetable farm, converting to hydroponics)

There was recurrent discourse around land leases in terms of length of tenure, rental rates, and payment schedules. Farmers had limited ability to negotiate with government agencies:

> Every 5 years it [the rent] can change … In 2016, they wanted to raise our rent by 42%, yes of course we fought back, so in the end it only raised by almost 40%, we fought back, we were successful in reducing a little bit but you know, we had to pay the lion's share basically despite after getting a valuer and everything, valuation done but this is … we are at the mercy of the land authority [SLA] because they are allowed to raise your rent up to 50%. (Interview 103, vegetable farm agroecology)

Part of the issue was due to multiple agencies at the government level and redundancies. Alignment and consolidation have been instigated among agencies (see Diehl et al., 2020); however, traditional farmers faced a complex and layered system required to navigate successfully to keep their business in operation:

> [W]hen you run a farm, every agency wants to regulate you, but no agency wants to help you, so your land is under SLA, your license is under AVA, your activities [are] under AVA and URA, your power is under PUB, your food is under NEA … and your road situation is under LTA. And then all the terrible accidents that happen around here is under traffic police. You basically deal with almost everyone—and wildlife is under NParks—so you basically deal with everyone. So, we have dealt with a whole host of agencies, but I will tell you the main ones that we normally deal with or that we should be dealing with of course before 1 April [2019] was AVA [now SFA], NParks, and, probably I would say, the SLA. (Interview 103, vegetable farm)

Among multiple and layered regulations was a productivity quota: 'They have a productivity quota for all the farms but it's very difficult to reach, so we just do what we can' (Interview 109, vegetable farm, raised-bed hydroponics). Another farmer elaborated: 'The government sets the quantity, but I know we are falling back, below the quantity that they bought up. Non-negotiable (quantity), it's a quota that is throughout all farms so matter what kind of produce' (Interview 107, vegetable farm).

Government agencies controlled farmland leases, length of tenure, what farmers could grow, and set minimum quantities for operation. Yet, traditional farmers mentioned several benefits including access to grants to experiment with high-tech systems (for example, the SFA's Agri-food Cluster Transformation Fund—website: https://www.sfa.gov.sg/food-farming/funding-schemes), monthly inspections and free tests for pesticide and plant health (documenting good practices and contributing to traceability and proof of high quality), recommendations and financial support to travel overseas and consult with knowledgeable farmers and researchers, and frequent events for farmers (traditional and high-tech) to network.

The macro network also included market agents. One farmer stated, 'You actually try to build [relationships with them] because you know that your goods are quite replaceable, so you have to give them the support they need.' Supermarkets selling the farmers' products included NTUC FairPrice (most frequently mentioned), Prime, and Sheng Siong. Farmers also sold online through RedMart, an online grocer similar to Amazon Fresh. Brick-and-mortar supermarkets require farmers to meet quantity, quality, and price standards. It was challenging to compete with imports from China and Malaysia—farmers had little power to negotiate because Singapore is a free port coupled with the small scale of Singapore farms; i.e., they could not produce volume at a discount. But it was not simply a case of David and Goliath (small-scale Singapore farms versus large-scale overseas farms); Singapore is a small place with intersecting histories that spill over to business relationships in positive ways. One farmer recalled when asked if he received any advice from the supermarket manager or if there was any support for local products:

> Some suggestions will be in the marketing side, for example Sheng Siong's boss, because he and I are old friends, we live in the same *kampung* [village], so we are more familiar with each other and he will give some suggestions on branding. Hence, I worked with him and came up with a heritage branding, he helped me with it. So, in this aspect, we work together, support local produce. (Interview 113, chicken egg farm)

There was less direct interaction with RedMart staff, yet farmers had more flexibility with the availability, quantity, cost, and pricing of products. One farmer stated,

> It's easier to actually use RedMart, because the people are tech people, so once you have something in mind that you want to carry out, it's easy for them to change the whole website, or change things ... We [don't] have to negotiate with them.

All but four (25%) farmers contracted with supermarkets and/or RedMart. In addition, many diversified for value-added benefits beyond income potential—nine (56%) participated in farmers' markets, and as one farmer explained:

> The other reason why we joined [the farmers' markets] is that we wanted to do it together with the other farmers because it's SG [Singapore] market ... we're not there to get a profit but to promote local farms, for the benefit of the whole industry. (Interview 115, vegetable farm, converting to hydroponics)

Traditional farmers could not compete with high-tech operations based solely on quantity; diversification and adding value through other avenues were essential. Some built social networks outside the Singapore food system to achieve this. For example, through tourism or skills-building:

> We've submitted [proposals] ... more recently about tourism ... to the Tourism Board. Also, to the Ministry of Education because I think [local] farms are so much more than primary producers, they should be more than primary producers, so we are hoping that each ministry can see the merit of farms from their perspective, then in that way we can raise the national value of farms, so they become much more than primary producers. (Interview 103, vegetable farm, agroecology)

> I had two consultants, one Malaysian and one, he is a New Zealander, he's a biodynamic consultant so because that time we practice biodynamic organic farming, and he taught us how to make compost and all that. So, he frequently came so we sort of applied for the fund and we gave him the money. (Interview 105, vegetable farm, food forest)

Yet, diversification and value-adding had tax implications:

> Our land use is agriculture, so it needs to be agriculture-based. But after we started to do the restaurant and the educational programs and things like that, we have coined the term agri-tainment and this term was then misappropriated by URA [Urban Redevelopment Authority], so when I build my restaurant and when I build my food museum and cooking school, I will have to pay a differential premium because the land use has changed from agriculture to commercial. (Interview 103, vegetable farm, agroecology)

In summary of the macro social network dimension, farms were losing government support and faced land lease expiration unless they chose to convert to high-tech operations, the small-scale production made it difficult to reliably meet market demands, and diversification and value-added improvements could have negative tax consequences.

BENEFITS AND BARRIERS

Traditional farmers described the benefits of farming in their current location, organizations they interacted with, pressing challenges, and what would make it easier to farm. One farmer stated that the three major common issues were land, labor, and market (e.g., the economic sustainability of sales). Lack of land ownership and expiring leases for traditional farmers was the single most challenging issue reported by all:

> [The renewal process] is rather complicated. We have to submit a lot of documents to them [SFA] and our productivity ... If [SFA] give us 10 years, we invest and renew everything. If not if [SFA] gives me 1 or 2 years, I cannot put in too much investments, I can only change whatever is damaged. (Interview 111, vegetable farm, agroecology)

Three reactions arose in response to SFA's plan to reallocate and develop the land where traditional farming was concentrated. First, farmers used social media to bring it to the public's attention (which resulted in one- to two-year extensions). Second, some farmers planned to leave the practice of farming:

> They told us that we have to move, and I said, 'If you want us to move, we will move, I'll just not farm anymore.' I have not recouped my investments and they want me to move to another place, reinvest and start over, it's impossible. If I have so much money, why should I continue working?

And, third, several acknowledged loss of natural and physical capital if relocated: 'The soil here is good, so the production ... it gets better, but the soil at the new farm is shit.' Another added, 'Our current resources are all here. As compared to a new vacant land, obviously the existing farm has all the resources as in the infrastructure, the machines ... but you compare to the new land you have to start all over again.' One farm, actively 'packing up' when interviewed, was featured in news media shortly after because of the cost of 'reinstating land.' When leases were not renewed, farmers were required to return the land in the state it was in prior to farming—to clear and replant according to government requirements, which in this case cost nearly 500,000 SGD (approx. 380,000 USD).

The debate over short versus long land leases becomes a moot point in the context of why farmers do not have ownership over the most critical asset for a sustainable productive livelihood. During interviews, it was framed in terms of economics:

> If you look at my piece of land, 20 years I've paid $3 million [SGD; approx. $2.28m USD], when 20 years is up, the $3 million is gone, for Malaysia, the land appreciates in value, now it's $1 million, after 10 years it becomes $10 million, we are $0, how are we to compete with others? (Interview 113, chicken egg farm)

> Land tenure [will make it easier/ more productive to farm] …We have workshops … overseas to see how wonderful our foreign partners are doing, how great they are, how technologically advanced they are, how first-first-first world they are … but why can they do so well? The land Ah Gong one [translated as: grandfather's land]—they own the land … it's freehold, it's 100 years or 200 years or 300 years lease, so whatever they dump in, they can get back. We are not dotcom companies, I take a long, long, longer time to get my money back. (Interview 102, goat dairy)

Economics is one layer. Land ownership is also about stewardship: long-term sustainability requires caring for soil, water, and other natural resources needed to produce healthy crops and animals. Land stewardship underpins what farmers expressed as key benefits to farming: lifestyle, working on the land, clean air, and healthy food:

> I was thinking that if we can farm sustainably and I can live in a farm, which in a way was good that they allowed us to, I'm getting my reward, in other ways not monetary but in the air we breathe, the vegetables we eat. (Interview 105, vegetable farm, food forest)

Another farmer elaborated:

> Maybe it's because we come from a farming background [that] we have a passion for farming, we have an attachment to it, so we know that as human being, we live to eat, so what do we eat? If no one does this [traditional farming], what we will be eating are the fake meats, fake eggs et cetera … I realized that even [some of] those … fake [things], they are also made from plants, you also need to grow them, the vegetables, the eggs, in order to be made into food. If no one grows them, where will you find food? (Interview 113, chicken egg farm)

High-tech farming has become the elephant in the room. Singapore's land scarcity 'problem' is being addressed by the promotion of landless (or at minimum soil-less) agriculture. Policies and incentives target and benefit high-tech entrepreneurs and traditional farms willing to convert. The consequence of the rapid shift was felt profoundly by traditional farmers:

> There's a great clarity for urban farms that are high-tech, there [is] a lot of support for them, so there's great clarity for them, so for them I'm sure it's very smooth sailing, there's a set of other problems, then for land-based peri-urban farms [like us], there's no clarity because despite the fact that [we] are the food basket, that [we] produce the bulk of the food affordably, [we] still not given any help, and … it shouldn't be the way. (Interview 103, vegetable farm, agroecology)

There was skepticism of high-tech farming to contribute to national food security. Traditional farmers speculated that high-tech farms needed long-lease tenures to recoup costs—efficiency afforded by technology was expensive. Others viewed high-tech as a trend to sell technology not food:

> Not dissing them, but the trend … is that a lot of them are going to high-tech farming, like indoor farms and stuff, but … if you look at the way they are operating, you know that what they are producing is not the priority, it's always to create a model farm with that technology and to sell that technology. So, in a way, I feel that is a result of the government's policy and the government's direction for farming because they are so fixated on Singapore's farming going towards more high-tech that they are kind of encouraging that and we've seen a lot of failed examples of farms come and go … And that one is very tricky because it's so hyped up and they get a lot of investment money so then you start to wonder what's going to happen? … So, there's a farm in … Kranji … they came, and they bought nutrients from us, and they bought seedlings from us as well, then, after a while, they folded also. A lot of other examples. So the reality of running a farm is not as simple as people think of it. (Interview 109, vegetable farm, raised-bed hydroponics)
>
> So, [at] the present moment, my 150,000 [quail farm], I only managed by four people, very efficient. But, of course, government is: 'Can four become two, or better one, then I give you the grant, I give you the money to automate, is it possible?' I said, 'Well of course it's possible but problem is that I don't have the land duration, I don't have the land time, so I pump in already, then after two years you want me to go, then I waste the taxpayers' money, I don't want to do that.' So, at [the] present moment I'm using all manual, all traditional methods. (Interview 110, quail egg farm)

Is there a disconnect between the high-level vision for a high-tech farm industry and a basic understanding of what it means to grow plants and raise animals? Several farmers lamented a loss of agricultural expertise within government agencies:

> We are supposed to contact AVA [now SFA] because they are [the] body that is supposed to be taking care of us. But I can tell you, every time we call, they don't respond, and they are not able to give us a solution … So they don't really help us and then also when you have diseases, fruit tree diseases, or your plants, you tell them 'Oh what can I do? I don't have the expertise, my workers also don't know what to do,' they say, 'Oh sorry, we have also lost the expertise,' and this is the agriculture department, because they no longer take care of these types of plants you see, it's all leafy greens now, you know and very specific types of crops, so now they look at the landscape like that, they don't know what the hell you're talking about, so they can't really help us. (Interview 103, vegetable farm, agroecology)

The loss of domestic knowledge has prompted farmers to look overseas at best practices suitable for climates very different from the tropical biome of Singapore:

> [I went to a workshop] in Bangkok, it's organised by NectarTek, which is a very big Israeli company. This was about 2 years ago. We later found out … they are a service provider, so you actually have to use their service … so that workshop is meant for their customers, but I think AVA just somehow invited us and we just went for it. So that was a very good experience because they went through the basics of farming, talk about things like greenhouses, different types of greenhouses, just nutrients, just … I think that was a very good … for me, a very good experience. (Interview 109, vegetable farm, raised-bed hydroponics)
>
>   They worked with agronomists from Israel, so Israel is one of the places in the world where … the first thought that comes into your mind is desert, so they found a way to grow food, so if they can grow food, I'm sure we would be able to as well, so a lot of technology is learnt from others and translated, it's not just from Israel, it's from Japan, they take all these technology know-how and they bring it here. (Interview 112, vegetable farm, traditional and high-tech)

Skills-building is immensely important for farmers to be able to improve their livelihoods; yet, what the farmer describes as a technological service package hints at the technological products marketed as part of the Green Revolution … solutions that are not designed to be adapted to local climatic, social, or economic conditions.

Plants are highly sensitive to climate conditions, and climate change is impacting agriculture globally at alarming rates. Only two traditional Singapore farmers mentioned the changing climate during interviews. According to data collected by the Meteorological Service Singapore since 1948, Singapore is trending above global averages for temperature increase rates—averaging 1–1.5°C above average temps compared to the global average of 0.5°C between 1995 and 2010. April is generally the hottest month with an average high of 32°C (low of 24.5°C). December is the coolest month with an average high of 30.5°C (low of 23.5°C). Most plants become stressed at a threshold of 32°C (90°F). Along with rising temperatures, recent records show heavier rainfall outside of monsoon months (which occurs from November to January) and some months drier than usual. One farmer described adaptation to changing weather conditions:

> [V]egetables require moisture, if the weather is hot, more watering is required to maintain its growth, so without you noticing you are actually increasing your cost. When it's hot, you water more times, and water requires money … [I plan to deal with this by] using more technology. For instance, automation in watering, because we are using manual now, automation will help us save a few workers. (Interview 111, vegetable farm, agroecology)

Several farmers thought technology could mitigate climate change risks. But there was the emergent question of what was 'organic'? As mentioned earlier in this chapter, the SFA developed the GAP program in 2021, which aligns with but is not the same as international 'organic certification' standards. For example, GAP allows produce grown hydroponically—technically grown using chemical fertilizers—to be eligible for approval. One traditional farmer in the process of converting to high-tech explained:

Having graduated from agriculture, our lecturer told us that for organic, it is grown on an undisturbed piece of land or forest land, you prepare it for farming, it is previously not farmed on before, then that is organic. You don't use any chemical fertilisers, or pesticides, the water is clean et cetera, that's organic. But for many people, their form of 'organic' means no chemical fertilisers and pesticides, for us who have studied agriculture, organic is not like that. (Interview 104, vegetable farm, converting to hydroponics)

An unintended consequence of broadening the 'organic' definition is that traditional soil-based farms compete for consumers on the shelf against hydroponic and other high-tech-produced products that would previously not be considered 'organic.'

Another repeated challenge to farming in Singapore was the lack of human capital in the Singapore population for the agricultural sector—the local people did not have the skills, wanted a salary greater than farmers could afford, or did not want to work outside in the hot, humid climate. This was described earlier in this chapter on migrant labor but warrants emphasis. One farmer exasperated:

[Hiring locally], it's so high cost, how much do they expect to make? A diploma holder, you want to pay [how much]? $1,000 [SGD; approx. $760 USD]? $1,200? Because production itself does not pay that demand you see, so that's what we experience, so we have graduates that comes here, they want a pay $3000, $2000 … how? How can the farmer live? (Interview 106, vegetable farm, converting to hydroponics)

Furthermore, the northwest of Singapore, where traditional commercial farms were located, is an anomaly in a city which boasts one of the best public transportation networks in the world. The northwest is poorly connected, making it difficult for Singaporeans, with low car ownership, to access work.

Disinterest extended beyond logistics and salary; one farmer pointed out a negative national identity because of the social construction of the 'farmer':

It's because of [the] national narrative, you know in the mainstream if you don't promote your farmers, you don't glorify them, then people do not aspire to be a farmer. If people aspire to be a farmer and agriculture is their first choice of degree, then it's different already, you get a different caliber of people. But you don't glorify your farmers [in the mainstream], you don't glorify your farming industry—in fact the minister of state himself say that farms are dirty, smelly and not desirable and hence we must go into urban high-tech farming. He told a room full of [traditional] farmers that what they did was dirty, smelly and … exactly, I'm surprised nobody stood up and scolded him … No, Koh Poh Koon … He was making an analogy why Singapore need to go into high-tech farming because to attract people, he said, you must change from the old way otherwise young people don't want to go to your farm because it's dirty, smelly and … okay … I mean that's how he talked about farmers; if this is the mindset of the leaders, then this is the mindset of the people. (Interview 103, vegetable farm, agroecology)

While some farmers reported a negative normative view of farming, others conceded invisibility. Yet, traditional farms had a mission—beyond meeting national food security targets—they wanted to connect local people back to the sources of their food … especially children. They wanted to instill indigenous knowledge through education, social media, conversations at farmers' markets, and branding and marketing:

Recently, I've got a bumper crop on moringa drumsticks, but because not many Singaporeans know how to eat it, so how to sell it? Marketing and making people want to eat because one challenge in Singapore is that we are so, don't know to say we are lucky or not, we have so much fruits and vegetables from all over the world that we are not … that the children are not familiar with what is grown here, that they are eating, many of the vegetables that are imported and that is not grown in this part of the climate, so they are not aware of what can be grown. And isn't it important for food security that we know what kind of vegetables that we must be eating? (Interview 105, vegetable farm, food forest)

Our role cannot be understated because connecting people that are inherently disconnected to farming, to the land, and to Singapore's very indigenous parts of Singapore's history and heritage is what we do, and no one else does that. (Interview 103, vegetable farm, agroecology)

Because I feel like Singaporeans don't know much about their own local produce, so I feel that Singaporeans should know more about local produce, that's why I printed the shirt, Support Local Produce. (Interview 115, vegetable farm, converting to hydroponics)

Three challenges converge here. First, traditional farmers pointed out that locals did not know local varieties—and wanted imports. Second, relatedly, locals preferred the cheapest option over the country of origin. Third, unique and diverse products were unwelcomed by markets because of low sales (due to cost), reinforced by government promotion of product uniformity (i.e., core products):

Singaporeans don't support our own product, out of 10 people, 8.5 to 8 people will say 'how much?' Ten years ago, 9 out of 10 people will say 'how much?' Now got improvement. (Interview 110, quail egg farm)

To create biodiversity, creating biodiversity is so important and the government should realise that there should be diversity in agriculture. (Interview 105, vegetable farm, food forest)

Several farmers wanted more financial support from the government to make it easier to compete with imports and ultimately contribute to national food security. One explained:

No matter how much you encourage the consumers to buy local produce, there is a limited amount of people that will support that, it's not across the board. I also don't wish for the locals to eat expensive produce, but my cost is so high, my land cost, my water cost, my property tax. It's scary. If I'm not wrong, he [government] charges me $870,000, how to count? $870,000 multiply by 12%, that's the amount I have to pay [annually for leasing the land]. Hello, why can't you factor this in for a special sector? We are supplying food for people's consumption, we don't have a high profit margin, but we have to pay this tax and that tax, and you want us to compete with others ... What I feel, the simplest method is that you reduce the amount tax, help the farmers. (Interview 113, chicken egg farm)

LIVELIHOOD ALTERNATIVES

Traditional Singapore farmers concluded interviews by describing livelihood alternatives including motivation to become a farmer, impacts on their dietary or other habits, other income sources, and what they would do if they could no longer farm (which seemed to be imminent for many who faced expiring land leases). Beginning with motivation, half the farmers were multi-generational and joined the farming business to support the family. Others were inspired by parents and exposure to nature; one reminisced, 'Growing up, my mum was a *kampung* girl, so growing up, I'm always out in the open.' Others were more socially oriented: 'We took an interest because there [were] a lot of articles about problems, food problems, and so this caught my eye and I just want to make a difference.' Beyond income, farmers were motivated to contribute to society—'Growing for other people to eat. If I don't grow, there will be no vegetables for consumption.' Farming was an alternative calling in a society where prestige and salary are prioritized:

I think farming is an experience that not many people can get ... there can be a lot of executives in Singapore, a lot of accounts, a lot of bankers, but when you talk about farming, I think not a lot will get, to try to join this industry, because also the government has been promoting food resilience, food security, what I think that it's actually an encouragement for us. Because you know that you are needed. You feel a responsibility to continue to do this. (Interview 116, beansprout farm)

Motivation to practice farming intersected with farmers' perception of their impact on society. Beyond contributing to national food security, themes of connecting people to food sources, educating children on how food is grown, promoting farm-to-table, producing high-quality safe food for people such as those with vulnerable immune systems, and resilience to global events—as one farmer reported (during an interview a few months prior to the global COVID-19 pandemic, which disrupted global markets):

I think the society become a bit more conscious of where their food comes from and the importance of it because Singapore is pretty safe in terms of food. Security-wise, [however] we're nowhere in which if suddenly areas around us close-up their doors to our food imports. What are we gonna do, right? So, I think it's really important that we are farming locally. (Interview 112, vegetable farm, mixed technology)

One farmer reflected on a conversation with another farmer about the passion to farm:

Actually, I was also saying this yesterday, if the government does not support our passion, it will slowly die off, we will lose our passion, and no one will continue farming anymore. Hence, I think that the government's support has its influence. (Interview 113, chicken egg farm)

Farmers also talked about how the practice of farming impacted their own diet and other habits. Responses ranged from eating more vegetables, plant-based diets and meat-free Mondays, and reduced food waste due to awareness of the energy invested to produce it—'I understand where the hard work is ... every vegetable has to be cherished, do not waste it.' Sensitivity to waste and lifecycle costs extended beyond food too: 'When you grow to be more appreciative [of the energy it takes to grow food], there is less wastage, order what you need. I suppose when you buy things, you also ask yourself where it comes from, you'll be a lot more educated.' In contrast, several farmers said their livelihood didn't impact them, but most were second- or third-generation farmers so it was possible the themes mentioned were already instilled in them as children.

Finally, farmers described what they would do if they were no longer able to farm—soon to become a reality for many. As a highly educated group, the options were diverse and thoughtful. There were five general pathways: expatriate and farm in another country (e.g., Malaysia, where a few already acquired land); go into processing and distribution (importing foodstuff) by leveraging existing market and customer social networks; go into accounting or marketing because of experience operating a business; a few nearing retirement age mentioned joining Grab (a car-sharing operation similar to Uber); and one planned to give farm tours and focus on agri-tourism. Rather than convert to high-tech operations (although some were doing just that), those who wanted to persist as traditional farmers were willing to move overseas where they had more control over their livelihood:

We are actually also leasing another area with another farmer [in Singapore], we're going to produce small amounts of vegetables, microgreens probably, and then [the rest of the] farming would have to shift to another country. We'll still be producing vegetables but not in Singapore, probably overseas to have maybe some contract farmers, we also have another farm which is tied down to us in the early 90s to produce vegetables that can't be grown in Singapore like spinach and cabbage, these kinds of items. Now we also prepared [an] extra piece of land in that area [overseas] to produce the bok choy, the choy sum that we're used to growing over here [in Singapore]. We will still be selling vegetables, but we won't be able to farm 100% in Singapore, maybe 5% in Singapore. (Interview 107, vegetable farm)

The irony of this quote is that the farmer planned to continue to produce for the Singapore market while growing overseas (most likely in Malaysia but not specified). Can this be the future of traditional farms in Singapore?

CONCLUSION

This chapter summarizes the perspective of traditional commercial Singapore farmers to better understand the sustainability of the local food system. Semi-structured qualitative interviews covered four topics: general farm practices, social networks, benefits and barriers, and livelihood alternatives. Interviewees were diverse in age; however, most were university educated, Singaporean born, had a family history of farming, and grew vegetables on 2- to 3-hectare farms. In addition to commercial orientation, some farms provided other services—specifically, public and educational farm tours. About half of the farms were family operations, which enabled the continuation of long-term business relationships and knowledge and skills. Farmers employed and often provided room and board for foreign workers who were already skilled in traditional farming practices; farmers knew many of the other traditional and new high-tech farmers through events convened through formal organizations; they had fixed contracts with vendors and markets; and made regular contact with customers through on-farm educational tours and farmers markets. At the same time, traditional farms were losing government support and faced land lease expiration unless they chose to convert to high-tech operations; small-scale production made it difficult to reliably meet market demands; and diversification and value-added improvements could have negative tax consequences.

Benefits to farming included the good lifestyle working on the land with access to clean air and food, connection back to sources of food, and exposure to farming as a public good for children. Barriers included government policies and incentives that targeted high-tech, lack of land ownership and expiring leases, markets that were not interested in unique or diverse products reinforced by product uniformity promoted by the government, and Singaporeans who did not know local varieties and wanted imports (they preferred the cheapest option). Labor was also an issue: Singaporeans lacked the skills, stamina, and motivation to join farming—the local population was too highly educated, and there was a need to hire foreign laborers with limited English. Only two farmers mentioned climate change despite rising temperatures already at the plant-stress threshold. Beyond income, farmers were motivated to contribute to society from contributing to national food security to connecting people to food sources and educating children on how food is grown. Finally, farmers described what they would do if they were no longer able to farm—which was soon to become a reality for many. As a highly educated group, the options were diverse and ranged from expatriating to farm overseas to going into accounting or marketing.

Overall, farmers were 'tight-knit' and many supported a family business. Yet, logistically, they faced a multitude of government agencies for approvals and regulations. An official document published by the SFA on starting a farm lists 'helpful' contacts: SFA, ESG, URA, NEA, BCA, PUB, SCDF, LTA, NParks, IRAS, and SLA (Singapore Food Agency, 2023). MOM is another contact required if a farmer wants to hire foreign laborers. So many agencies! Farmers relied on the grassroots-organized Kranji Farmers Association and government-backed SAFEF to mediate with government agencies including how to meet future targets and grant support for improvements. Farmers received government support to go overseas for precedents and technologies but found they were designed for climates different from Singapore. Conversion to high-tech was misaligned with motivations to practice farming and overshadowed the indigenous knowledge base growing domestically as traditional farmers experimented with food forests and other forms of agroecology. They questioned why everything had to move to high-tech, which was not perceived to be sustainable. There were stories of high-tech farms with high costs, unable to recoup investments in a short timeframe, and closing down operations without anyone knowing about it. Traditional farming was experienced as a livelihood—a lifestyle, an identity, a history—not a job.

In the end, all of this will be the driver of Singapore's food future. Arable land in Singapore is increasingly scarce—soil quality is poor, and daytime temps are rising. Traditional in-soil farms are at the end of land leases; some plan to relocate overseas but continue to sell to the Singapore market. High-tech farming requires a different skill set; traditional knowledge is not easily transferable. Over a few generations there has been a loss of agricultural expertise at all levels: from citizens to government agencies. With less than 1% of land in agricultural use, the high-density, city-state of Singapore is testing integrative approaches to where and how food can be grown in the city. However, policies support high-tech intensification and promotion of industrial and commercial land uses toward efficient and value-added activities that exclude traditional farms. There is a need to resolve regulatory and legal constraints that enable high- *and* low-tech farms to produce substantially more food in the city. As urbanizing cities like Singapore face the double threats of urban food insecurity and land scarcity, diverse low- to high-tech multi-functional agriculture could be a critical adaptation for the sustainability of future cities. As one farmer said succinctly:

> Government says that we need to feed our people, but don't put all your eggs into one basket, okay you can go to high-tech, maybe high-tech can supplement a bit of the leafy vegetable but also don't forget the natural way of farming: I am now graduating into a food forest. (Interview 105, vegetable farm, food forest)

ACKNOWLEDGMENTS

The author acknowledges research assistants Ching Sian Sia, who conducted interviews, Pei Yun Tay, who translated and transcribed interviews, and Lixia Bao, who assisted with qualitative coding. At the time of research, they were graduate students in the Department of Architecture at the National University of Singapore. This research was funded by the Singapore Ministry of Education (MOE) Start-up Grant for New Faculty, funding # R-295-000-141-133. An early version of this chapter was presented virtually under the title, 'Roots versus Reach: A Political Ecology of Traditional Farmers in Singapore' at ASFS/AFHVS/CAFS/SAFN 2021, a joint conference hosted by the Association for the Study of Food and Society (ASFS), Agriculture, Food & Human Values Society (AFHVS) held June 9–15, 2021, in New York City, USA. The National University of Singapore (NUS) Institutional Review Board for Social, Behavioural and Educational Research (IRB-SBER) approved this study.

REFERENCES

Agri-Food & Veterinary Authority of Singapore. (2017). *Handled with Care*: *Annual Report 2016/2017*. Retrieved from Singapore: https://www.sfa.gov.sg/docs/default-source/publication/annual-report/ava-ar-2016-17.pdf

Brashears, M. E., & Quintane, E. (2015). The microstructures of network recall: How social networks are encoded and represented in human memory. *Social Networks*, *41*, 113–126. doi:10.1016/j.socnet.2014.11.003

Crooks, A., Pfoser, D., Jenkins, A., Croitoru, A., Stefanidis, A., Smith, D.,.... Lamprianidis, G. (2015). Crowdsourcing Urban form and function. *International Journal of Geographical Information Science*, *29*(5), 720–741.

Diehl, J. A. (2020). Growing for Sydney: Exploring the Urban food system through farmers' social networks. *Sustainability*, *12*(8), 3346.

Diehl, J. A., & Bose, M. (2023). A sustainable livelihoods approach to measuring mobilization of resources through social networks among vulnerable populations: A case study of Delhi farmers. *Social Sciences & Humanities Open*, *8*(1), 100689.

Diehl, J. A., & Oviatt, K. (2019). Productive Urban landscapes: Emerging typologies of form and function in Urban landscapes. In B. M. Rinaldi & P. Y. Tan (Eds.), *High Density Cities*. Basel: Birkhäuser Verlag.

Diehl, J. A., Sweeney, E., Wong, B., Sia, C. S., Yao, H., & Prabhudesai, M. (2020). Feeding cities: Singapore's approach to land use planning for urban agriculture. *Global Food Security*, *26*, 100377.

Fong, L. S., Leng, M. J., & Taylor, D. (2020). A century of anthropogenic environmental change in tropical Asia: Multi-proxy palaeolimnological evidence from Singapore's Central Catchment. *The Holocene*, *30*(1), 162–177.

Knoke, D., & Yang, S. (2008). *Social Network Analysis* (2nd ed.). Los Angeles: SAGE Publications.

Lepoer, B. L. (Ed.). (1991). *Singapore*: *A country study*. (Vol. 550, No. 184). Claitor's Pub Division.

Lin, N. (2001). Building a network theory of social capital. In N. Lin, K. Cook, & R. S. Burt (Eds.), *Social Capital*: *Theory and Research* (pp. 3–29). New York: Aldine De Gruyter.

Lin, N., Fu, Y.-c., & Hsung, R.-M. (2001). The position generator: Measurement techniques for investigations of social capital. In N. Lin, K. Cook, & R. S. Burt (Eds.), *Social Capital*: *Theory and Research* (pp. 57–81). New York: Aldine De Gruyter.

O'Dempsey, T., Emmanuel, M., van Wyhe, J., Taylor, N. P., Tan, F. L., Chou, C.,.... Heng, C. (2014). *Nature contained*: *Environmental histories of* Singapore: NUS Press.

Osborne, P. L. (2012). *Tropical Ecosystems and Ecological Concepts* (2nd ed.). Cambridge University Press.

Perreault, T. A., Bridge, G., & McCarthy, J. (2015). *The Routledge Handbook of Political Ecology*. London: Routledge.

Scoones, I. (2009). Livelihoods perspectives and rural development. *Journal of Peasant Studies*, *36*(1), 171–196. doi:10.1080/03066150902820503

Scott, J. (1991). *Social Network Analysis*: *A Handbook*. London: Sage Publications.

Shakya, H. B., Christakis, N. A., & Fowler, J. H. (2017). An exploratory comparison of name generator content: Data from rural India. *Soc Networks*, *48*, 157–168. doi:10.1016/j.socnet.2016.08.008

Shim, K. F. (1987). *Part IV country papers (Contd)-Singapore*. *Animal feed resources in Asia and the Pacific*. Retrieved from https://www.fao.org/3/AC153E/AC153E13.htm#ch4.11

Singapore Food Agency. (2023). *Starting a Farm: An Industry Guide*. Retrieved from https://www.sfa.gov.sg/docs/default-source/food-farming/sfa-farming-guide_fa-spread-high-res.pdf

Singapore Government Agency. (2023). *Singapore Green Plan 2030*. Retrieved from https://www.greenplan. gov.sg/

Swee-Hock, S. (2012). *The population of Singapore*. Institute of Southeast Asian Studies.

The Economist Group. (2019). Global Food Security Index. Retrieved from https://foodsecurityindex.eiu.com/ Country

The World Bank. (2019a). *Agriculture land (% of land area)*. Retrieved from https://data.worldbank.org/indicator/AG.LND.AGRI.ZS

The World Bank. (2019b). *GPD per capita (current US$)*. Retrieved from https://data.worldbank.org/indicator/ NY.GDP.PCAP.CD

Wolf, E. R. (1990). Distinguished lecture: Facing power - Old insights, new questions. *American Anthropologist*, *92*(3), 586–596.

Wolf, E. R. (2001). *Pathways of Power*. Berkeley: University of California Press.

Worldometer. (2023). *Singapore Population based on United Nations Population data for 2019*. Retrieved from https://www.worldometers.info/world-population/singapore-population/

7 Growing Cities, Changing Demands
Scope of Urban Agriculture as a Sustainable Agricultural Intensification Strategy in India

Sreejith Aravindakshan, Hage Aku, and Michi Tani

INTRODUCTION

Addressing one of the paramount challenges of the 21st century involves meeting the needs of a projected global population of 11 billion humans while simultaneously reducing the agricultural footprint of humanity (Lal, 2016). More than half of the world's population resides in urban areas, accounting for approximately 56% or 4.4 billion people (World Bank, 2023; Figure 7.1). The World Urban Forum reports that the world's cities have surpassed 10,000 in number as of 2023 (PopulationU, 2023). Among these cities, several stand out due to their substantial populations, with over 20 million inhabitants each (World Bank, 2023). Notable examples include Tokyo, Delhi, Seoul, Shanghai,

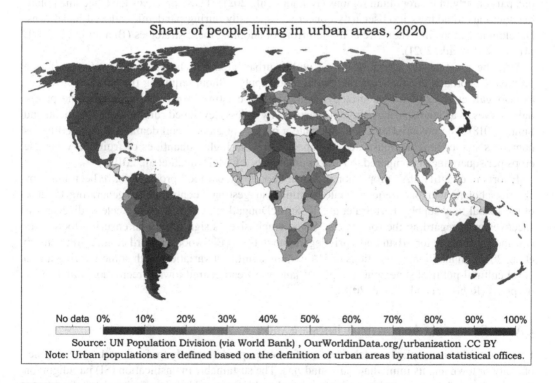

Share of people living in urban areas, 2020

No data 0% 10% 20% 30% 40% 50% 60% 70% 80% 90% 100%

Source: UN Population Division (via World Bank) , OurWorldinData.org/urbanization .CC BY
Note: Urban populations are defined based on the definition of urban areas by national statistical offices.

FIGURE 7.1 Share of the population living in urban areas as of the year 2020.

DOI: 10.1201/9781003359425-7

New York, São Paulo, Mexico City, Cairo, Mumbai, Beijing, and Dhaka. Moreover, recent trends indicate that numerous cities within India have experienced rapid growth between 2012 and 2023 (PopulationU, 2023). This trend toward urbanization is projected to persist, with approximately 68% (6.7 billion individuals) expected to inhabit urban regions by the year 2050 (UN, 2018).

In light of these developments, urbanization presents challenges and opportunities for sustainable land use. As the world experiences unprecedented urbanization, with more people living in cities than ever, land use dynamics are undergoing significant shifts. More than 80% of the global gross domestic product is generated within urban centers (Seto et al., 2014), signifying the potential for urbanization to drive sustainable growth through heightened productivity and innovation. The rapid pace of urban development has resulted in the extensive conversion of vegetated zones into built environments, precipitating the loss of cultivable lands and forests. This ongoing shift has significant implications, given that urbanization and human activities exert transformative impacts on landscapes, influencing their ecological dynamics (McDougall et al., 2019) and food security. Urbanization has introduced a myriad of challenges for food security, including the restricted supply of agricultural produce due to the loss of arable land, a growing demand for fresh and locally sourced products, and increased environmental pressures. Traditional approaches that help to increase crop production often entail either converting natural vegetation to crop cultivation or intensively using existing arable areas, but both strategies are closely tied to biodiversity loss and increased greenhouse gas emissions (GHGs) (Smith et al., 2013). In situations where cropland availability is limited, particularly in urban settings where alternative land use is prioritized, urban agriculture (UA) emerges as a potential avenue to partially meet the consumption needs of urban residents. UA could potentially address the arable land constraints by utilizing available open spaces and cultivable areas within the built environment and adopting innovative sustainable intensification (SI) techniques like soil-less culture, vertical farming, and aeroponics (Al-Kodmany, 2018; Yuan et al., 2022). The escalating trend of urbanization and increasing food demands in expanding cities has fueled the proliferation of UA practices, including kitchen gardening and vertical farming, particularly in metropolitan regions (Bannor et al., 2021). These practices have become potential solutions to address food security concerns, especially during pandemic-induced lockdowns. Developing nations like India are at the forefront of adopting these strategies (Bannor et al., 2021; Mishra & Pattnaik, 2021).

UA, the practice of cultivating crops within urban areas for human consumption, offers a promising approach to enhancing the global food supply without requiring further land clearance (McDougall et al., 2019). This strategy capitalizes on existing urban land resources and potentially serves as a "land-sparing" strategy, particularly in less-developed rural regions (Wilhelm and Smith, 2018). The potential of UA to alleviate a portion of global food demand is compelling, as numerous studies have demonstrated its capacity to yield higher quantities of fruit and vegetable crops per square meter compared to rural counterparts (cf. McDougall et al., 2019).

Efforts to quantify UA's proportional contribution to global food provision have been made by several scholars, with one frequently cited estimate suggesting it contributes to generating 15–20% of the global food supply (Ferreira et al., 2018; McDougall et al., 2019). Nevertheless, diverse perspectives exist regarding the role of UA. Some emphasize its significance in ensuring food security and subsistence for urban underprivileged populations (Bellwood-Howard et al., 2018; Schwab et al., 2018). In addition, perceptions of UA exhibit significant variation, with some viewing it as a novel cultural-political statement or a global land use trend geared toward recreational and social purposes (Robineau and Dugué, 2018).

UA FOR SUSTAINABLE AGRICULTURAL INTENSIFICATION

UA is the cultivation of plants and the raising of animals, predominantly for food, within the confines of a city or town and its immediate surroundings. The sustainable intensification (SI) paradigm, on the other hand, focuses on enhancing agricultural yields while addressing environmental concerns and restraining the expansion of cultivation into ecologically fragile environments (Godfray and Garnett, 2014). The SI approach is deeply rooted in the fundamental objectives of ensuring food security while preserving environmental sustainability. The SI paradigm emerged as a response to recognizing the adverse impacts associated with conventional agriculture. Agriculture, encompassing a substantial 37.1% of the Earth's land area and supporting the livelihoods of approximately 5.6 billion individuals, is responsible for a significant portion of the alarming 5,326.0 million tons of GHGs (FAOSTAT, 2019). The broader context of agriculture, forestry, and land use is equally disconcerting, contributing roughly 25% of global CO_2 emissions (Smith et al., 2014), which rises further when considering emissions from machinery and other sources. Given these environmental challenges, exploring the role of UA within the framework of urban resilience and global sustainability strategies is promising.

UA has been heralded as a potential catalyst for enhancing urban resilience and advancing global sustainability, primarily by offering diverse ecosystem services. UA contributes to various functions, including soil fixation, flood mitigation, stormwater runoff control, pollination support, and microclimate regulation (Edmondson et al., 2020; Nicholls et al., 2023). Moreover, UA has been lauded for its capacity to promote social inclusion, empowerment, and community ownership, particularly within marginalized communities (Orsini et al., 2013).

Modern UA practices emphasize large-scale, standardized, intensive production with reduced pesticide and fertilizer usage (Orsini et al., 2013). While this approach offers safer and more environmentally sustainable food than traditional agriculture, the reality can be more complex. Modern UA and traditional agriculture encounter pest and disease management challenges. Excessive pesticide use in traditional agriculture affects food quality and safety, while UA aims to address this issue by adopting green and organic practices (Aubry and Manouchehri, 2019). Nevertheless, transitioning away from pesticides in modern UA remains an ongoing process (Montiel-León et al., 2019). Recent findings by Ping et al. (2022) highlight persistent concerns, as banned pesticides like Chlorpyrifos and Phorate were detected in vegetables from urban greenhouses, exceeding maximum residue limits. Although the health risks were deemed limited, this underscores the ongoing struggle to ensure pesticide-free produce.

However, despite the optimism surrounding UA, it is imperative to approach the topic critically and scrutinize its claims of sustainability and possible alternatives to sustainably intensify food production in urban areas. One particular aspect raising questions about the sustainability of UA is its carbon footprint. Various factors, including infrastructure, field management practices, and distribution methods influence carbon emissions (CEs) associated with UA. For instance, heated greenhouse production systems tend to produce higher CEs than unheated greenhouses and open-field cultivation (Kumar and Aravindakshan, 2022). The choice of energy sources for heating, which ranges from electricity to fossil fuels or renewable options, significantly impacts emissions. Material inputs, such as fertilizer application rates and the preference for mineral fertilizers or manure, also play a pivotal role in determining CEs. Beyond on-farm production, the complexities of supply chains further contribute to the CE landscape. Shortening supply chains is often advocated to reduce emissions compared to long-distance transportation, and even transportation methods influence emissions (Stoessel et al., 2012).

While UA may offer potential benefits, many sustainability claims are marred by the complexities of addressing pests, disease management, and CEs in real-world situations outside the experimental stations. To fully understand UA's impact and viability, it's essential to critically analyze the practical challenges and potential trade-offs of integrating UA into the broader context of sustainable agricultural intensification. That is, the optimistic consideration of UA's potential for SI must be grounded firmly in empirical studies and evidence that substantiate its claims of augmented food production and concurrent mitigation of deleterious environmental impacts.

While the potential advantages of UA are widely claimed, a compelling need exists for robust empirical evidence and comprehensive, long-term studies that can validate these assertions. The current gaps in available data and the absence of extensive trials hinder a holistic comprehension of the multifaceted impacts of UA. In response to this imperative, urgent attention should be directed toward scrutinizing the sustainability of UA, considering essential social, environmental, and economic indicators through the lens of case studies and primary data. As the momentum of urbanization persists and cities progressively become the focal points of sustainability initiatives, the comprehensive integration of UA into land use planning emerges as an inescapable necessity. In these collective endeavors, relying on empirical data will undoubtedly serve as the guiding beacon, empowering UA to fully realize its potential as a strategic instrument for sustainable agricultural intensification. This realization would bolster urban resilience and global sustainability and foster holistic well-being across human societies. Constrained by the limitations of open-field cultivation in urban areas, UA embraces space-saving agricultural innovations that occupy minimal space. The rationale behind investigating UA innovations such as vertical farming, soil-less agriculture, and similar others mainly involves pilot efforts rather than broader-scale adoption. Consequently, progress is gradual, although this phenomenon, still in its infancy, garners attention from many quarters. A research gap exists concerning UA's potential to function as an alternative to conventional extensive cultivation. This chapter aims to address this gap by delving into the sustainability of UA. This will be accomplished by analyzing primary data collected from urban gardens in two distinct Indian urban ecologies: Arunachal Pradesh in northeastern India and Kerala in the southern region.

UA FOR SI: A CONCEPTUAL FRAMEWORK

A conceptual framework has been developed here to understand the potential of UA as a strategy for SI and facilitate its sustainability assessment. To dive deeper into sustainable UA, exploring its intrinsic attributes is crucial. These range from understanding local soil–plant interactions to deciphering domestic market dynamics and distribution mechanisms for UA products (Adam et al., 2020). Biophysical elements like soil fertility, climate conditions, and pest management shape UA (Bedeke, 2023). A thorough analysis of diverse management practices and their alignment with environmental conditions is also pivotal in understanding their impact on crop yields (Aravindakshan et al., 2022). Examining farmers' decision-making, biophysical productivity, and economic, environmental, and technical efficiency is critical within the realm of agricultural sustainability and intensification (Aravindakshan et al., 2018; Aravindakshan et al., 2020; Aravindakshan et al., 2021a). From an economic perspective, UA serves as both a local agricultural enterprise and an integral part of the regional or national economy. Economic sustainability relies on evaluating production costs and long-term viability projections amid evolving environmental, social, and economic dynamics (Rachael et al., 2021). Zooming out, UA addresses the growing demand for food and fiber in urban settings. Sustainable UA revolves around meeting local and global food and fiber needs, ensuring food quality and security, facilitating technology transfer, and improving food distribution systems' efficiency and inclusivity (Aravindakshan et al., 2020; Bathaei and Štreimikienė, 2023). Sustainability perspectives change with the scale of observation. At the micro-level, urban farmers focus on soil health, nutrients, water, and crop growth (Schulthess et al., 2019). Operational aspects span crop and livestock production, management practices, and overall system viability. UA also influences land use patterns, guiding responsible regional resource utilization. In the literature, sustainable UA involves three key dimensions: economic, environmental, and social, as highlighted by Paganini et al. (2018).

Achieving sustainability in UA requires a balance between these dimensions, as emphasized by Ghosh (2021), Ana et al. (2021), and Zhang et al. (2021). It focuses on mitigating environmental and health impacts, leveraging local ecosystem resources, and preserving biodiversity. Environmental sustainability in UA covers factors like topography, slope, and soil quality (Abhishek et al., 2021). Economic indicators encompass agricultural productivity, income generation, and economic efficiency considerations. The participation of farmers, their satisfaction, and their technical expertise are integral to holistic, sustainable UA (Khanh, 2022).

It may be argued that the conventional sustainability paradigm, with its three primary dimensions (environmental, economic, and social), falls short in comprehensively evaluating and sustainably managing UA. It fails to incorporate a crucial component: cultural and traditional aspects of society. The cultural dimension was formally introduced as the fourth pillar of sustainable development in 2002 and further emphasized in 2010 (Appendino, 2017). The role of culture and political indicators in sustainability, including UA, has become a multidimensional area of research (Alqahtany and Aravindakshan, 2022). Integrating these dimensions into analysis is challenging due to their intricate interplay. The following section aims to identify and distill quantifiable indicators relevant to UA sustainability.

Environmental Dimension of UA

In the context of the environmental dimension of UA sustainability, substantial efforts have been made in the past two decades to address growing concerns regarding sustainability and environmental challenges (Vrolijk et al., 2020). In a thorough examination of environmental indicators drawn from existing literature, this study has categorized them into nine distinct environmental themes or topics. These topics encompass both observable physical aspects of the environment and human activities with substantial environmental implications. They encompass aspects like overall environmental efficiency (EE), soil quality, biodiversity, nutrient management, pesticide usage (Aravindakshan et al., 2022), non-renewable resource utilization (e.g., energy and water), land management practices, GHGs (Aravindakshan et al., 2015; Rahman et al., 2021; Nayak et al., 2023), and substances contributing to acidification.

Economic Dimension of UA

UA, as envisioned by van Cauwenbergh et al. (2007), has the potential to bring prosperity to urban farming communities. The capacity of UA systems (UASs) to withstand economic shifts, which is closely linked to economic viability, aligns well with the principles of economic sustainability (van Cauwenbergh et al., 2007). Changes influence economic stability in this context in production and input costs, crop yields, available markets for produce, and crop prices. The concept of long-term economic viability extends beyond a single farmer's lifetime and may span multiple generations, entwined with the notions of durability and the intergenerational transition of UA. Economic viability encompasses key aspects such as profitability, liquidity, stability, and productivity (van Cauwenbergh et al., 2007).

Profitability hinges on the balance between revenue and costs, while liquidity pertains to the ability to meet short-term financial obligations. Stability encompasses factors like equity and growth, while productivity assesses how efficiently input resources are converted into outputs. Indicators related to income, such as gross income, and those associated with profitability and productivity are primarily quantitative, often expressed in monetary terms or as ratios. These quantitative indicators are widely used for evaluating economic sustainability. In some cases, reference scales are also employed alongside these quantitative measures (Tauqir and Bhatti, 2020). While there have been proposals for a broader range of indicators to encompass various economic aspects of farming systems in the context of sustainability, it is important to note that the assessment of economic sustainability frequently relies heavily on these quantitative indicators, with less emphasis on comprehensive measures (Aravindakshan et al., 2020; Mariia et al., 2019; van Cauwenbergh et al., 2007).

Social Dimension of UA

The exploration of social dimensions within the realm of sustainable UA spans a spectrum of levels, sectors, and conceptual frameworks. The realm of research traverses domains such as gender, development studies, and sociological analyses (Slater, 2001; Orsini et al., 2013; Reynolds, 2015), and, within the sphere of UA, it encompasses participatory approaches (Aravindakshan and Sherief, 2015), social learning within farming communities (Aravindakshan et al., 2021b), and even the perspectives of urban consumers (Notarnicola et al., 2017).

The well-being of urban farmers and their households assumes paramount importance within the context of social sustainability, encompassing facets such as contentment, education, employment generated (quantified by hours, workload, and workforce), and the overall quality of life (gauged by isolation and social engagement) (Lebacq et al., 2013). The ambit of social sustainability extends beyond individual well-being to encompass broader societal considerations molded by values and societal requisites. The spectrum of indicators encompasses multifunctionality, acceptable agricultural practices, and food safety and quality (Renting et al., 2009; Aravindakshan and Sherief, 2015; Lebacq et al., 2013). The synergy between environmental, economic, and social alignment finds expression in the integration of social development with sustainability principles (Hallett et al., 2016). Over the past couple of decades, the schism between social and natural sciences has gradually faded, culminating in a more unified perception of sustainability (Mooney et al., 2013).

CULTURAL DIMENSION OF UA

Within the multifaceted tapestry described in the previous sections, UA resonates with several cultural dimensions, intricately woven by an array of determinants. These cultural facets are shaped by the historical tapestry and traditions of the community, the local tapestry of climate and geography, and the socio-economic fabric that envelops the residents (Ilieva et al., 2022).

A salient cultural dimension of UA lies in its capacity to foster community cohesion, as expounded by Artmann and Sartison (2018). UA serves as a communal canvas where individuals from disparate backgrounds converge to pursue a shared objective. It forms the crucible for forging relationships among neighbors and nurturing a profound sense of community. For instance, the narrative of community gardens often unfolds with communal workdays and vibrant events, acting as crucibles for social interaction and mutual learning.

Another significant cultural facet of UA is its role in preserving and promoting traditional food cultures (Zail, 2023). This is exemplified through the production and marketing of traditional beverages like Phalap (smoked tea) by the Singpho tribe in Arunachal Pradesh, India, as well as the branding and sale of banana chips, a traditional snack, in Kerala. UA provides a conduit through which individuals can cultivate and save foods that hold profound significance within their cultural heritage. Notably, in many immigrant enclaves, UA emerges as the vanguard for nurturing traditional crops and herbs that might be scarce or unavailable within mainstream commercial channels. Beyond the mere act of sustenance, UA thus becomes a torchbearer, facilitating the intergenerational transmission of traditional food wisdom and practices.

The cultural impact of UA extends further, leaving an indelible mark on the urban landscape itself, a phenomenon highlighted by Zail (2023). UA has the transformative potential to introduce new green spaces, enhancing the visual and environmental aesthetics of neighborhoods. Simultaneously, it exerts a magnetic pull on tourists and businesses, catalyzing the renaissance of declining urban communities. For instance, consider the transformation of parts of Pallichal, Trivandrum, Kerala—a once marshy expanse now vibrant with marigold fields, a product of collaboration between Kudumbasree workers and the state agriculture department. This showcases UA's urban revitalization prowess, drawing tourists and breathing new life into the surroundings.

POLITICAL DIMENSION OF UA

UA can be used to challenge the status quo and promote political change. For example, UA can be used to promote food security, especially in marginalized communities. One important political dimension of UA is its role in promoting food sovereignty (Chihambakwe et al., 2018). Food sovereignty is the right of people to define their own food systems and to produce, distribute, and

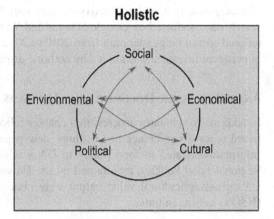

FIGURE 7.2 Theoretical framework of sustainability: a *reductionistic* approach to *holistic* (adapted from Alqahtany and Aravindakshan, 2022).

consume food in a way that is healthy, culturally appropriate, and ecologically sound. UA can help promote food sovereignty by giving people control over their food production and reducing their reliance on industrial agriculture.

Another important political dimension of UA is its role in promoting food security and food self-sufficiency as political goals through various policies and programs (Frayne et al., 2014). Food security is the state of having reliable access to a sufficient quantity of affordable, nutritious food, while food self-sufficiency is the ability to produce one's food. UA can help promote food security by giving people access to fresh, healthy food, even in low-income communities. UA can also help to reduce food insecurity during times of crisis, such as natural disasters or economic recessions.

Drawing from the theoretical insights presented above, a comprehensive theoretical framework has been constructed (depicted in Figure 7.2). This framework serves as the epistemological foundation for the examination of UA in India and its sustainability assessment. It establishes a connection among three crucial elements. Initially, the study examines the status, demand, and supply dynamics of UA in India. Subsequently, it delves into the evaluation of the gross production value of UA in the country. Finally, based on environmental, economic, social, cultural, and political indicators, the sustainability assessment and the impact of UA are expounded, drawing insights from the study of urban households in Kerala and North East India.

It is imperative to highlight that the theoretical framework presented herein, as applied in this specific case, is designed to offer a probabilistic comprehension of UA rather than providing deterministic predictions.

METHODOLOGY

UA is a multifaceted phenomenon encompassing economic, environmental, social, and political change processes. Given its interdisciplinary nature, assessing the potential for SI in UA poses significant methodological challenges. To address these challenges, the current study employed a mixed-method approach that included (1) literature search and review, (2) analysis of public documents and databases, (3) survey and primary data collection, (4) expert assessment, and (5) sustainability index construction and statistical analysis.

LITERARY SEARCH AND REVIEW

The methodology employed in this chapter encompasses a comprehensive global-level inquiry into articles pertaining to UA, with a focus on India aimed at discerning the patterns observed in terms of area, production, and productivity across various regions in India. To achieve this, an extensive search was conducted on Google Scholar and Web of Science platforms on April 18, 2022, with a tailored search range spanning from 2010 to 2022. In contrast to a systematic review, the selection of pertinent literature was guided by authors' discretion and judgment.

ANALYSIS OF PUBLIC DOCUMENTS AND DATABASES

In addition to scholarly articles, this chapter relied on a diverse range of sources, including published reports, conference proceedings, newspaper articles, and government institution websites. Information related to food policy in UA was sourced from the Global Database for City and Regional Food Policies, maintained by the University of Buffalo in the USA. Data pertaining to UA's gross agricultural value output were obtained from the National Statistical Organization's (NSO's) website in India.

SURVEY AND PRIMARY DATA COLLECTION

Concurrently, a demand and sustainability analysis of UA was executed, utilizing primary data collected from urban vegetable gardens situated in two distinct regions of India: Kerala in South India and Arunachal Pradesh in the North East India. The sampling methodology employed was non-probability purposive sampling, whereby urban households engaged in cultivating vegetables within their premises were identified based on lists furnished by the respective Agriculture or Horticulture offices within the two regions.

From Kerala, four prominent cities were selected for study: Trivandrum, Kochi, Thrissur, and Calicut. Similarly, from Arunachal Pradesh, three locations were chosen for investigation, namely Itanagar, Namsai, and Ziro. The sampling approach yielded varying sample sizes across the two states (Table 7.1). In the case of Arunachal Pradesh, a total of 11 urban households from each of the two cities were surveyed. Conversely, in Kerala, 21 urban households were surveyed from each of the four selected cities.

TABLE 7.1
Sampling Framework—Urban Agricultural Systems (UAS) in Surveyed Cities and Corresponding Crops

State	UAS	Cities	Crops	Sample Size
Kerala	Kitchen garden	Trivandrum (6), Kochi (3), Trissur (2)	Ash gourd, bitter gourd, bottle gourd, cabbage, brinjal, chilli, cowpea, pumpkin, radish, snake gourd	11
	Rooftop farming	Trivandrum (2), Kochi (3), Trissur (1)	Cauliflower, cucumber, okra	6
	Polyhouse	Trivandrum (1), Kochi (1), Trissur (1), Calicut (1)	Tomato	4
Arunachal Pradesh	Kitchen garden	Itanagar (3), Namsai (2), Ziro (6)	Beans, brinjal, broccoli, cabbage, cauliflower, cherry tomato, chilli, coriander, king chilli, leaf lettuce, spinach	11

Note: The values in parentheses next to the cities represent the sample size corresponding to each city.

EXPERT ASSESSMENT

Two teams comprised of 15 agricultural experts were established—one from Kerala and the other from Arunachal Pradesh. The first team's purpose was to assess UA households in Kerala, while the second team was tasked with assessing UA households in Arunachal Pradesh. These teams independently visited the UA households that were previously surveyed in "survey and primary data collection". The visits occurred at various times according to the teams' convenience, spanning from June 2022 to June 2023.

Their primary objective during these visits was to evaluate and rate the performance and sustainability of the UA farms in both regions. To conduct these assessments, the experts used a set of five criteria: (1) EE, (2) economic efficiency, (3) positive cultural impact, (4) positive social impact (PSI), and (5) the effectiveness of households in implementing government schemes related to UA. They assigned scores on a scale from 0% to 100% for their evaluations, where "0" indicated the lowest rating and "100" signified the highest.

SUSTAINABILITY INDEX CONSTRUCTION AND STATISTICAL ANALYSIS

The expert assessments, which considered the five criteria for UA as outlined in "expert assessment" along with the indicators of UA sustainability, were utilized to formulate the sustainability index. To estimate CEs from the surveyed UA farms, we utilized the Cool Farm Tool (https://app.coolfarm-tool.org/), a web-based program developed in collaboration between the University of Aberdeen, Unilever, and the Sustainable Food Lab. This tool is widely recognized as a carbon footprint calculator used in 118 countries (Cool Farm Alliance, 2020). The emissions calculations were based on comprehensive farm-level data collected from the surveyed farms, including information on inputs, outputs, and relevant biophysical and management data specific to UA. In line with the methodology demonstrated by Aravindakshan et al. (2021a), principal component analysis to integrate the various variables was employed, and dimensions were used in constructing the composite sustainability index for UA. The survey data and the numerical data obtained from statistical databases and expert assessment scores were analyzed using R statistical software, version 4.2.1.

RESULTS AND DISCUSSION

STATUS, DEMAND, AND SUPPLY DYNAMICS OF UA IN INDIA

India is experiencing rapid urbanization, with its urban population projected to increase from 497 million in 2020 to 881 million in 2050, making it one of the most urbanized countries globally (UN, 2023). This urbanization trend is putting significant pressure on food supply chains, resulting in a heightened demand for fresh, locally sourced produce. Furthermore, India's expanding middle class is witnessing rising incomes and evolving dietary preferences, leading to increased consumption of fruits, vegetables, and other nutritious foods (Sharma & Singh, 2020). The demand for organic, fresh, local food is expected to grow by ~15–25% annually (Cumulative annual growth rate) over the next decade (IBEF, 2023). By 2030, it is projected that India's agriculture and allied sectors will generate USD 813 billion in revenue, creating 152 million jobs and impacting 1.1 billion lives (Aspire Impact, 2021). This shifting demand landscape offers new opportunities for UA to cultivate and market high-value crops.

However, this growing demand comes with its own set of challenges. India's urban areas occupy a relatively small portion of its vast geographical expanse, approximately 6.77% (The World Bank, 2010), while accommodating around 35% of the population, roughly 500 million people (The World Bank, 2018), excluding peri-urban regions. Peri-urban areas, while not clearly delineated, offer fertile ground for UA initiatives. In fact, approximately 65% of the produce found in urban markets in India originates from peri-urban production (Bhat and Paschapur, 2020).

To provide context, if Indian cities were to designate 10% of their urban space for green infrastructure, as suggested in the Urban & Regional Development Plans Formulation & Implementation guidelines (Ministry of Urban Development, Government of India), it would amount to around 22,268 km^2 of urban land. Even if only half of this area (11,134 km^2) were repurposed for UA instead of traditional parks, gardens, playgrounds, and horticulture, it would constitute a mere 5% of the total urban area and a minuscule 0.56% of the nation's agricultural land (Awasthi, 2013). These figures underscore the spatial constraints faced by UA while emphasizing its significant potential, especially when combined with modern technologies and sustainable practices. Despite these spatial challenges, UA remains poised to play a crucial role in India's food security and sustainable urban development. Realizing this potential necessitates the adoption of innovative techniques and region-specific urban planning approaches (Awasthi, 2013).

Yet, urban farmers in India encounter a host of challenges, including limited land supply, restricted access to essential resources, and a lack of policy support (Sharma & Singh, 2020). Only 4% of India's urban land is allocated for agriculture (Jha, 2022). Additionally, urban farmers often face hurdles in accessing water, fertilizers, and quality seeds. Water supply is particularly scarce in many Indian cities, necessitating reliance on rainwater harvesting or reclaimed water for irrigation (Sahasranaman, 2016). Additionally, obtaining affordable and high-quality agricultural inputs can be a significant hurdle.

Despite these challenges, the UA sector in India is experiencing rapid growth. A study by the National Bank for Agriculture and Rural Development in 2020 reported a 50% increase in the UA area in India over the previous five years. Notably, during the COVID-19 pandemic, nearly 30% of the urban population was engaged in UA directly or indirectly, including food sales (Chandrasekharam, 2020). The 2020–2021 Agriculture Census revealed the presence of over 60 million kitchen gardens in India, covering an area exceeding 1.5 million hectares (or 15,000 km^2). According to the analysis, surprisingly, more than 50% of the gross value output (GVO) from kitchen gardens originated from Indian cities. Moreover, the GVO of kitchen gardens in urban areas surpassed that in rural regions, with an average GVO of approximately USD 135 (Rs 10,000) per hectare in urban areas compared to about USD 68 (Rs 5,000) per hectare in rural areas. In specific cities like Chennai, Tamil Nadu, a 2021 study by the Indian Institute of Technology Madras identified over 12,000 hectares (or 120 km^2) of kitchen gardens. These gardens were found to be maintained by the average household across Chennai, covering approximately 100 square feet. Similarly, in Ahmedabad, Gujarat, a 2019 study by the Gujarat Agricultural University reported over 6,000 hectares (or 60 km^2) of kitchen gardens, with the average household cultivating an area of around 25 square feet.

Delhi, the bustling capital, serves as a prominent exemplar of how UA can significantly bolster a city's food supply. As reported by Cook et al. (2015), the fertile expanse of the Yamuna floodplain in Delhi accommodates around 2,500 small-scale urban farms. While there was some variation in the sizes of farm plots, the average land area under cultivation amounted to slightly less than 0.81 hectares (or 8,100 square meters), as documented by Cook et al. in 2015. Impressively, approximately 60% of Delhi's meat demand, 25% of its milk, and 15% of its vegetable requirements are fulfilled by locally grown produce (Nitnaware, 2023). Conversely, in the state of Kerala, with its unique geography and climate, UA takes on a distinctive form. Here, the vibrant tradition of kitchen gardening flourishes, often encompassing 50% of rooftop areas or an equivalent of 50 square meters. At the household level, UA in the state typically covers around 0.02 hectares (or 200 square meters). Turning to our primary data on UA farms, it indicates that household vegetable demand is met by 4–64% in Kerala and 25–40% in North East India. Almost 60–70% of the produce from UA farms in North East India and Kerala is sold.

Within the metropolis of Bangalore, India, the total agricultural land encompasses a significant 11,463 hectares (or 114.63 km^2), providing a snapshot of the UA landscape in this dynamic urban center. A recent study in 2020 by the University of Agricultural Sciences, Bangalore reported over 3,237 hectares (or 32.37 km^2) solely for kitchen gardens. The study also revealed that the average household in Bengaluru tends to have a kitchen garden of around 0.05 hectares (or 500 square meters).

In Bihar, a distinct proclivity toward UA is evident, with households in the major cities typically dedicating approximately 1 hectare of land to these endeavors (Bannor et al., 2021). This inclination reflects a preference for larger-scale UA initiatives in this region, emphasizing the diversity in approaches to UA across India.

Urban Agricultural Gross Production Value in India

The state-wise and item-wise GVO of India's agriculture, forestry, and fishing sectors from 2011–2012 to 2020–2021 were analyzed to understand the growth and dynamics of UA. The data from the NSO (2023) was used to investigate the intricate dynamics of UA, with a focus on the contribution of kitchen gardens to the nation's GVO. Kitchen gardens, a fundamental facet of UA in India, have emerged as significant contributors to the country's GVO. It was found that the GVO from kitchen gardens in India has increased steadily over the period, from 1.07 billion USD in 2011–2012 to 1.66 billion USD in 2020–2021. This represents an overall increase of 56.0%. The five leading states in terms of GVO from kitchen gardens in 2020–2021 were Madhya Pradesh, Rajasthan, Gujarat, Maharashtra, and Uttar Pradesh (Table 7.2). These states accounted for over 50% of the total GVO from kitchen gardens in India. The five lagging states in terms of GVO from kitchen gardens in 2020–2021 were Arunachal Pradesh, Manipur, Meghalaya, Mizoram, and Nagaland (Table 7.2). These states accounted for less than 1% of the total GVO from kitchen gardens in India.

The percentage change in GVO from kitchen gardens between 2011–2012 and 2020–2021 varied widely across states. The highest percentage increase was in Mizoram (154.8%), followed by Nagaland (77.6%), Meghalaya (82.3%), and Manipur (69.0%). The lowest percentage increase was in Arunachal Pradesh (–19.5%). The GVO from kitchen gardens in India increased by 10.2% during the complete COVID-19 lockdown from March 25, 2020, to May 31, 2020. This increase can be attributed to a number of factors, including people spending more time at home and having more time to garden, a desire to grow their own food and reduce reliance on grocery stores, and government initiatives to promote kitchen gardens during the lockdown. The GVO from kitchen gardens in India continued to increase during the partial lockdowns from mid-2020 to early 2021, with an overall increase of 16.7% for the year 2020–2021.

TABLE 7.2
Leading and Lagging States in GVO from Kitchen Gardens in India (2011–2021)

State	Gross Value Output (GVO) in Million USD (2011–2012)	GVO in Million USD (2020–2021)	Percentage Change (%)
Leading states			
Madhya Pradesh	172.50	261.40	52.00
Rajasthan	68.04	144.25	112.00
Gujarat	96.64	128.54	33.00
Maharashtra	105.89	154.11	46.00
Uttar Pradesh	125.26	195.34	56.00
Lagging states			
Arunachal Pradesh	1.46	1.18	−19.50
Manipur	2.32	3.92	69.00
Meghalaya	1.50	2.74	82.00
Mizoram	0.47	1.18	154.8
Nagaland	1.81	3.22	77.60

Source of data for analysis: NSO (2023). The average exchange rates of the US dollar to INR, obtained from https://www.ceicdata.com/ for the years under analysis, were utilized for the calculations.

ENVIRONMENTAL IMPACT OF UA IN SELECT INDIAN CITIES

The summary statistics presented in Table 7.3 offer valuable insights into the performance and efficiency of three distinct UASs: kitchen garden, polyhouse, and rooftop garden. These systems exhibit notable variations across several key variables, shedding light on their sustainability and productivity profiles.

Crop Irrigation Water Productivity

Polyhouses demonstrate exceptional efficiency in crop irrigation water productivity (CIWP) with a mean value of 0.07. This signifies that polyhouses are remarkably adept at maximizing agricultural productivity while minimizing water usage. Their controlled environments and advanced irrigation systems likely play a crucial role in optimizing water distribution. In contrast, kitchen gardens and rooftop gardens maintain similar but comparatively lower CIWP values at 0.01. This suggests that these UASs may benefit from more efficient water management strategies, potentially by adopting water-saving technologies or optimizing irrigation schedules.

TABLE 7.3
Sustainability Indicators and Holistic Sustainability of Urban Agricultural Systems

Variable	Kitchen Garden	Rooftop Garden	Polyhouse Agriculture
A. Environmental indicators			
Crop irrigation water productivity (CIWP)	0.01*** (0.01)	0.01** (0.00)	0.07 (0.01)
Crop nitrogen use productivity (CNUP)	0.24** (0.16)	0.13* (0.05)	0.74 (0.13)
Carbon emissions (kgCO$_2$e per ha)	3,030.38 (1,881.93)	2,836.83 (717.75)	3,852.79 (643.22)
Environmental efficiency (0–100%)	64.34 (13.25)	65.91 (11.14)	67.72 (5.45)
B. Economic indicators			
Annual Gross income (USD/ha)	3,055.82* (1,106.78)	1,443.31** (548.51)	8,629.59 (1,043.65)
Crop yield (tons/ha)	19.83** (11.16)	16.00** (6.03)	106.88 (20.73)
Economic efficiency (0–100%)	49.29** (13.80)	65.99* (11.56)	89.05 (1.84)
C. Social indicators			
Employment created (man-days of hired labor)	19.47 (6.73)	18.17 (8.33)	17.50 (5.45)
Food security provision (share of crop produce for own consumption)	29.59* (14.86)	27.67 (20.29)	12.00 (6.48)
Social impact rating (0–100%)	56.58 (10.95)	66.06 (11.45)	89.77 (1.63)
D. Cultural indicator			
Cultural impact rating (0–100%)	57.22* (11.14)	66.51* (9.93)	41.65 (0.39)
E. Political/policy indicator			
Efficiency in implementing urban agriculture schemes (0–100%)	51.46* (11.64)	66.71* (10.88)	89.80 (1.71)
F. Sustainability			
Composite sustainability index score	−0.89** (1.25)	−0.50** (0.72)	5.66 (0.87)

Note: Mean values are provided. Values in parentheses represent standard deviations. Significance levels (*, **, and ***) indicate differences carried out using the Kruskal–Wallis test between kitchen garden and rooftop agriculture against polyhouse culture and are significant at the 10%, 5%, and 1% levels, respectively. Composite sustainability index scores for each of the samples are calculated based on indicators A to E mentioned above. The average exchange rates of the US dollar to INR, obtained from https://www.ceicdata.com/ for the years under analysis, were utilized for the calculations.

Crop Nitrogen Use Productivity

Polyhouses consistently outperform the other UASs in crop nitrogen use productivity (CNUP), boasting a mean value of 0.74. This indicates that polyhouses effectively utilize nitrogen fertilizers to enhance crop yields while minimizing excess application. Their ability to precisely control nutrient inputs within enclosed environments likely contributes to this efficiency. Conversely, kitchen gardens and rooftop gardens display lower CNUP values of 0.24 and 0.13, respectively. This suggests that these UASs have room for improvement in nitrogen fertilizer management. Implementing more precise nutrient application methods or adopting organic farming practices could help enhance CNUP.

Carbon Emissions

Polyhouses exhibit the highest average CEs at 3,852.79 kgCO$_2$e per hectare. This finding is attributed to the energy-intensive nature of polyhouses, which often require artificial lighting, heating, and cooling systems. These energy demands contribute to increased CEs. In contrast, kitchen gardens emit 3,030.38 kgCO$_2$e per hectare. While they may also require energy inputs for irrigation, the scale and complexity of their operations generally result in lower emissions compared to polyhouses. Rooftop gardens display relatively lower emissions at 2,836.83 kgCO$_2$e per hectare, which can be attributed to their smaller scale and potentially more environmentally friendly designs. Rooftop gardens often utilize available sunlight, reducing the need for artificial lighting and heating.

Environmental Efficiency

Polyhouses stand out once again in terms of EE, with a mean value of 89.82%. This high score suggests that polyhouses effectively utilize resources for sustainable practices, encompassing efficient water and nutrient management and reduced waste. Rooftop gardens also perform well, with a mean EE of 65.91%, indicating their positive contribution to sustainable UA. While maintaining a respectable score of 64.34%, kitchen gardens have room for improvement to further enhance their EE. This could involve the adoption of more sustainable farming practices, such as organic farming or efficient use of resources like composting and rainwater harvesting.

ANALYZING THE ECONOMIC DIMENSIONS OF UA

When evaluating economic indicators (Table 7.3), it was observed that there are significant variations among UASs. The Annual Gross income (USD/ha) portrays substantial differences between polyhouses and the other UASs. Polyhouses boast an impressive mean income of 8,629.59 USD/ha, significantly higher than kitchen gardens (3,055.82 USD/ha) and rooftop agriculture (1,443.31 USD/ha). This divergence can be attributed to the high-value crops typically cultivated within polyhouses and controlled environments that enhance crop yields and market value. Despite spatial constraints in urban settings, rooftop agriculture manages to secure a respectable income, while kitchen gardens yield gross returns higher than rooftop agriculture.

Crop yield (tons/ha) demonstrates a similar pattern. Polyhouses exhibit remarkably high crop yields, averaging 106.88 tons/ha, substantially greater than kitchen gardens' yields (19.83 tons/ha) and rooftop agriculture (16.00 tons/ha). The advanced technology, optimized conditions, and crop selection within polyhouses contribute to this substantial output. Kitchen gardens and rooftop agriculture, often reliant on traditional methods and facing space limitations, achieve comparatively lower crop yields.

Economic efficiency (0–100%) reflects the overall economic sustainability of each UAS. Polyhouses stand out with an average efficiency score of 89.05%, signifying efficient resource utilization and high economic sustainability. Rooftop agriculture also showcases commendable efficiency, scoring an average of 65.99%. While kitchen gardens yielded a higher gross income (3,055.82 USD/ha) compared to rooftop gardens (1,443.31 USD/ha), it is noteworthy that kitchen gardens demonstrate lower economic efficiency (49.29%). This disparity may be attributed to factors such as diminished net returns, spatial constraints, and diverse practices and input utilization among households practicing kitchen gardens.

EXPLORING THE SOCIETAL IMPACT OF UA

Turning to social indicators (Table 7.3), insights into the socio-cultural dimensions of UA in different settings were uncovered. Employment created (man-days of hired labor) reveals that, on average, all three UASs contribute to job creation. Kitchen gardens, with an average of 19.47 man-days, and rooftop agriculture, with 18.17 man-days, provide relatively similar employment opportunities. However, polyhouses generate slightly fewer man-days at 17.50, possibly due to greater mechanization and limited labor requirements.

Food security provision (share of crop produce for own consumption) demonstrates intriguing variations. Kitchen gardens, predominantly found in North East India, display a higher mean share of crop produce for own consumption at 29.59%, potentially reflecting the self-sufficiency culture prevalent in the region. Rooftop agriculture follows with 27.67%, indicating a substantial portion of household food supply. In contrast, polyhouses allocate a lower proportion for their own consumption, averaging 12.00%, possibly due to their focus on commercial high-value crops.

Social impact rating (0–100%) unveils significant differences in how these UASs contribute to the broader social well-being of communities. Polyhouses, with an average rating of 89.77%, exhibit the highest social impact, indicating their significant contribution to the community's socio-economic development. Rooftop agriculture, with an average rating of 66.06%, also plays a substantial role in social welfare, whereas kitchen gardens, with a rating of 56.58%, lag slightly behind regarding social impact.

These observations underscore the multifaceted nature of UA, where economic and social outcomes depend on various factors, including geography, local practices, and resource availability. These indicators illuminate the diverse contributions of different UASs to urban communities' economic and social fabric.

EXPLORING THE CULTURAL SIGNIFICANCE OF UA

Exploring the cultural dimension of UA, it was found that the cultural impact rating (0–100%) provides exciting insights into how these practices influence the cultural fabric of communities. Rooftop agriculture emerges as a leader in this aspect, boasting an average rating of 66.51% (Table 7.3). The proximity of these gardens to households and the active participation of residents in their maintenance likely contribute to their cultural significance. Kitchen gardens follow closely with a mean rating of 57.22%, reflecting their role in preserving traditional agricultural practices and community engagement.

In contrast, polyhouses, with a significantly lower mean rating of 41.65%, appear to have a comparatively lower cultural impact. This might be attributed to their commercial orientation, resulting in a diminished connection to traditional farming practices and cultural aspects.

EFFICIENCY IN IMPLEMENTING UA SCHEMES

The assessment of the efficiency in implementing UA schemes (0–100%) offers a critical lens through which to evaluate the effectiveness of policy measures and governance structures in promoting and supporting UA. Notably, rooftop agriculture emerges as the most proficiently implemented UAS, boasting a mean efficiency score of 66.71%. This observation underscores the presence of a conducive policy environment and the effectiveness of governance mechanisms, which foster the proliferation of rooftop gardens within urban domains.

Following closely, kitchen gardens exhibit a mean efficiency score of 51.46%, indicating a moderate level of policy support and functional governance structures. This suggests a favorable environment for establishing and maintaining kitchen gardens in urban settings.

Intriguingly, polyhouses present the lowest mean efficiency score at 89.80% despite their economic success and remarkable productivity. This finding is particularly noteworthy, suggesting that while polyhouses excel in various aspects, there is potential for developing more supportive policies and streamlining governance processes to further enhance the effectiveness of UA schemes involving polyhouses.

Several Indian states have proactively taken steps to foster UA by initiating programs and subsidies aimed at its promotion. For instance, the government of Kerala actively champions rooftop vegetable gardening through various schemes. These initiatives encompass incentives such as providing 20 clay or high-quality plastic pots at a total cost of USD 50.00. Furthermore, the government actively promotes the utilization of irrigation units for terrace cultivation, offering a family drip unit at USD 85.00, which includes 25 drip systems, and wick irrigation units of 25 numbers at USD 30 per unit.

A visionary pilot project by the government of Kerala aspires to stimulate vegetable cultivation across all homesteads within five major cities in the state (Trivandrum, Kochi, Calicut, Kollam, and Trichur). This initiative extends need-based support to residence associations, contingent upon project submissions to expand vegetable cultivation within these urban areas (Government of Kerala, 2022). This multifaceted approach highlights the government's commitment to bolstering UA and underscores the importance of tailored policy measures and governance structures in achieving sustainable UA objectives.

Assessing Holistic Sustainability and Potential for SI in UASs

Five crucial dimensions were assessed in evaluating SI potential among different UASs, each rated on a scale from 0% to 100%. These dimensions included EE (EnvEff), economic efficiency (EconEff), PSI, positive cultural impact, and efficiency in implementing government schemes, complemented by additional variables detailed in Table 7.3. The radar plot (Figure 7.3) illustrates that among the UASs under scrutiny, polyhouses emerged as robust contenders, exhibiting the highest economic efficiency (89.05%) and PSI (89.8%). This implies that polyhouses excel economically and possess significant cultural significance within the surveyed regions. Rooftop gardening also displayed noteworthy performance, particularly in terms of EE (65.91%), PSI (66.06%), and efficiency in implementing government schemes (66.71%). Despite kitchen gardens demonstrating relatively lower economic efficiency (49.29%) and PSI (56.58%), they still made positive contributions to EE (64.34%) and positive cultural impact (~60%).

The composite sustainability index scores, constructed from these multidimensional assessments, offer a comprehensive perspective on the sustainability potential of each UAS. Notably, polyhouses and rooftop gardening systems exhibit a higher potential for SI, supported by their substantial economic and environmental performance, PSI, and efficient utilization of government initiatives. Kitchen gardens, while contributing positively to various aspects, could benefit from economic and EE enhancements, as well as the implementation of government schemes. In summary, these findings highlight the diverse strengths and areas for improvement across different UASs, underscoring the necessity for customized approaches to enhance sustainability in UA and promote SI within varying urban contexts. Lastly, the composite sustainability index score provides a holistic assessment of the overall sustainability of each UAS. Polyhouses stand out significantly, with an average score of 5.66. Their focus on high-value crops, advanced technology, and efficient resource use contribute to their superior sustainability. rooftop agriculture attained a mean sustainability score of −0.50, a satisfactory result considering the space constraints and the utilization of low-level crop management practices, along with minimal input resources in rooftop gardens. In contrast, kitchen gardens presented a mean score of −0.89, indicating challenges in environmental

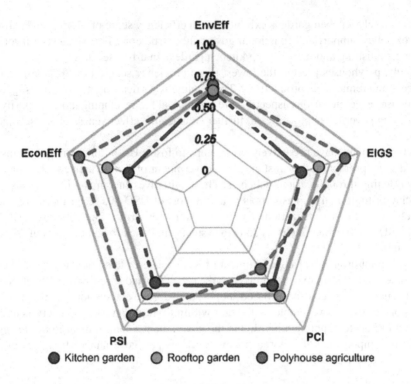

FIGURE 7.3 Radar plot illustrating the five sustainability indicators and their relative positions across the various urban agricultural systems (UAS).

impact and resource management. These difficulties may be associated with traditional farming practices and variations in household-level sustainability efforts.

CONCLUSION

An extensive assessment of the potential of three distinct urban agricultural systems (UASs) (kitchen gardens, rooftop farming, and polyhouses) to serve as a viable strategy for sustainable agricultural intensification in India was conducted using a comprehensive mixed-method approach, which involved analyzing peer-reviewed literature, official documents, publicly available secondary data, and primary data collected from UA households. Despite limitations in the availability of reliable secondary data and a relatively small sample size of 32 urban farms, robust statistical methods were employed, and innovative tools like the "Cool Farm Tool" were utilized in this study, resulting in a rigorous analysis.

The findings indicate that polyhouses show considerable promise in optimizing water and nutrient utilization. However, they tend to have higher CEs due to their energy-intensive nature. On the contrary, kitchen gardens and rooftop farming systems demonstrate potential for improvement in water and nutrient management while maintaining more favorable EE scores. These variations in environmental indicators among the UASs underscore the importance of customizing agricultural approaches to local conditions, fostering technological adoption, and considering the scale of operations to enhance sustainability and minimize environmental impacts.

The comparative analysis of GVO from kitchen gardens in South Indian and North East Indian states highlights significant regional disparities. South Indian states exhibit higher GVO figures and have recorded more substantial percentage increases in recent years. Nevertheless, North East Indian states, particularly Mizoram, Nagaland, and Meghalaya, are making commendable progress, demonstrating substantial percentage growth in GVO from kitchen gardens between 2011–2012 and 2020–2021.

Despite the relatively lower sustainability scores observed in rooftop and kitchen gardens, these systems can still achieve sustainability if managed meticulously. Implementing efficient irrigation and fertilization practices and adopting best-bet technologies contribute to their sustainability.

The findings shed light on the intricate dynamics of UA across diverse geographical contexts. Notably, the impact of the COVID-19 pandemic on the GVO of kitchen gardens is evident in the NSO data for 2020–2021. The GVO of kitchen gardens in North East India increased by 10% in 2020–2021, compared to an average annual increase of 5% over the previous decade. In South India, the GVO of kitchen gardens saw a 20% increase in 2020–2021, compared to an average annual increase of 3% over the previous decade.

In the post-COVID era, several Indian states have taken proactive measures by introducing programs and subsidies to promote UA. A noteworthy example is the Kerala government, which actively encourages rooftop vegetable gardening through various schemes. These initiatives include incentives such as providing clay or high-quality plastic pots and irrigation units. Their forward-thinking pilot project is designed to stimulate vegetable cultivation in major cities across Kerala, extending support to residence associations to expand vegetable cultivation within these urban areas.

These regional variations underscore the significant influence of geography, climate, local practices, and policy support on shaping UASs' economic, social, cultural, and sustainability dimensions. It is crucial to acknowledge these diversities to tailor interventions and policies effectively, thereby maximizing the benefits of UA while addressing specific challenges inherent to different urban settings.

Further research is essential to understand the factors driving the expansion of UAS in India. However, UA holds significant potential for enhancing food security and improving regional nutritional outcomes. To realize this potential, it is imperative to conduct participatory experimental research to enhance these systems' productivity, efficiency, and sustainability. Equally critical is the need for robust agricultural extension support to effectively communicate research findings to urban growers. Presently, urban agricultural research in India remains underdeveloped, and the availability of data on Indian UA is limited, primarily due to the limited engagement of agricultural scientists in scholarly and public discourse surrounding UA. By making strategic investments in research and development, providing training and support to urban farmers, and promoting sustainable management practices, the full potential of the UAS to contribute significantly to sustainable agricultural intensification in urban areas can be unlocked.

ACKNOWLEDGMENTS

The first author acknowledges with appreciation the research support received from the One CGIAR Regional Integrated Initiative, Transforming Agrifood Systems in South Asia (TAFSSA; https://www.cgiar.org/initiative/20-transforming-agrifood-systems-in-south-asia-tafssa/), and CIMMYT during the data analysis and manuscript preparation. The views expressed in this chapter are solely those of the authors and do not necessarily represent those of One CGIAR, CIMMYT, and TAFSSA, or their funders, and should not be used for advertising purposes.

REFERENCES

Abhishek, R., Jhariya, M. K., Khan, N., Banerjee, A., & Meena, R. S. (2021). Ecological intensification for sustainable development. In Ecological Intensification of Natural Resources for Sustainable Agriculture (pp. 137–170). Springer.

Adam, J. V., Aizen, M., Cordeau, S., Garibaldi, L., Garratt, M. P., Kovács-Hostyánszki, A., Lecuyer, L., Ngo, H., & Potts, S. (2020). Transformation of agricultural landscapes in the anthropocene: Nature's contributions to people, agriculture and food security. In *Advances in Ecological Research* (pp. 193–253). Elsevier.

Al-Kodmany, K. (2018). The vertical farm: A review of developments and implications for the vertical city. *Buildings*, 8(2), 24. https://doi.org/10.3390/buildings8020024

Alqahtany, A., & Aravindakshan, S. (2022). Urbanization in Saudi Arabia and sustainability challenges of cities and heritage sites: Heuristical insights. *Journal of Cultural Heritage Management and Sustainable Development*, 12(4), 408–425.

Ana, T., Marta-Costa, A., & Fragoso, R. (2021). Principles of sustainable agriculture: Defining standardized reference points. *Sustainability*, 13, 4086.

Appendino, F. (2017), October. Balancing heritage conservation and sustainable development-the case of Bordeaux, *IOP Conference Series: Materials Science and Engineering*, Vol. 245 No. 6, 62002. Bristol, UK. https://doi.org/10.1088/1757-899X/245/6/062002

Aravindakshan, S., & Sherief, A. K. (2015). Collective action on improving environmental and economic performance of vegetable production: Exploring pesticides safety in India. In *Biodiversity Conservation-Challenges for the Future* (pp. 127–135). Bentham Science Publishers.

Aravindakshan, S., Rossi, F. J., & Krupnik, T. J. (2015). What does benchmarking of wheat farmers practicing conservation tillage in the eastern Indo-Gangetic Plains tell us about energy use efficiency? An application of slack-based data envelopment analysis. *Energy*, 90, 483–493.

Aravindakshan, S., Rossi, F., Amjath-Babu, T. S., Veettil, P. C., & Krupnik, T. J. (2018). Application of a bias-corrected meta-frontier approach and an endogenous switching regression to analyze the technical efficiency of conservation tillage for wheat in South Asia. *Journal of productivity analysis*, 49(2–3), 153–171.

Aravindakshan, S., Krupnik, T. J., Groot, J. C., Speelman, E. N., Amjath-Babu, T. S., & Tittonell, P. (2020). Multi-level socioecological drivers of agrarian change: Longitudinal evidence from mixed rice-livestock-aquaculture farming systems of Bangladesh. *Agricultural Systems*, 177, 102695.

Aravindakshan, S., Krupnik, T. J., Amjath-Babu, T. S., Speelman, S., Tur-Cardona, J., Tittonell, P., & Groot, J. C. (2021a). Quantifying farmers' preferences for cropping systems intensification: A choice experiment approach applied in coastal Bangladesh's risk prone farming systems. *Agricultural Systems*, 189, 103069.

Aravindakshan, S., Krupnik, T. J., Shahrin, S., Tittonell, P., Siddique, K. H., Ditzler, L., & Groot, J. C. (2021b). Socio-cognitive constraints and opportunities for sustainable intensification in South Asia: Insights from fuzzy cognitive mapping in coastal Bangladesh. *Environment, Development and Sustainability*, 23(11), 16588–16616.

Aravindakshan, S., AlQahtany, A., Arshad, M., Manjunatha, A. V., & Krupnik, T. J. (2022). A metafrontier approach and fractional regression model to analyze the environmental efficiency of alternative tillage practices for wheat in Bangladesh. *Environmental Science and Pollution Research*, 29(27), 41231–41246.

Artmann, M., & Sartison, K. (2018). The role of urban agriculture as a nature-based solution: A review for developing a systemic assessment framework. *Sustainability*, 10(6), 1937.

Aspire Impact. (2021). *"Food, Agri and AgriTech" Report*. Aspire Impact Ratings Pvt Ltd. Gurugram. https://aspireimpact.in/impact-research/ ; accessed 23rd September 2023.

Aubry, C., & Manouchehri, N. (2019). Urban agriculture and health: Assessing risks and overseeing practices. Field Actions Science Reports. *The journal of field actions*, (Special Issue 20), 108–111.

Awasthi, P. (2013). Urban agriculture in India and its challenges. *International Journal of Environmental Science*, 4(2). Retrieved from https://www.ripublication.com/ijesdmspl/ijesdmv4n2_12.pdf

Bannor, R. K., Sharma, M., & Oppong-Kyeremeh, H. (2021). Extent of urban agriculture and food security: Evidence from Ghana and India. *International Journal of Social Economics*, 48(3), 437–455.

Bathaei, A., & Štreimikienė, D. (2023). A systematic review of agricultural sustainability indicators. *Agriculture*, 13(2), 241.

Bedeke, S. B. (2023). Climate change vulnerability and adaptation of crop producers in sub-Saharan Africa: A review on concepts, approaches and methods. *Environment, Development and Sustainability*, 25(2), 1017–1051.

Bellwood-Howard, I., Shakya, M., Korbeogo, G., & Schlesinger, J. (2018). The role of backyard farms in two West African urban landscapes. *Landscape and Urban Planning*, 170, 34–47.

Bhat, C. & Paschapur, A. (2020). Urban agriculture: The saviour of rapid urbanization. *Indian Farmer* 7(01): 1–9.

Chandrasekharam, D. (2020, December 7). *Urban Agriculture. Dornadula C, India, politics, TOI*. https://timesofindia.indiatimes.com/blogs/dornadula-c/urban-agriculture/; accessed 30th August 2022.

Chihambakwe, M., Mafongoya, P., & Jiri, O. (2018). Urban and peri-urban agriculture as a pathway to food security: A review mapping the use of food sovereignty. *Challenges*, 10(1), 6.

Cook, J., Oviatt, K., Main, D. S., Kaur, H., & Brett, J. (2015). Re-conceptualizing urban agriculture: An exploration of farming along the banks of the Yamuna River in Delhi, India. *Agriculture and Human Values*, 32, 265–279.

Cool Farm Alliance. (2020). *Cool Farm Impact Report*, October 2020. https://coolfarmtool.org/wp-content/uploads/2020/10/Cool-Farm-Impact-Report.pdf; accessed 15 July 2023.

Edmondson, J. L., Cunningham, H., Densley Tingley, D. O., Dobson, M. C., Grafius, D. R., Leake, J. R., McHugh, N., Nickles, J., Phoenix, G. K., Ryan, A. J., & Stovin, V. (2020). The hidden potential of urban horticulture. *Nature Food*, 1(3), 155–159. https://doi.org/10.1038/s43016-020-0045-6

FAOSTAT (2019) *Statistical Database*. Food and Agriculture Organization of the United Nations, Rome. https://www.fao.org/faostat/en/#data; accessed 20th September 2023.

Ferreira, A. J. D., Guilherme, R. I. M. M., & Ferreira, C. S. S. (2018). Urban agriculture, a tool towards more resilient urban communities? *Current Opinion in Environmental Science & Health*, 5, 93–97.

Frayne, B., McCordic, C., & Shilomboleni, H. (2014). Growing out of poverty: Does Urban agriculture contribute to household food security in Southern African Cities?. *Urban Forum*, 25, 177–189.

Ghosh, S. (2021). Urban agriculture potential of home gardens in residential land uses: A case study of regional City of Dubbo, Australia. *Land Use Policy*, 109, 105686.

Godfray, H. C. J., & Garnett, T. (2014). Food security and sustainable intensification. *Philosophical Transactions of the Royal Society B: Biological Sciences*, 369(1639), 20120273.

Government of Kerala. (2022). *How can Kerala double its vegetable production in the next five years: Report of the working group for the Fourteenth Five-Year Plan (2022-2027)*. Agriculture Division, Kerala State Planning Board. https://spb.kerala.gov.in/sites/default/files/inline-files/How%20can%20Kerala%20double%20its%20vegetable%20production%20in%20the%20next%20five%20years.pdf

Hallett, S., Hoagland, L., & Toner, E. (2016). Urban agriculture: Environmental, economic, and social perspectives. *Horticultural Reviews* 44, 65–120.

Ilieva, R. T., Cohen, N., Israel, M., Specht, K., Fox-Kämper, R., Fargue-Lelièvre, A., Poniży, L., Schoen, V., Caputo, S., Kirby, C. K., & Goldstein, B. (2022). The socio-cultural benefits of urban agriculture: A review of the literature. *Land*, 11(5), 622.

India Brand Equity Foundation. (2023). *Agriculture and Allied Industries Industry Report*. India Brand Equity Foundation. https://www.ibef.org/industry/agriculture-india; accessed 23rd September, 2023.

Jha, R. (2022). *Optimizing Urban agriculture: A pathway to food security in India*. ORF Issue Brief No. 590. Observer Research Foundation, Mumbai.

Khanh, C. N. T. (2022). Driving factors for green innovation in agricultural production: An empirical study in an emerging economy. *Journal of Cleaner Production*, 368, 132965.

Kumar, B. M., & Aravindakshan, S. (2022). Carbon footprints of the Indian AFOLU (Agriculture, Forestry, and Other Land Use) sector: A review. *Carbon Footprints*, 2, 1B.

Lal, R. (2016). Feeding 11 billion on 0.5 billion hectare of area under cereal crops. *Food and Energy Security*, 5(4), 239–251.

Lebacq, T., Baret, P. V., & Stilmant, D. (2013). Sustainability indicators for livestock farming. *A Review. Agronomy for Sustainable Development*, 33, 311–327.

Mariia, K., Pigosso, D. C., & McAloone, T. (2019). Towards the ex-ante sustainability screening of circular economy initiatives in manufacturing companies: Consolidation of leading sustainability-related performance indicators. *Journal of Cleaner Production*, 241, 118318.

McDougall, R., Kristiansen, P., & Rader, R. (2019). Small-scale urban agriculture results in high yields but requires judicious management of inputs to achieve sustainability. *Proceedings of the National Academy of Sciences*, 116(1), 129–134.

Mishra, A., & Pattnaik, D. (2021). Urban agriculture during and post Covid-19 Pandemic. *Biotica Research Today*, 3(1), 62–64.

Montiel-León, J. M., Duy, S. V., Munoz, G., Verner, M. A., Hendawi, M. Y., Moya, H., ... & Sauvé, S. (2019). Occurrence of pesticides in fruits and vegetables from organic and conventional agriculture by QuEChERS extraction liquid chromatography tandem mass spectrometry. *Food Control*, 104, 74–82.

Mooney, H. A., Duraiappah, A., & Larigauderie, A. (2013). Evolution of natural and social science interactions in global change research programs. *Proceedings of the National Academy of Sciences*, 110(supplement_1), 3665–3672.

Nayak, H. S., Parihar, C. M., Aravindakshan, S., Silva, J. V., Krupnik, T. J., McDonald, A. J., ... & Sapkota, T. B. (2023). Pathways and determinants of sustainable energy use for rice farms in India. *Energy*, 272, 126986.

Nicholls, E., Griffiths-Lee, J., Basu, P., Chatterjee, S. and Goulson, D., (2023). Crop-pollinator interactions in urban and peri-urban farms in the United Kingdom. *Plants, People, Planet*, 5:759–775.

Nitnaware H. (2023). Cultivated idea: Urban farming in India requires holistic policy support; here is why. *Down to Earth.* https://www.downtoearth.org.in/news/agriculture/cultivated-idea-urban-farming-in-india-requires-holistic-policy-support-here-is-why-85450; accessed 11th August 2023.

Notarnicola, B., Sala, S., Anton, A., McLaren, S. J., Saouter, E., & Sonesson, U. (2017). The role of life cycle assessment in supporting sustainable agri-food systems: A review of the challenges. *Journal of Cleaner Production*, 140, 399–409.

NSO (National Statistical Office). (2023). State-Wise and Item-Wise Value of Output from Agriculture, Forestry, and Fishing Year: 2011-12 to 2020-21 with Base Year: 2011-2012. *Ministry of Statistics and Programme Implementation, Government of India.* https://www.mospi.gov.in/sites/default/files/publication_reports/Agr_forestry-fishingBrochure2023N_0.pdf ; accessed 8th June 2023.

Orsini, F., Kahane, R., Nono-Womdim, R., & Gianquinto, G. (2013). Urban agriculture in the developing world: A review. *Agronomy for sustainable development*, 33, 695–720.

Paganini, N., Raimundo, I., & Lemke, S. (2018). The potential of urban agriculture towards a more sustainable urban food system in food-insecure neighborhoods in Cape Town and Maputo. *Economia agro-alimentare / Food Economy*, 20(3), 401–423. https://doi.org/10.3280/ECAG2018-003008.

Ping, H., Wang, B., Li, C., Li, Y., Ha, X., Jia, W., ... & Ma, Z. (2022). Potential health risk of pesticide residues in greenhouse vegetables under modern urban agriculture: A case study in Beijing, China. *Journal of Food Composition and Analysis*, 105, 104222.

PopulationU (2023). Research papers on Population dynamics and Social affairs. Available at: https://www.populationu.com/world-cities; accessed 2nd July 2023.

Rachael, D. G., Cammelli, F., Ferreira, J., Levy, S., Valentim, J., & Vieira, I. (2021). Forests and sustainable development in the Brazilian Amazon: History, trends, and future prospects. *Annual Review of Environmental Resources*, 46, 625–652.

Rahman, M. M., Aravindakshan, S., Hoque, M. A., Rahman, M. A., Gulandaz, M. A., Rahman, J., & Islam, M. T. (2021). Conservation tillage (CT) for climate-smart sustainable intensification: Assessing the impact of CT on soil organic carbon accumulation, greenhouse gas emission and water footprint of wheat cultivation in Bangladesh. *Environmental and Sustainability Indicators*, 10, 100106.

Renting, H., Rossing, W. A., Groot, J. C., Van der Ploeg, J. D., Laurent, C., Perraud, D., Stobbelaar, D. J., & Van Ittersum, M. K. (2009). Exploring multifunctional agriculture. A review of conceptual approaches and prospects for an integrative transitional framework. *Journal of Environmental Management*, 90, S112–S123.

Reynolds, K. (2015). Disparity despite diversity: Social injustice in New York City's urban agriculture system. *Antipode*, 47(1), 240–259.

Robineau, O., & Dugué, P. (2018). A socio-geographical approach to the diversity of urban agriculture in a West African city. *Landscape and Urban Planning*, 170, 48–58.

Sahasranaman, M. (2016). Future of urban agriculture in India. Institute for resource analysis and policy. *Occasional Paper No.10-1216*, 2(10), 1–24.

Schulthess, U., Ahmed, Z. U., Aravindakshan, S., Rokon, G. M., Kurishi, A. A., & Krupnik, T. J. (2019). Farming on the fringe: Shallow groundwater dynamics and irrigation scheduling for maize and wheat in Bangladesh's coastal delta. *Field Crops Research*, 239, 135–148.

Schwab, E., Caputo, S., & Hernández-García, J. (2018). Urban agriculture: Models-in-circulation from a critical transnational perspective. *Landscape and Urban Planning*, 170, 15–23.

Seto K.C., S. Dhakal, A. Bigio, H. Blanco, G.C. Delgado, D. Dewar, L. Huang, A. Inaba, A. Kansal, S. Lwasa, J.E. McMahon, D.B. Müller, J. Murakami, H. Nagendra, and A. Ramaswami, 2014: Human Settlements, Infrastructure and Spatial Planning. In: *Climate Change 2014: Mitigation of Climate Change. Contribution of Working Group III to the Fifth Assessment Report of the Intergovernmental Panel on Climate Change* [Edenhofer, O., R. Pichs-Madruga, Y. Sokona, E. Farahani, S. Kadner, K. Seyboth, A. Adler, I. Baum, S. Brunner, P. Eickemeier, B. Kriemann, J. Savolainen, S. Schlömer, C. von Stechow, T. Zwickel and J.C. Minx (eds.)]. Cambridge University Press, Cambridge, United Kingdom and New York, NY, USA.

Sharma, P., & Singh, A. K. (2020). Urban farming in India: A review of drivers, opportunities, and challenges. *Journal of Agroecology and Rural Development*, 11(1), 92–101.

Slater, R. J. (2001). Urban agriculture, gender and empowerment: An alternative view. *Development Southern Africa*, 18(5), 635–650.

Smith, P, Clark H, Dong H, et al. (2014). Chapter 11 - Agriculture, forestry and other land use (AFOLU). In: *climate change: mitigation of climate change. IPCC Working Group III Contribution to AR5*. NY, Cambridge University Press, p. 811–922.

Smith, P., Haberl, H., Popp, A., Erb, K.H., Lauk, C., Harper, R., Tubiello, F.N., de Siqueira Pinto, A., Jafari, M., Sohi, S. and Masera, O. (2013). How much land-based greenhouse gas mitigation can be achieved without compromising food security and environmental goals?. *Global change biology*, 19(8), pp.2285–2302.

Stoessel, F., Juraske, R., Pfister, S., & Hellweg, S. (2012). Life cycle inventory and carbon and water footprint of fruits and vegetables: Application to a Swiss retailer. *Environmental science & technology*, 46(6), 3253–3262.

Tauqir, A., & Bhatti, A. A. (2020). Measurement and determinants of multi-factor productivity: A survey of literature. *Journal of Economic Surveys*, 34, 293–319.

The World Bank (2023). Urban Development. Retrieved from https://www.worldbank.org/en/topic/urbandevelopment/overview#:~:text=Today%2C%20some%2056%25%20of%20the,people%20will%20live%20in%20cities; accessed 27th June 2023.

The World Bank. (2010). Urban land area (sq km) - India. Retrieved from https://data.worldbank.org/indicator/AG.LND.TOTL.UR.K2?locations=IN

The World Bank. (2018). Urban population (% of total population) - India. Retrieved from https://data.worldbank.org/indicator/SP.URB.TOTL.IN.ZS?locations=IN

UN (United Nations). (2023). Department of economic and social affairs, population division. World Population Prospects 2022. Retrieved from https://www.un.org/development/desa/pd/sites/www.un.org.development.desa.pd/files/wpp2022_summary_of_results.pdf; accessed 11th January, 2023.

UN (United Nations). (2018). Department of Economic and Social Affairs. 68% of the world population projected to live in urban areas by 2050, says UN 2018. Available from: https://www.un.org/development/desa/en/news/population/2018-revision-of-world-urbanization-prospects.html; accessed 20th August 2022.

Van Cauwenbergh, Nora, et al. (2007) "SAFE-A hierarchical framework for assessing the sustainability of agricultural systems." *Agriculture, Ecosystems & Environment*, 120(2-4),: 229–242.

Vrolijk, H., Reijs, J., & Dijkshoorn-Dekker, M. (2020). Towards sustainable and circular farming in the Netherlands: Lessons from the socio-economic perspective. *Wageningen Economic Research.* https://edepot.wur.nl/533842; accessed 23rd September 2023.

Wilhelm, J. A., & Smith, R. G. (2018). Ecosystem services and land sparing potential of urban and peri-urban agriculture: A review. *Renewable Agriculture and Food Systems*, 33(5), 481–494.

Yuan, G. N., Marquez, G. P. B., Deng, H., Iu, A., Fabella, M., Salonga, R. B., Ashardiono, F., & Cartagena, J. A. (2022). *A Review on Urban Agriculture: Technology, Socio-Economy, and Policy.* Heliyon.

Zail, D. B. (2023). Growing culturally relevant food at the Urban farm: An examination of sovereign foodways, place-making practices, and autonomous identity-shaping. *Pitzer Senior Theses*, 152. https://scholarship.claremont.edu/pitzer_theses/152; accessed 17th August 2023.

Zhang X, Yao G, Vishwakarma S, Dalin C, Komarek AM, Kanter DR, Davis KF, Pfeifer K, Zhao J, Zou T, D'Odorico P. (2021). Quantitative assessment of agricultural sustainability reveals divergent priorities among nations. *One Earth*, 4(9), 1262–1277.

8 The Social, Economic, and Policy Impacts of Urban and Peri-Urban Agriculture
Farmer's Experience from Dar es Salaam City and Morogoro Municipality, Tanzania

Betty Mntambo

INTRODUCTION

Worldwide, the percentage of urban residents rose from 30% to 55% between the 1950s and 2018 (United Nations, 2019), and more people are anticipated to move into cities. The population of Tanzania increased from 44 million in 2012 to 61 million in 2022 (NBS, 2022). According to projections, Tanzania's urban population would rise from 19,959,000 in 2018 to 35,529,000 in 2030, and more to 76,542,000 in 2050 (George, 2022). This also means that by 2050, metropolitan regions will be home to half of the population (Wolf et al., 2018). With 5.3 million residents, Dar es Salaam City has Tanzania's largest urban population (NBS, 2022). Additionally, Morogoro Municipality has seen a huge population increase from 74,114 in 1978 to 622,000 in 2016 (Dayoub et al., 2019; Sumari et al., 2019). Urban poverty and rising food and employment demand are indicators of urbanization. Urban and peri-urban agriculture (UPA) has emerged as one of the answers to urban food insecurity, income, and employment in this environment of urbanization (FAO, 2012; Mntambo, 2017, 2021;Rao et al., 2022). UPA has become one of the methods for boosting the availability of fresh and nutritious food locally (Kennard and Bamford, 2020)

Globally, research has shown that UPA improves the lives of urban residents and that people are becoming more involved in it. In Pune, India, home gardens provide socio-cultural and environmental benefits (Zasada et al., 2020). Farming in towns is said to be widespread in sub-Saharan Africa. For example, in Zambia, half of the people used UPA as a food source (Simatele and Binns, 2018), and it was one of the strategies used to address economic challenges (Theresa and Pride, 2017). Livestock production, such as poultry and dairy cattle, is widespread in Uganda (Lee-Smith, 2010). In Nakuru, Kenya, approximately 35% of the population participated in UPA (Lee-Smith, 2010), whereas UPA empowered women in South Africa (Slater, 2001). With an estimated 20 million urban dwellers, backyard gardening is a common practice in West Africa (Drechsel et al., 2005). COVID-19, on the contrary, has increased the scale and scope of UPA. COVID-19 increased urban food insecurity from 135 million in January 2020 to 265 million by the end of 2020 (Lal, 2020). The pandemic has had an impact on urban food sources and supply chains. This has resulted in people recognizing and participating in UPA, mainly to obtain healthy and fresh vegetables to boost their immunity. In Tanzania, urban agriculture has grown in importance to supply urban residents with food and boost their immunity against the virus. The potential of UPA has expanded thanks to easy access to fresh produce and herbs. However, in most developing nations, UPA is seen as one of the expanding employment and income-generating opportunities (Mkwela, 2013). According to estimates, UPA employs around 40%

 DOI: 10.1201/9781003359425-8

of Africa's urban population (FAO, 2012). For instance, the FAO (2012) estimates that approximately 650 hectares of land are used for agricultural purposes in Dar es Salaam, Tanzania and that 32% of the population of Morogoro Municipality works in agriculture (URT, 2012).

In addition to tackling sustainable production and consumption (SDGs 12), food and nutrition security (SDGs 2), and poverty reduction (SDGs 1), UPA has contributed to the development of sustainable cities and communities (SDGs 11 and New Urban Agenda). Because it addresses social, economic, and environmental challenges in urban areas, this demonstrates the potential of UPA. UPA has gained popularity for urban residents to raise their household income, expand their food options, and improve their nutrition. Due to access to fresh dairy, meat, fruits, and vegetables, people who practice UPA are more likely to have a varied diet (Wagner and Tasciotti, 2018; Blakstad et al., 2022). The nutritional status of the household members is subsequently improved. In addition, they raise household income by selling their surplus production. Women are also empowered through the income they receive from UPA, which improves their agency and decision-making. Additionally, it encourages neighborhood farmers to connect socially. Putting flowers and trees throughout the city protects and enhances the urban environment. Numerous studies in Tanzania have emphasized the significance of UPA in dietary diversity (Wagner and Tasciotti, 2018), home gardening as a means of enhancing nutrition (Blakstad et al., 2022), boosting household income (Foeken et al., 2004; Mntambo, 2017, 2021; Zella, 2018), role in food security (Malekela and Nyomora, 2018), empowering women (Flynn, 2001; Mntambo, 2012, 2017, 2021), and UPA's negative aspects, such as heavy metal contamination in agricultural soils (Mwegoha and Kihampa, 2010), environmental degradation (Mlozi, 1997), and policy perspectives in UPA (Mkwela, 2013). However, no studies have addressed the UPA's social, economic, and policy impacts while focusing on Dar es Salaam City and Morogoro Municipality. As a result, this chapter investigates urban agriculture's social, economic, and policy implications by drawing on farmers' experiences in Dar es Salaam City and Morogoro Municipality.

POLICY FRAMEWORKS INFLUENCING UPA

Past Policies That Promoted UPA in Tanzania

The British colonial government enacted the first local authority by-laws in 1928, which prohibited agricultural activities in urban areas to separate urban and rural areas (Mlozi, 2003; Halloran and Magid, 2013). Following the economic crisis and the need for self-sufficiency, the post-colonial government invalidated the by-laws in the 1970s. During the economic downturn, the government encouraged urban residents to grow food in their backyards and open spaces (Mlozi, 2003). The following policies encouraged urban farmers to grow their food: "Kilimo cha Kufa na Kupona" (Agriculture for Life and Death) in 1974/75, and Siasa ni Kilimo (Politics is Agriculture) in 1972 (Mlozi, 2003). This was a plan to improve city dwellers' food security and cash income. The aforementioned policies provided opportunities for urban residents to begin agricultural activities to address income, food, and employment issues. Low-income earners, for example, began raising local chickens and growing vegetables to achieve food self-sufficiency. Furthermore, Tripp (1989) stated that most urban people in Dar es Salaam turned to vegetable cultivation or livestock, increasing household income and food security. People's participation in UPA has grown; for example, between 1967 and 1991, urban farmers in Dar es Salaam City increased from 18% to 67% (see Halloran and Magid, 2013). This indicates that UPA will continue to grow despite farmers' challenges, as discussed in the forthcoming sections.

CURRENT POLICIES IN UPA

Various national policies recognize the importance of UPA in urban livelihoods. Despite the UPA's efforts to address economic, social, and environmental issues, Tanzania lacks a specific policy on urban agriculture. Many policies recognize its contribution to sustaining city dwellers' livelihoods by providing employment, income, and food. However, most people oppose UPA because of its negative impact on the environment (Mlozi, 1997). The following policy statements acknowledge UPA activities:

S. 3.23.2. Promoting peri-urban livestock farming to provide employment improves household income and food security (Livestock Policy, 2006)

S. 3.1.3. Commercial poultry production is mostly practiced in urban and peri-urban areas (Livestock Policy, 2006)

S. 3.3.2. (i) Agriculture is not a principal function of towns, but when properly organized urban agriculture has the potential to provide employment, income and is a supplementary source of food supply (Agriculture and Livestock policy, 1997)

S. 4.3.7.2. (iii) Review the existing laws to facilitate planned urban agriculture (National Human Settlements Development Policy, 2000)

Although the above policy statements recognize the role of UPA in addressing income, food, and employment challenges, UPA is described in the 1997 Agriculture and Livestock Policy as an unorganized activity that is not part of town functions. As a result, according to Mkwela (2013), UPA is an unregulated informal activity, and there are only a few demarcated areas for livestock and crop activities in Dar es Salaam City and Morogoro Municipality. This is consistent with Halloran and Magid (2013), who stated that urban authorities still regard UPA as a minor land use and urban activity. As a result, farmers are landless and face challenges that limit their ability to reach their full potential in the UPA, as discussed in the following section. The following policies restrict UPA activities:

S. 6.7.1: The government will continue to regulate the conduct of urban agriculture and will ensure that it does not disrupt planned urban development (Agricultural and Livestock Policy, 1997)

S.3.23.3. Increased livestock populations and human activities related to livestock production in some areas of the country have resulted in over-exploitation of natural resources. Efforts will be undertaken to promote proper land use planning for livestock production (Livestock Policy, 2006)

S.77.1.4.5. No person shall be allowed to practice urban farming activities within built-up areas which are not zoned for urban farming purposes (Urban Planning Act, 2018)

S. 78. 3 (1): No person shall occupy or use more than three acres of land for urban farming (Town and Country Planning Regulations, 1992)

S. 6.7.0 (ii): In their present form, agricultural activities often conflict with the proper planning of urban land uses (National Land Policy, 1997)

The above policy statements demonstrate the opposing views on the UPA. For example, the Town and Country Planning Regulations 1992 seek to limit the size of land that an urban farmer can occupy, whereas the Livestock Policy 2006 encourages urban commercial poultry production. Regarding regulating UPA practices, each policy has its own focus. This implies that municipal policies are not integrated and localized. Some measures do not meet the needs of UPA's smallholder farmers and thus cannot be implemented locally. This implies that conflicting viewpoints between or within policies make implementing UPA challenging. The draft 2013 Agricultural Policy has the potential to promote and advance UPA practices by highlighting some of the challenges that impede UPA production and advocating for three-pronged interventions to increase UPA production, productivity, and profitability. The 2013 Agricultural Policy included measures to promote UPA, such as creating a regulatory framework for UPA and promoting good agricultural practices for the long-term viability of UPA. Meanwhile, the 2013 Agricultural Policy is still in draft form and has yet to be implemented. It acknowledges, however, that the expansion of UPA practices in Tanzania parallels the expansion of the urban population and that the lack of a clear regulatory framework and guidelines for UPA poses challenges to urban producers. The policy review in this section implies that policy gaps have resulted in several challenges in the UPA, as discussed in the coming section. As a result, closing these policy gaps is critical for promoting a more sustainable UPA.

METHODOLOGY

This chapter results from the British Academy-funded project "Urban and Peri-urban Agriculture as Green Infrastructure: Implications for Well-being and Sustainability in the Global South (India and Tanzania)". The project investigated how UPA is perceived as a green infrastructure and how it contributes to well-being and urban sustainability in Dar es Salaam and Morogoro Municipality. The current chapter investigates the social, economic, and policy implications of UPA through the lens of farmers' experiences in the two locations. The two regions were chosen not for comparison but to investigate UPA's social and economic impacts in each. Dar es Salaam is a large city in Tanzania with a high population, so the demand for food is high. Although her urbanites are involved in agricultural activities in production and food supply, the supply is insufficient due to the city's vastness, necessitating reliance on other nearby regions such as Morogoro. Morogoro Municipality is located 192 kilometers from Dar es Salaam and is one of eight districts in the Morogoro region. The Morogoro region is agriculturally based, popular for food and horticultural crops, and is regarded as the country's food granary. The high production of food and horticultural products, as well as its proximity to Dar es Salaam, makes the municipality one of the greater suppliers of food to Dar es Salaam City and also Morogoro depending on the market from Dar es Salaam, thus creating a food synergy between the two regions.

To address the UPA's social and economic impact on farmers' lives in Dar es Salaam City and Morogoro Municipality, we conducted 415 farmer household surveys, 20 farmer case studies, and field observations. Farmers were asked questions in household surveys about UPA's socio-economic benefits and challenges. SPSS was used to analyze the data, which resulted in the main discourse's content. During case study visits, photographs were also taken. Farmers' experiences and perceptions of UPA's contribution and challenges were gathered qualitatively. We also produced farmer's testimonies that address UPA's social and economic impact on their daily lives. For policy impact, we examined various national policies related to UPA to determine how they recognize, support, restrict, and promote UPA and identify policy gaps that limit the UPA's potential.

CHARACTERIZATION OF FARMERS AND UPA

UPA is defined as crop cultivation and livestock raising within intra-urban and peri-urban areas (FAO, 2007). It is regarded as an activity carried out by employed and unemployed individuals and educated and uneducated individuals. Employed people use UPA to supplement their income, while unemployed people use it as their primary source of income (Sawio, 1994; McLees, 2011; Simiyu, 2012; Mntambo, 2017). As a result, the primary debates on UPA are as follows: first, UPA is regarded as a survival strategy for low-income earners (Rakodi, 1988); second, UPA is done for people other than the urban poor (survivors) (Sawio, 1994; FAO, 2007; Lee-Smith, 2010; McLees, 2011; Simiyu 2012). That is, different groups of people, such as the employed, urban poor, unemployed, and landless, engage in it as a primary or secondary source of income. According to the current study, different types of farmers participate in UPA, including urban poor, middle-class, and economically well-off residents. Government officials, retired officers (such as professors, lecturers, nurses, and teachers), married men and women, single mothers, widowed, part-time and full-time farmers, and others are among the urban farmers of Morogoro Municipality and Dar es Salaam City.

Finally, UPA has adverse effects on human health and the urban environment. For example, livestock keeping harms the environment (Mlozi, 1995), agricultural input spill-over effects in the soil Mwegoha and Kihampa, 2010), and livestock bacteria transmission to humans and the environment

(Lupindu et al., 2015). This demonstrates that, despite UPA's positive contribution, there are negative consequences for the environment and human health. Understanding these opposing discourses is essential for investigating UPA's social, economic, and policy impact because it influences how policymakers and other stakeholders perceive UPA. It impacts how UPA is supported and/or integrated into urban planning in some way.

SOCIO-DEMOGRAPHIC CHARACTERISTICS OF THE RESPONDENTS

A total of 187 farmers from Morogoro and 228 farmers from Dar es Salaam participated in the study; 45% were women and 55% were men. Only 0.5% of farmers were under 18 (child labor), according to Table 8.1, whereas the majority (29%) were in the 41–50 age range. According to Table 8.1, most farmers (56%) had completed their primary education, while only 0.7% had completed their religious education. Additionally, 6.7% of farmers had less than a year of experience, whereas the majority (30%) had been farming for 15 years or longer (Table 8.1).

Farmers typically ranged in age from 30 to 60 years old, with the youngest farmer having 32 years and the oldest having 57 years (Table 8.1). All the farmers had at least 10 years of farming experience. Most farmers from Morogoro Municipality and Dar es Salaam City had a primary education level, in contrast.

DIFFERENT TYPES OF UPA ACTIVITIES

There are various UPAs activities in the current study, including aquaculture, floriculture, fruit trees, urban forests, and raising livestock (dairy cattle, chickens, goats, and pigs), as well as growing horticultural crops (leafy vegetables, tomatoes, and green peppers) and food crops (maize, cassava, and rice) (Table 8.2).

The survey and case study visits showed that livestock husbandry and vegetable and crop farming were the most popular practices in both regions. The results are comparable to those published by Halloran and Magid (2013), who stated that 60% and 90% of the eggs and green vegetables produced in the city of Dar es Salaam were produced within the intra- and peri-urban areas. This is because UPA has the potential to fill the void left by some food items that are perishable, such as dairy milk, leafy vegetables, horticulture crops, and poultry eggs, which are in great demand. The "Methodology" section indicates that the municipality of Morogoro reflects the agricultural nature of the Morogoro region. In Morogoro, as opposed to Dar es Salaam City, more food crops and livestock are kept, according to Table 8.2. The cultivation of green vegetables is more prevalent in Dar es Salaam. In the city of Dar es Salaam, leafy vegetable cultivation is widespread. However, there

---

TABLE 8.1
Socio-demographic Characteristics of the Respondents

Demographic Characteristics					
Age		**Education**		**Years of Farming**	
<18	2	None	11	<1 year	28
>70	15	Primary school	232	1–4 years	121
18–30	65	Religious teachings	3	10–14 years	60
31–40	70	Secondary school	83	5–9 years	81
41–50	122	Tertiary (college/university)	77	15 years and above	125
51–60	88	Vocational training	9		
61–70	53				

TABLE 8.2

Types of Urban and Peri-urban Agriculture Activities

Types of Urban and Peri-Urban Agriculture	Dar es Salaam (N=228)		Morogoro (N=187)	
	N	Percentage	N	Percentage
Vegetable cultivation	128	30.8	95	22.9
Livestock keeping	85	20.5	89	21.4
Horticultural crops	50	12.0	35	8.4
Food crops	57	13.7	70	16.9
Flowers	20	4.8	22	5.3
Tree planting	12	2.9	12	2.9
Fishing	7	1.7	13	3.1

FIGURE 8.1 Case Study visits, 2021. Photos show vegetable cultivation in the open spaces and flower cultivation at home.

is a significant need for leafy vegetables because of the city's dense population. As a result, several farmers in Morogoro Municipality supply vegetables to the city, particularly during the rainy season. When it rains in Dar es Salaam, parts of the open spaces used for farming become submerged in water, forcing farmers to stop their work. Farmers in Morogoro Municipality take advantage of this market opportunity to meet the need. Additionally, Table 8.2 demonstrates that both regions regularly keep livestock. From small to large-scale producers, we have included many types of livestock keepers, for instance, from 10 to 50,400 to 200,000 chicken owners (Figure 8.1).

LOCATION OF UPA AND LAND OWNERSHIP

In both regions, UPA is practiced in intra-urban (home-based and open spaces) and peri-urban settings. UPA activities vary depending on location; for example, home-based is primarily for dairy cattle, poultry, and home gardens (McLees, 2011; Mntambo, 2017). Open space cultivation is primarily for leafy vegetables, whereas horticultural and food crops (such as maize, rice, and tomatoes) are common in peri-urban areas. It has also been discovered that land size and ownership impact the production scale. Home-based farmers, for example, typically own the land because the land tenure is attached to the household; thus, the land size can be small, medium, or large. Farmers rent plots of land in open spaces, with an estimated size of 80 m² per plot, depending on the location. Farmers in the peri-urban area rent, borrow, or own land, and the size of the farm can range from 1 to 5 acres depending on location. In general, 6.3% of farmers own or rent less than 20 m² of land, 46.5% own or rent between 101 and 2,023 m² of land, 2.2% own or rent between 8,095 and 12,140 m² of land, and 7.2% own or rent more than 12,140 m² of land. Land ownership varies according to capital, location, and the relationship between a farmer and a landlord. Because UPA is considered an informal activity, social relationships are important in gaining access to resources such as land (see McLees, 2011; Mntambo, 2021). Land ownership influences production scale; that is, regardless of other production factors such as inputs and water, the larger the land size, the larger the production (see McLees 2011; Mntambo, 2021).

MOTIVATION TO START UPA

The motivation for farmers to participate in UPA varies according to their socioeconomic status. It was discovered that income was a major motivator for 36.2% of those who participated in UPA, and 24.8% were motivated to grow their food to diversify their diet and save money on food. About 18.1% began UPA as a hobby, primarily among retired employees. Approximately 12.7% were motivated for nutritional reasons, i.e., they have access to fresh food and can grow their foods. The final two categories are home-based farmers who cultivate vegetables in small gardens or buckets and flower cultivation to beautify households. Different motivations of urban farmers indicate that it may impact the scale of production and time spent in the UPA. It is noted that UPA is a transitory activity because it competes with urban resources such as land and water. They contended that UPA activities would halt as urbanization progressed due to limited resources. However, according to Halloran and Magid (2013), urban farmers practice shifting cultivation despite limited resources. When evicted from one location, they relocate to another; this is a common practice among open space farmers. This implies that people will continue to practice UPA because their motivations for starting UPA differ. Due to increasing urbanization and demand for food, income, and employment, people will continue to engage in UPA to solve income, food, and employment challenges. According to the findings of this study, several people have been involved in UPA for a long time (more than five years), as indicated in Table 8.1. This suggests that, despite the difficulties, more people are sticking with UPA. In this case, UPA is a more permanent than transitory activity, so it requires government support and integration in urban planning to address challenges. The dynamics of UPA activities and urban farmers were revealed in the previous sections. The reasons for starting UPA, the production scale, and the types of activities vary. Different types of UPA activities and production scales demonstrate UPA's potential to address food challenges in both regions.

ECONOMIC IMPACT OF UPA

According to FAO's 2007 report, self-employment, income from surplus sales, food savings, and the exchange of agricultural products for other economic goods are direct economic benefits and costs for agricultural production in urban areas. UPA can potentially produce marketable perishable goods such as leafy vegetables, milk, eggs, and so on. Given the growing population, these products are in high demand and have a comparative advantage over rural production (FAO, 2007). Mlozi (2004) found that livestock keepers in Dar es Salaam made an average of TSh. 240,000 (US $500) per year, and livestock keepers selling milk earned TSh. 64 244 (US $75.80) per month, more than the government minimum wage of TSh. 50,000 (US $58.80). Jacobi (1997) discovered that a farmer with a vegetable garden can earn more than the minimum government wage. An urban farmer engaged in mixed vegetable cultivation earns a monthly net income of 60 USD, whereas Tanzania's monthly net income per capita is 24 USD. Other research shreds of evidence have also confirmed the economic benefits of UPA in Tanzania, such as contributing to household income in Morogoro Municipality (Mntambo, 2017, 2021), contributing to poverty alleviation in Dodoma Municipality (Zella, 2018), and contributing to household income in Morogoro Municipality and Mbeya Municipality (Foeken et al., 2004).

ECONOMIC BENEFITS OF UPA

We discovered that some farmers do not keep records of their daily operations, making it difficult to calculate the costs and benefits of their activities. This supports Moustier's (2001) argument that the diversity of farmer profiles, seasonality of crops, scattered crops, and continuous harvesting make it difficult for farmers to estimate actual income figures. Furthermore, farmers in open spaces are so insecure about their land that they move from one location to another (see FAO, 2007), and these patterns affect the estimation of the actual figure on income. Although we did not ask farmers about the amount of income they generate from UPA activities in the current study, we focused on the economic benefits farmers receive after selling UPA produce. As a result, as shown in Table 8.3, we requested various UPA benefits for farmers.

Table 8.3 shows that income and food are the major benefits of UPA in both regions, this concurs with different research shreds of evidence in Tanzania, economic benefits (Foeken et al., 2004; Mntambo, 2017; Zella, 2018); food security (Malekela and Nyomora, 2018); and dietary diversity (Wagner and Tasciotti, 2018; Jacobi et al., 2000; Blakstad et al., 2022). Table 8.3 supports the previous findings that income and food are the primary reasons farmers started UPA in both Morogoro and Dar es Salaam. This is also true for those who engage in UPA as a hobby or for physical exercise, as discussed in the coming sections. This implies that, in addition to material benefits, other non-material benefits are essential for entry into UPA. It also implies that UPA's benefits in urban people's lives extend beyond economic benefits because UPA engages different profiles of farmers with varying motives, interests, and scales of production. Thus, UPA should not be dismissed as a mere informal activity but rather as one that, if properly promoted and advanced, can significantly contribute to economic, social, and environmental sustainability outcomes. Table 8.4 shows the economic benefits of UPA.

The previous sections described the various profiles of urban farmers in both regions, including employed and unemployed individuals. The former use UPA to supplement their income, whereas the latter use UPA as their primary source of income. For some farmers, UPA provides additional income because they are employed or have another source of income, whereas for others, UPA is their sole source of income. Table 8.4 shows that, in Dar es Salaam City, 53.9% and 21.1%, respectively, and 37.4% and 47.6% strongly agree and agree that UPA provides extra income. This implies that UPA contributes to household income, whereas those who disagree and/or strongly disagree regard UPA as a secondary source of income. During case study visits, we discovered that some farmers have diversified their income sources with UPA, such as small welding workshops, retail shops, and brick manufacturing, among other things. The sales of UPA produce provided the start-up capital for other income-generating activities (IGAs). Farmers' IGAs in Morogoro Municipality and Dar es Salaam City are depicted in the photos in Figure 8.2.

TABLE 8.3

Benefits of Urban Agriculture

Dar es Salaam	Frequency	Percentage	Morogoro	Frequency	Percentage
Income	208	34.0	Income	165	27.7
Food	198	32.4	Food	174	29.2
Hobby	81	13.3	Hobby	88	14.8
Diversify diet	46	7.5	Diversify diet	89	14.9
Environmental concern	28	4.6	Environmental concern	14	2.3
Physical exercise	50	8.2	Physical exercise	66	11.1
Total	611	100.0	Total	596	100.0

NB: This was a multiple-response question.

TABLE 8.4
UPA Provides Extra Income

Dar Es Salaam		Frequency	Percentage	Morogoro	Frequency	Percentage
Valid	Strongly agree	123	53.9	Strongly agree	70	37.4
	Agree	48	21.1	Agree	89	47.6
	Neutral	13	5.7	Neutral	6	3.2
	Disagree	7	3.1	Disagree	12	6.4
	Strongly disagree	37	16.2	Strongly disagree	10	5.3
	Total	228	100.0	Total	187	100.0

NB: This was a multiple-response question.

FIGURE 8.2   Case Study visits, 2021. Photos show brick-making and welding activities.

ECONOMIC SAVINGS ON FOOD

Urban food insecurity has increased due to several factors, including poor purchasing power, insufficient and erratic access to food, growing supply costs, and food distribution from rural to urban regions (FAO, 2007). As a response, UPA is addressing these issues. According to a farmer in Morogoro, "Growing your own food in urban settings saves money because the food produced is also consumed within the household".

According to Table 8.5, 54.4% and 16.7% of respondents in Dar es Salaam City and 42.2% and 49.7% of respondents in Morogoro Municipality strongly agree and concur that they save money on food since they also eat what they produce. This result is comparable to Simatele and Binns' (2008)

TABLE 8.5
UPA Saves Money Spent on Vegetables

Dar Es Salaam		Frequency	Percentage	Morogoro	Frequency	Percentage
Valid	Strongly agree	124	54.4	Strongly agree	79	42.2
	Agree	38	16.7	Agree	93	49.7
	Neutral	20	8.8	Neutral	5	2.7
	Disagree	6	2.6	Disagree	8	4.3
	Strongly disagree	40	17.5	Strongly disagree	2	1.1
	Total	228	100.0	Total	187	100.0

NB: This was a multiple-response question.

findings from Zambia, where farmers saved at least $100 a month since their families ate 65% of the vegetables they grew in their gardens. The household produces and consumes its food, saving money that would have otherwise been spent on food. The allotted funds can then be used for other household expenses. Farmers were said to benefit from this since they could use the money they saved to pay other expenses like home rent, medical expenses, and more. Various food products are made from horticultural and food crops, as well as animal products such as leafy vegetables, chicken eggs, and dairy milk, which are easily consumable perishables in the home. Farmers frequently use a percentage of their produce for food to diversify their diet, improve family nutrition, and save money on food for other household expenses to lower living costs. The testimony listed below demonstrates UPA's economic contribution:

> "*I began farming to make money. Before I retired, I used to have a small number of animals. I expanded my farm with the funds I received from my retirement pension in order to increase income*". Female, Morogoro
>
> "*Because we are near to the market, urban agriculture is more profitable than rural agriculture. For instance, I produced up to 1,500 buckets of tomatoes last season, but they were only used for farm sales*". Male, Morogoro
>
> "*I've been farming for 20 years, and it's a profitable activity. I built a 6-room house and purchased 2 acres of land and a small plot. When business is booming, I make about TShs. 600,000–700,000 (255–297 USD) per season, and when sales are down, I make about TShs. 200,000–300,000 (85–128 USD). These amounts are comparable to a person's monthly pay if they were employed, however I make my living through farming*". Male, Morogoro
>
> "*People should participate in UPA activities since, in my experience, they pay significantly more than jobs do. No human being can survive without food, thus if you work in agriculture you will definitely make money*". Female, Dar es Salaam

The aforementioned claims demonstrate that financial gain was one of the main motivations for starting UPA. Some of the farmers began UPA while still working. Therefore, UPA added to household income. UPA is one of the activities that individuals continue to practice because they have direct access to and social relationships with customers, and it has commercial potential. Table 8.5 showed that income was the central advantage farmers mentioned. This is because farmers can sell their products on farms, in their neighborhoods, or at local marketplaces. For instance, a farmer in the Morogoro Municipality's peri-urban area said he could produce up to 1,500 buckets of tomatoes to sell on the farm. In general, income from UPA is used for a variety of expenses. For instance, building a house is one of the advantages farmers have mentioned, and it is a sign of financial success. Other household expenses paid from UPA revenue include household bills, child school fees, and start-up money for IGAs, among others. Even though we were unable to obtain precise costs and benefits of UPA, survey results and farmer testimonies have shown that UPA is economically significant in their households, which is consistent with several studies in Tanzania (Foeken et al., 2004; Mntambo, 2017; Zella, 2018).

JOB CREATION

Urban residents are known to benefit from peri-urban and urban agriculture as a source of employment. UPA made up 60% of the informal sector in Dar es Salaam; according to research by Sawio (1994), it was the second-largest urban employer. According to Dongus (2000), 4,000 or more urban residents of Dar es Salaam participated in UPA. In Morogoro Municipality, 32% of the population worked in agriculture in 2011 (URT, 2012). Farmers who participated in the current study attested that, depending on the season, they might hire casual laborers through UPA to help with various tasks. Irrigation, planting, milking cows, pesticide application, cleaning livestock shade, harvesting, and many other tasks fall within this category. As a result, workers may provide for their family's fundamental requirements by receiving payment from UPA operations. According to the survey, 55% and 45% of farmers in the municipalities of Morogoro and Dar es Salaam employed laborers for UPA tasks. Some farmers make seasonal hiring based on agricultural activities like irrigation, plowing, harvesting, and collecting animal wastes from the shade, among others. Some of the farmers employed workers all year long. For instance, we came across a case of a farmer who kept chickens in Dar es Salaam and hired 107 male and female workers in chicken units, as well as a human resource officer, veterinarian assistant, and marketing assistant. His employees receive monthly pay and a bonus of one tray of chicken eggs each month. This story demonstrates how, if properly promoted and developed to reach its full potential, UPA can employ more urban residents. The UPA has also given street hawkers work, allowing them to make money by purchasing vegetables from farmers and selling them to homes on the street. Both regions engage in this practice frequently. In this situation, particularly in Morogoro Municipality, more women than males work in the hawking industry. Since different kinds of leafy vegetables are grown in open spaces, hawking business occurs year-round.

SOCIAL IMPACT OF URBAN AGRICULTURE

To comprehend the social effects of the UPA, we concentrated on interpersonal interactions and relationships, as Golden (2013) described. The main factors used to analyze the social effects of the UPA were gender, networking, and physical and mental health.

According to Table 8.6, 45.2% and 19.3% of respondents in Dar es Salaam City and 41.2% and 41.2% of respondents in Morogoro Municipality strongly agree and concur that UPA has aided them in creating social networks among farmers and other UPA stakeholders, including customers, extension agents, NGO officials, and input suppliers, among others. Farmers in Dar es Salaam City and Morogoro Municipality indicated good relationships with other farmers based on their experiences in UPA. Urban farmers' socio-economic and cultural characteristics vary, and, as a result, their understanding of agriculture and level of productivity also vary. Networking with other farmers is an opportunity to learn from one another and share agricultural resources like input (pesticides, seeds, fertilizers) and skills:

> *"My wife distributes the seedlings of orange-fleshed sweet potatoes with the neighbors from the garden. She also instructs the women's group in agriculture on how to grow sweet potatoes".* Male, Morogoro

TABLE 8.6
UPA Has Helped Build My Social Network

Dar Es Salaam		Frequency	Percentage	Morogoro	Frequency	Percentage
Valid	Strongly agree	103	45.2	Strongly agree	77	41.2
	Agree	44	19.3	Agree	77	41.2
	Neutral	45	19.7	Neutral	14	7.5
	Disagree	18	7.9	Disagree	17	9.1
	Strongly disagree	18	7.9	Strongly disagree	2	1.1
	Total	228	100.0	Total	187	100.0

NB: This was a multiple-response question.

TABLE 8.7

UPA Provides Physical Exercise for Me and My Family

Dar Es Salaam		Frequency	Percentage	Morogoro	Frequency	Percentage
Valid	Strongly agree	84	36.8	Strongly agree	76	40.6
	Agree	69	30.3	Agree	89	47.6
	Neutral	42	18.4	Neutral	11	5.9
	Disagree	9	3.9	Disagree	7	3.7
	Strongly disagree	24	10.5	Strongly disagree	4	2.1
	Total	228	100.0	Total	187	100.0

NB: This was a multiple-response question.

> "*In the event that I don't have the required products, I typically connect consumers to another farmer, and other farmers do the same. Despite the fact that we all produce the same goods, sharing consumers has helped us keep and strengthen our bond*". Male, Dar es Salaam

If there aren't enough or no available products at the time, they also share customers. In addition to social networks, UPA improves human health, as indicated in Table 8.7.

UPA activities are linked to socializing and physical and mental relaxation (Bellows et al., 2008). Although health professionals undervalue the benefits of exercise-related urban agriculture (Bellows et al., 2008), the current study did not find a direct link between UPA and health improvement through exercise. However, farmers view UPA as a form of exercise. According to Table 8.7, UPA counts as exercise for 36.8% and 30.3% of respondents in Dar es Salaam City and 40.6% and 47.6% of respondents in Morogoro Municipality. Farmers claimed that, even though they employed casual laborers, they still had to perform various tasks on the farm, which dramatically impacted their physical and mental health. As noted by the following comments, because they are physically fit, they lower their risk of contracting diseases:

> "*Urban agriculture has been useful to me as I become older. I have never had an illness identified. This is because I work hard at every task on my farm; to me, it's like exercising.*" Male, Dar es Salaam

The testimonials demonstrate how UPA fosters better social ties among farmers by encouraging resource sharing. Farmers believe participating in UPA activities helps their physical and mental health because they keep them active, busy, and financially stable. This demonstrates that farmers also value the social benefits of UPA, which makes them stick with UPA. This implies that decision-makers need to know that the UPA has advantages for farmers beyond its financial gains. This suggests that a shift in policymakers' perception toward UPA will encourage its inclusion in urban planning.

WOMEN IN UPA

The positive contribution of women in UPA has been demonstrated through research in Kenya (Simiyu and Foeken, 2014; Simiyu, 2012, 2015), South Africa (Slater, 2001), and Botswana (Hovorka, 2005). Women play a crucial role in UPA in Tanzania as well, and their participation increases their capacity to provide for the household financially and promotes empowerment (Mntambo, 2012; Flynn, 2001). Women are crucial in UPA, but their contributions are frequently underestimated because they are frequently associated with traditional female roles like cooking, cleaning, and washing. Therefore, their labor-related contributions are frequently considered part of their gendered responsibilities (see Hovorka, 2005; Mntambo, 2017, 2021). However, women use UPA as a source of income to support themselves and their families. As a result, the gender component in UPA is vital because both men and women are involved, and their experiences and challenges differ. According to the current study, there are 186 female farmers out of 415 in Dar es Salaam City and Morogoro Municipality. Women in both regions engage in poultry keeping (chicken), home gardening, fish keeping, and open-air cultivation of leafy vegetables, to name a few activities. Most rely on the UPA to meet their and their families' needs. UPA has a positive impact on the lives of women, as evidenced by the following female farmers:

> *"Urban and peri-urban agriculture has had a significant impact on my life and that of my family. It has allowed me to pay my children's school fees as well as other household expenses. It has become the primary source of income in my family. I am not financially dependent"*. Female, Dar es Salaam
> *"I have been a widow for over 30 years. I was a secondary school teacher who decided to keep poultry to supplement my income. I've been keeping chickens full-time since I retired. I paid for my children's school fees and supported them until they graduated from high school. I built my own house, bought a car, and can meet my basic needs"*. Female, Morogoro

Financial independence and the ability to meet the needs of household members boost women's spirits and fulfillment. UPA economically and socially empowers women. They are satisfied because they can meet their own and their families' needs. Furthermore, income from UPA empowers women to make their own decisions and achieve financial independence.

CHALLENGES AND OPPORTUNITIES OF URBAN AGRICULTURE

According to the survey and case study findings, farmers face a variety of challenges, including the presence of pests and diseases, limited access to extension services and government support, and thus lack of access to various agricultural inputs, such as rural farmers, land tenure insecurity, price variations of agricultural inputs (seeds and fertilizers), and water scarcity. The latter was attributed to climate change, which has altered rainfall patterns. Water scarcity, theft, and animal slaughter are among the others According to Mwalukasa (2000) and Foeken et al. (2004), despite UPA's contribution to the urban economy and households, it is still characterized by structural and policy challenges such as the neglect of small urban livestock keepers and crop growers, the failure of relevant authorities to designate and allocate land for urban agriculture, and limited access to agricultural inputs and extension services. Land tenure insecurity is another challenge mentioned by farmers in this study, particularly farmers in open spaces and peri-urban areas. Access to ward extension officers is vital so that farmers can obtain agricultural knowledge and information such as fertilizer and pesticide application, market information, and treatment of animal diseases, among other things. In this study, 202 (49%) of the 415 total respondents in Dar es Salaam and Morogoro reported having access to extension services. Further discussions revealed that farmers have access to a variety of extension services, the majority of which do not involve ward extension officers. Farmers instead obtain extension services from various sources, including NGOs affiliated with UPA, their own experience, the internet, friends and fellow farmers, experts from universities such as SUA, and others. The following statements indicate access to extension services:

> *"Our extension services are limited. They only suggested purchasing pesticides. However, I chose to learn from the internet. As an officer in Information and Technology, I learn from my efforts. Very rare extension officers have time for us, we are so many farmers in the municipality"*. Male, Morogoro

"An extension officer can come and get information on the number of livestock you have for their records. In the past when I had a challenge with my livestock I used to call a private veterinary doctor, but now I have hired my doctor". Male, Morogoro.

"Extension officers are useless to us. If I have a problem, I go to the AGROVET shop for medication advice. Very rarely you call them and they come". Female, Dar es Salaam

The challenges described here result from a lack of government support and a negative perception of the UPA. As a result, UPA is uncontrolled and informal. Hence, the UPA's chances of increasing production and sustaining cities are limited. To address these issues, an interdisciplinary approach is unavoidable. Municipal departments such as agriculture, environment, and land, for example, can collaborate to address land tenure insecurity while also protecting the environment and increasing agricultural production. Farmers have demonstrated individual strategies and efforts to deal with various challenges. Sharing resources, learning from the internet, hiring private veterinary doctors, and connecting with NGO officials dealing with agricultural issues are just a few examples. Investing in various strategies to sustain UPA activities demonstrates that farmers are unwilling to abandon UPA.

CONCLUSIONS

Using farmers' experiences, this chapter examined UPA's social, economic, and policy implications. The chapter demonstrated how UPA helps to improve the economic and social well-being of urban farmers in both regions. Farmers are classified according to their reasons for starting in UPA, characteristics, and type of activities. Income and food are essential benefits of the UPA for farmers, but other benefits, such as social benefits, are also significant. We have also demonstrated that there is recognition of UPA as well as contradictions within and between policies, making UPA challenging to implement. Despite this, farmers employ various strategies to sustain UPA activities and earn a living. We discovered farmers who have been practicing UPA for more than 15 years. This implies that, as urbanization increases, UPA activities will also increase despite limited resources. Because UPA is dynamic regarding farmers' socio-economic profiles and scale of production, any policy intervention should focus on the dynamics of urban farmers because one size does not fit all. Despite its negative image, we argue that UPA is a long-term and significant activity that will continue to address income, food, and employment issues. As a result, it must be considered in the political, economic, and social aspects of urban planning and development.

THE WAY FORWARD

Given that UPA addresses income, food, and employment issues, specific mechanisms should be put in place to recognize UPA's contribution to the economy. Opening new avenues for political and institutional support for urban farmers through land allocation for cultivation to ensure land security, as well as encouraging intensification and diversification of vegetable cultivation, is unavoidable. As a result, educating farmers on sustainable agriculture to protect the environment, improving credit access, forming farmers' small cooperatives, supplying agricultural inputs, and providing extension services are some of the immediate and long-term measures that would move UPA forward more sustainably. Furthermore, targeting the UPA is critical, given the informal means of access to resources for any assistance. This is because any formalization of resources without recognizing the informal means may distort the available means of access to poor farmers, resulting in inequity among farmers, particularly among women.

ACKNOWLEDGEMENT

This chapter derives from the project titled "Urban and Peri-urban Agriculture as Green Infrastructure: Implications on Well-Being and Sustainability in the Global South (Tanzania and India)", funded by British Academy, Grant No: UWB190091. The opinions and interpretations presented in this paper are from the author.

REFERENCES

Bellows, A. C., Brown, K., & Smit, J., (2008). *Health Benefits of Urban Agriculture*. A Paper from Members of the Community Food Security Coalition's North American Initiative on Urban Agriculture.

Blakstad, M, M., Mosha D., Bliznashka, L., Bellows, A, L., Canavan, C, R., Yussuf, M, H., Mlalama, K., Madzorera, I., Chen, J, T., Noor, R, A., Kinabo, J., Masanja, H., Fawzi, W, W., (2022). Are home gardening programs a sustainable way to improve nutrition? Lessons from a cluster-randomized controlled trial in Rufiji, Tanzania. *Food Policy*, 109:102248

Dayoub, M., Thonglor, O., & Korpela, T., (2019). Trends of Precipitation and Temperature in Morogoro Region in Tanzania. *Proceeding of the 8th International Conference on Integration of Science and Technology for Sustainable Development (8th ICIST)*, In November 19-22 at Huiyuan International Hotel, Jingde, Anhui Province, P.R. China.

Dongus, S. (2000). Dar es Salaam Urban Agriculture. [online] www.cityfarmer.org. Available at: https://www.cityfarmer.org/daressalaam.html

Drechsel, P., Graefe, S., Danso, G.,, Keraita, B., Obuobie E., Amoah P., Cofie, O.O., Gyiele, L. & Kunze, D., (2005). *Informal Irrigation in Urban West Africa*. IWMI Research Report, No. 102.

FAO, (2007). *Profitability and Sustainability of Urban and Peri-Urban Agriculture*. Agricultural *Management, Marketing and Finance Occasional Paper* No. 19. FAO, Rome.

FAO, (2012). *Growing Greener Cities in Africa. First Status Report on Urban and Peri-Urban Horticulture in Africa*. Rome, Food and Agriculture Organization of the United Nations.

Flynn, K. C., (2001). Urban Agriculture in Mwanza, Tanzania. *Journal of International African Institute*, 71 (4), 666–691.

Foeken, D., Sofer, M., & Mlozi, M., (2004). Urban agriculture in Tanzania: Issues of sustainability. African Studies Centre, Research Report 75/2004

George, C., (2022). The transformation of cities in Tanzania: An overview. *Dispatch* 1. September, 2022. Fes 2022

Golden, S., (2013). Urban Agriculture Impacts: Social, Health, and Economic: A Literature Review. US Sustainable Agriculture Research and Education Program Agricultural Sustainability Institute at UC Davis

Halloran, A., & Magid, J., (2013). Planning the unplanned: Incorporating agriculture as an Urban land use into the Dar Es Salaam master plan and beyond. *Environment & Urbanization*, 25 (2), 541–558.

Hovorka, A. J., (2005). The (Re) production of gendered positionality in Botswana's commercial Urban agriculture sector. *Annals of the Association of American Geographers*, 95 (2), 294–313.

Jacobi, P., (1997). *Importance of Vegetable Production in Dar Es Salaam, Tanzania*. Dar es Salaam: Urban Vegetable Promotion Project

Jacobi, P., Amend, J., & Kiango, S., (2000). Urban agriculture in Dar es Salaam: Providing an indispensable part of the diet. City case study Dar es Salaam. In: Growing Cities, Growing Food: Urban Agriculture on the Policy Agenda. A Reader on Urban Agriculture. Bakker, N. Dubbeling, M. Gundel, S. Sabel-Koschella, U and De Zeeuw, H. (Eds) *Food and Agriculture Development Centre*. Germany

Kennard, N.J., & Bamford, R.H., (2020). Urban agriculture: Opportunities and challenges for sustainable development. In: Leal Filho, W., Azul, A., Brandli, L., Özuyar, P. and Wall, T. (Eds) *Zero Hunger. Encyclopedia of the UnSustainable Development Goals*. Springer, Cham

Lal, R., (2020). Home gardening and urban agriculture for advancing food and nutritional security in response to the Covid-19 pandemic. *Food Security*, 12, 871–876.

Lee-Smith, D., (2010). Cities feeding people: An update on urban agriculture in equatorial Africa. *Environment & Urbanization*: 22(2), 483–499.

Lupindu, A. M., Dalsgaard, A., Msoffe, P.L.M., Ngowi, H. A., Mtambo, M.M & Olsen, J.E., (2015). Transmission of antibiotic-Resistant escherichia coli between cattle, humans and the environment in peri-Urban livestock keeping communities in Morogoro, Tanzania. *Preventive Veterinary Medicine*, 118, 477–482

Malekela, A. A., & Nyomora, A., (2018). Food security: The role of Urban and peri-Urban agriculture. A case of Dar es Salaam city, Tanzania. *International Journal of Agronomy and Agricultural Research* (IJAAR), 13(2), 50–62

Mclees, L., (2011). Access to land for Urban agriculture in dar Es Salaam, Tanzania: Histories, benefits and insecure Tenure. *Journal of African Studies*, 49 (4), 610- 624

Mlozi, M.R.S., (1995). '*Information and the Problems of UA in Tanzania: Intentions and Realizations*. PhD Thesis. Department of Educational Studies, Faculty of Graduate Studies. University of British Columbia.

Mlozi, M.R.S., (1997). Urban agriculture: Ethnicity, cattle raising and some environmental implications in the city of Dar Es Salaam, Tanzania. *African Studies Review* 40 (3), 1–28

Mlozi, M.R.S., (2003). *Legal and Policy Aspects of Urban Agriculture in Tanzania. Urban Agriculture Magazine* No 11, RUAF, Wageningen

Mlozi, M. R. S (2004). Urban Agriculture in Tanzania: Its Role Toward Income Poverty Alleviation and Resources for its Rersistence. UNISWA Research Journal 7(1).

Mkwela, H. S., (2013). Urban agriculture in Dar Es Salaam: A dream or reality? *Sustainable Development and Planning*, 173, 161–172.

Mntambo, B., (2012). How does gardening activities fit into the lives and gendered responsibilities of women? The case of vegetable cultivation in Morogoro municipality. *Journal of Education, Humanities And Social Sciences*, 10 (2), 53–68

Mntambo, B., (2021). Land tenure security and urban agriculture: Focusing on the vegetable cultivation in morogoro municipality. *Huria Journal*, 28 (1), 1–19.

Mntambo, B. D., (2017). *Intra-Household Gender Relations and Urban Agriculture: The Case of Vegetable Cultivation in Morogoro Municipality, Tanzania.* PhD Thesis Submitted to the School of International Development, University Of East Anglia, UK.

Moustier P., (2001). *Assessing the Socio-Economic Impact. UA Magazine*, 5 December 2001, Appropriate Methods For Urban Agriculture, Leusden RUAF.

Mwalukasa, M (2000). Institutional Aspects of Urban Agriculture in the City of Dar es Salaam Thematic Paper 6. In: Bakker, N. Dubbeling, M. Gundel, S. Sabel-Koschella, U and De Zeeuw, H. (Eds) *Growing Cities, Growing Food: Urban agriculture on the Policy Agenda. A Reader on Urban Agriculture.* Food and Agriculture Development Centre, Germany.

Mwegoha, W. J. S., & Kihampa, C. (2010). Heavy metal contamination in agricultural soils and water in Dar es Salaam city, Tanzania. *African Journal of Environmental Science and Technology*, 4(11): 763–769.

NBS, (2022). *National Bureau of Statistics and Housing Report.* United Republic of Tanzania.

Rakodi, C., (1988). Urban agriculture: Research questions and Zambian Evidence. *The Journal of Modern African Studies*, 26 (3), 495–515

Rao, N., Patil, S., Singh, C., Roy, P., Pryor, C., Poonacha, P & Genes, M., (2022). Cultivating sustainable and healthy cities: A Systematic literature review of the outcomes of Urban and peri-Urban agriculture. *Sustainable Cities and Society*, 85: 104063

Sawio, C. J., (1994). Who are farmers of dar Es Salaam?: In: Egziabher, A. G., Lee- Smith, D., Maxwell, D. G, Memon, P. A, Mougeot, L. J.A, & Sawio, C.J. *Cities Feeding People: An Examination of Urban Agriculture in East Africa.* International Development Research Centre 1994. Ottawa, On, IDRC, 1994.

Simatele D. M & Binns T., (2018). Motivation and marginalization in African Urban agriculture: The case of Lusaka, Zambia. *Urban Forum*, 19 (1), 1–21.

Simiyu, R. R., (2012). *I Don't Tell my Husband about Vegetable Sales: Gender Dynamics in Urban Agriculture in Eldoret, Kenya.* African Studies Collection, 46. African Studies Centre.

Simiyu, R, R and Foeken, D. (2014). Gendered divisions of labour in urban crop cultivation in a Kenyan town: implications for livelihood outcomes. *Gender, Place and Culture*, 21 (6):768–784.

Simiyu, R. R. (2015). I know how to handle my husband: Intra-household decision making and urban food production in Kenya. *Eastern Africa Social Science Research Review*, 31 (2): 63–81.

Slater, R. J., (2001). Urban agriculture, gender and empowerment: An alternative view. *Development Southern Africa*, 18 (5), 635–650

Sumari, N.S., Xu, G., Ujoh, F., Korah, P. I., Ebohon, J. O And Lyimo, N. N., (2019). A geospatial approach to sustainable urban planning: Lessons for morogoro municipal council, Tanzania. *Sustainability*, 11 (22), 6508.

Theresa, K., & Pride, C., (2017). The social, economic and health impacts of urban agriculture in Zambia. *Asian Journal of Advances in Agricultural Research*, 3(1), 1–8.

The Town and Country Planning Ordinance (Cap. 378), The Town and Country Planning (Urban Farming) Regulations (1992), Government Printer, Dar es Salaam.

The Urban Planning Act. (ACT NO. 8 OF 2007): Regulations. Urban Planning (Urban Farming). Government Notice No. 90 Published on 9/3/2018

Tripp, A. M. (1989). Women and the changing urban household economy in Tanzania. *Modern African Studies*, 27 (4): 601–623

United Nations, (2019). *World Urbanization prospects: The 2018 revision (St/Esa/Ser.A/420).* New York: United Nations. Department of Economic and Social Affairs, Population Division.

United Republic of Tanzania (URT), (2012). *Morogoro Municipal Council Socio-Economic Profile* 2010. *Jointly Prepared by Ministry of Finance*, National Bureau of Statistics and Morogoro Municipal, Tanzania

URT, (1997). Agricultural and Livestock Policy. Ministry of Agriculture and Co-operative Development. Dar es Salaam.

URT, (1997). National Land Policy. 2nd Edition. The Ministry of Lands and Human Settlements. Dar es Salaam, Tanzania.

URT, (2006). National Livestock Policy. The United Republic of Tanzania. Ministry of Livestock Development

URT, (2000). The National Human Settlements Development Policy 2000. Ministry of Lands, Housing and Human Settlements Development (MLHHSD)

Wagner, T & Tasciotti, L., (2018). Urban agriculture, dietary diversity and child health in a sample of Tanzanian Town Folk. *Canadian Journal Of Development Studies*, 39(2), 234–251

Wolf, S. M., Kuch, A., & Chipman, J., (2018). *Urban Land Governance in Dar Es Salaam: Actors, Processes and Ownership Documentation*. Working Paper Ref No C-40412-TZA-1. International Growth Centre. February 2018.

Zasada, I., Weltin, M., Zoll, F., & Benninger, S. L., (2020). Home gardening in Pune (India), the role of communities, Urban environment and the contribution to Urban sustainability. *Urban Ecosystems*, 23 (4): 403–417

Zella, A. Y., (2018). Effects of Urban Farming Practices on Income Poverty Reduction in Dodoma Municipality, Tanzania. *Current Investigations in Agriculture and Current Research* 3(3), 354–367.

9 In the Face of Climate Change and Food Insecurity in the Middle East and North African Regions
Are Urban and Peri-Urban Agriculture Viable Options?

Tarek Ben Hassen and Hamid El Bilali

INTRODUCTION

Today, urban areas house nearly 60% of the world's population. Further, the global population is anticipated to reach 9.7 billion by 2050, with 70% living in cities, mostly in poor and middle-income African and Asian nations (UN Department of Economic and Social Affairs, 2018). As cities expand, they face colossal challenges in providing crucial services such as transportation, healthcare, education, and food to all residents (Acharya et al., 2021). Additionally, the urbanization process is intimately linked to rising urban poverty, food insecurity, and malnutrition, particularly among low-income urban residents (FAO, 2020; WFP, 2022), which became apparent during the COVID-19 pandemic (disruptions in the food supply chain, aggravation of physical and economic obstacles to food access, etc.) (Lal, 2020). In many developing countries, the ongoing migration of rural populations to urban areas often leads to the relocation of poverty, hunger, and malnutrition to the cities, commonly referred to as the "urbanization of poverty" (FAO et al., 2022; Liddle, 2017). These challenges are exacerbated by fluctuations in food prices and financial, energy, and economic crises, disproportionately affecting urban consumers who rely heavily on purchased food. Among them, low-income urban residents are especially vulnerable and face significant barriers to accessing nutritious and affordable food. In response to these challenges, there is growing interest in developing more resilient and sustainable food systems (El Bilali et al., 2021). One solution that has gained attention is the promotion of urban and peri-urban agriculture (UPA), which involves cultivating food in and around cities. According to FAO et al. (2022),

> Urban and peri-urban agriculture (UPA) can be defined as practices that yield food and other outputs through agricultural production and related processes (transformation, distribution, marketing, recycling ...), taking place on land and other spaces within cities and surrounding regions. It involves urban and peri-urban actors, communities, methods, places, policies, institutions, systems, ecologies and economies, largely using and regenerating local resources to meet changing needs of local populations while serving multiple goals and functions.

> *(p. 11)*

In many developing countries, UPA supplies a relatively significant amount of the food eaten in cities by utilizing the available urban and peri-urban land for food production. The production of crops and animal products in urban and peri-urban areas frequently constitutes a significant portion of urban food needs. For instance, in Nakuru (Kenya), it accounts for as much as 8%; in Dakar (Senegal), it represents 10%; and in Hanoi (Vietnam), it contributes to 44% of the total food requirements. Agri-food products from UPA are often fresher, more nutritious, and more varied than food purchased from stores, marketplaces, or street vendors (World Bank, 2013). Therefore, UPA seems a viable option that has the potential to enhance nutritional variety, urban space quality, and community action and empowerment.

Over the past few decades, the significance of UPA has increased. It has been gradually acknowledged as a crucial factor in providing safe and nutritious food, feeding expanding urban populations, and contributing to all aspects of urban food systems (FAO et al., 2022). In the face of the pandemic's catastrophic consequences, for example, home gardens offered nutritious and healthful food supplies as well as ecological services (Lal, 2020). Globally, UPA is changing due to political, economic, environmental, and technological developments (Smit et al., 1996). UPA includes a broad range of production systems within urban and surrounding areas. Crops, including fruits and vegetables, fish and cattle production, and herbal, medicinal, and decorative plants for household consumption and sale, are part of these systems. Accordingly, UPA activities may range in size and direction from subsistence-oriented farming to micro-scale, more leisurely forms of agriculture to medium- and large-scale commercial enterprises (World Bank, 2013). UPA involves a diverse array of actors and encompasses various growing techniques. The products of UPA are also highly diverse, and the practice takes place in a wide range of locations. Additionally, UPA employs different organizational arrangements and serves multiple functions (FAO et al., 2022). One of the most significant characteristics that set UPA apart is its integration within the economic, social, and ecological systems of urban areas. Moreover, UPA is significantly impacted by urban policies, competition for land, and market conditions such as prices and has a considerable influence on the overall urban system (Tawk et al., 2014). Despite the multiple benefits of UPA, there is a lack of knowledge on its development state and perspectives in developing countries, such as those of the Middle East and North Africa (MENA) region.

The MENA region refers to a geographic area that includes countries situated in Western Asia, the Arabian Peninsula, and North Africa, such as Algeria, Bahrain, Djibouti, Egypt, Iran, Iraq, Israel, Jordan, Kuwait, Lebanon, Libya, Mauritania, Morocco, Oman, Palestine, Qatar, Saudi Arabia, Syria, Tunisia, the United Arab Emirates (UAE), and Yemen. The region has significant economic diversity, including countries with varying wealth levels, including high-income countries (e.g., Qatar and UAE) and lower-income ones (e.g., Yemen). The MENA region is experiencing some of the highest rates of population growth globally, with Lebanon and Jordan being among the top five countries with growing populations due to the impact of the refugee influx (UNICEF, 2021). According to the World Population Prospects 2017 report by the United Nations (2017a), the highest population growth rates between 2010 and 2015 were observed in countries located in the MENA region and Africa. The population growth in the region is expected to continue in the coming decades. This trend will further increase the water demand, making it even more challenging for countries in the region to meet their water needs, exacerbating the already-severe water scarcity issues. In addition, the growing population will require more food, increasing the pressure on the agricultural sector and further straining water resources (FAO, 2022).

Furthermore, the urban population in the region has been steadily increasing in recent years. Today, approximately 58% of the total population in the region lives in urban areas. This trend is expected to continue, with the urban population projected to increase to 70% by 2050 (United Nations, 2020). Further, the MENA region is characterized by a harsh environment dominated by deserts and, hence, limited arable land, with less than 5% of the total land being arable, half of the global average, poor soils, and a high level of aridity and water stress (FAO, 2022). Indeed, with low freshwater resources, average water availability of just roughly 1200 m^3/person/year, compared to a global average of 6500, and high water demand owing to population increase and economic development, the MENA region is one of the world's most water-scarce regions (FAO, 2013). Except for Iraq and Mauritania, all 17 Near East and North Africa countries[1] have values below the 1000 m³/ year water scarcity level. For instance, Syria, Sudan, Morocco, Lebanon, and Egypt are experiencing water stress (between 500 and 1000 m³/capita/year). In contrast, almost two-thirds of MENA countries are experiencing extreme water shortage (less than 500 m³/capita/year) (FAO 2022).

Accordingly, to meet their food needs, the countries of the region are structurally reliant on global markets (OECD & FAO, 2018). Due to its significant dependence on food imports, the MENA region is regarded as one of the world's most food insecure and, consequently, one of the most vulnerable to price volatility (Tawk et al., 2014). Due to the growing population, fast urbanization, and rising per capita earnings, some Middle Eastern nations started to experience an expanding "food gap" in the 1970s. Meanwhile, rising incomes changed local consumption patterns, increasing demand for grains, protein, and various fruit and vegetable sources. As a result, food self-sufficiency for several countries in the region has declined substantially over the previous few decades, and a growing reliance on food imports has become the norm (Babar & Kamrava, 2014). In addition, the MENA region is projected to be severely impacted by climate change, with many increasingly evident adverse effects, such as droughts, floods, and extreme heat waves (Bayoumi et al., 2022; Lewis et al., 2018). Currently, 70% of the region's agricultural output is rain-fed, making the region very vulnerable to temperature and precipitation fluctuations, resulting in ramifications for food security, social security, and rural livelihoods. Indeed, the impact on the region will be significant under both 2°C and 4°C warming scenarios, primarily due to the high rise in predicted heat extremes, a notable decrease in water availability, and anticipated implications for regional food security. In some countries, there is a possibility of a reduction in agricultural productivity by a maximum of 30% when the temperature rises by 1.5–2°C and a potential fall of almost 60% at 3–4°C in certain areas within the region (World Bank, 2014).

UPA is a longstanding practice in the MENA region, with roots dating back thousands of years. Moreover, despite facing negative pressures and growing demands for land and water resources, crop cultivation and animal husbandry continue to be prevalent practices in the cities of the MENA region (Nasr & Padilla, 2004). Cultivating vegetables (including tomatoes, cucumbers, and radishes) and herbs (such as mint, parsley, and basil) are also prevalent practices (Al-Asad & Zureikat, 2018). In recent years, UPA has been gaining significance in addressing the pressing issue of food security and environmental sustainability (Tawk et al., 2014). UPA in the MENA region is implemented in various forms, including community gardens, rooftop gardens, urban farms, and aquaponics systems. These initiatives aim to promote sustainable and healthy food production while providing job opportunities and improving the quality of life in urban areas. UPA has several benefits in the MENA region. First, the increasing urbanization in the region has resulted in the decline of agricultural land, leading to greater dependence on food imports. Hence, UPA helps reduce the dependency on food imports, which can be costly and also subject to market volatility, as observed with the ongoing war in Ukraine (Al-Saidi, 2023; Ben Hassen & El Bilali, 2022a; Rabbi et al., 2023). Second, it provides fresh and locally produced food to urban residents, which is more nutritious and sustainable. Third, UPA creates job opportunities, particularly for women and youth, often marginalized in the formal economy. Finally, by using sustainable and organic farming practices, UPA helps mitigate environmental degradation, such as air pollution.

Despite the numerous benefits of UPA, several challenges need to be addressed. These include limited access to water and soil, lack of infrastructure and technology, and limited access to markets, with urban and peri-urban land use being redirected toward more financially viable purposes. In many cities in the region, the lack of government initiatives combined with issues related to land tenure and high land prices make UPA less prevalent and viable. Indeed, in many countries, public policies haven't caught up with UPA and don't support it.

Accordingly, this chapter aims to explore the state of UPA in the MENA region and the main challenges the region faces in developing it. The chapter is divided into two sections that (i) present the main issues of agriculture in the MENA region in relation to climate change and food insecurity, and (ii) discuss the main factors that encourage or hinder the development of UPA in the region. The chapter combines bibliographical and topical analyses of the scholarly and gray literature and includes case studies from various MENA countries as well as some policy recommendations to develop UPA.

ADVANTAGES OF UPA IN THE MENA REGION

A growing body of evidence shows that UPA can have multiple and multifaceted advantages, and the MENA region is no exception. Indeed, UPA seems a viable option to address many challenges the region faces toward sustainability and sustainable development. In this regard, referring to Egypt, Kamel and El Bilali (2022) state that "UPA can perform a wide range of socio-economic and environmental roles, including aesthetic urban design, waste management, circular economy, energy use efficiency, microclimate control, preservation of cultural heritage, biodiversity conservation, and health and well-being promotion" (p. 48). However, one of the most critical roles of UPA seems to be its contribution to food and nutrition security.

Food insecurity and water scarcity are significant challenges in the MENA region. Further, today, more than 58% of the region's population lives in urban areas (United Nations, 2020). However, urban agriculture (UA) could solve some of these challenges, offering sustainable food production, employment, and community development opportunities. Indeed, UA in the MENA region can contribute to food security by increasing the availability of fresh and nutritious food, improving livelihoods, and enhancing social capital (Tawk et al., 2014). Indeed, numerous studies have indicated that incorporating agricultural practices in urban and peri-urban areas can help tackle issues such as urban food insecurity, poverty, malnutrition, and health problems, as well as alleviate various environmental and social challenges (Zezza & Tasciotti, 2010). In times of crisis, it plays a crucial role in ensuring an adequate supply of emergency food.

First, one of the most significant advantages of UPA in the MENA region is its potential to enhance food security. Studies have indicated that the agricultural productivity in the region, while low compared to other countries (due to climate change, soil degradation, water scarcity, etc.), is not keeping up with the region's increasing population (FAO et al., 2023). In the MENA region, the annual population growth rate over the past decade has been nearly twice the world average. This rapid population growth, coupled with rapid urbanization, limited water and land resources, and low domestic agricultural productivity, has led to a high level of dependence on imports to meet the demand for food. Through imports, the region meets over 50% of its food requirements (FAO, 2014). As the population is expected to grow, policy and decision-makers are increasingly concerned about the region's reliance on global markets and trade to meet its food needs (FAO et al., 2023). This puts its countries at risk of food insecurity, as global food prices can be volatile, and trade disruptions can occur.

Consequently, MENA countries have been adversely impacted by the recent global food shocks like the Covid-19 pandemic and the war in Ukraine due to their reliance on food imports (Al-Saidi, 2023; Ben Hassen et al., 2022a; Ben Hassen & El Bilali, 2022b). In 2023, most countries in the region still encounter substantial obstacles in accomplishing Sustainable Development Goal (SDG) target 2.1, which is to guarantee consistent access to adequate, safe, and reasonably priced nutritious food for all individuals, as well as SDG target 2.2, which is to eradicate all types of malnutrition (FAO et al., 2023). Further, MENA is the most water-stressed region in the world. According to a forecast from the World Bank, the MENA region is likely to face the most significant economic losses due to climate-related water scarcity compared to other regions worldwide. The projected losses could account for 6–14% of the region's GDP by 2050 (World Bank, 2018). According to a global analysis by Martellozzo et al. (2014), UA can play a significant role in meeting the demand for vegetables in urban areas, especially in regions like MENA, where limited resources present challenges to traditional and rural agriculture. Likewise, UPA could contribute to household food security by providing access to fresh, healthy produce at lower costs and income-generating opportunities.

Second, UPA may also help reduce post-harvest food losses. In many MENA countries, most fresh produce is transported from rural areas to urban centers, often over long distances, resulting in significant food losses due to inadequate infrastructure and storage facilities. Food loss and waste are prevalent issues in the MENA region, exacerbating food insecurity, water scarcity, and environmental degradation. Moreover, it increases the region's dependence on food imports, which is already high (Abiad & Meho, 2018; El Bilali & Ben Hassen, 2020). According to FAO (2011), approximately 34% of food is lost or wasted throughout the region. The MENA region discards a considerable amount of food, with estimates suggesting up to 250 kg per individual annually, which exceeds the worldwide average (FAO, 2014). UPA can provide an alternative to this system by bringing the production of fresh produce closer to urban consumers (cf. short food supply chains), thus reducing transportation and storage costs and minimizing food loss due to spoilage. It can also help reduce food contamination risks and increase fresh and locally produced food supply.

Third, UA can also provide a more diversified food supply by growing a variety of foods, thereby enhancing the nutritional value of the foods and diets. The region faces a dual challenge of malnutrition, where issues of undernutrition co-occur with over-nutrition and obesity (FAO, 2015). As of 2021, the region has an estimated 54.3 million undernourished people, equivalent to 12.2% of the population (FAO et al., 2023). Consequently, most countries in the region still encounter substantial obstacles in accomplishing SDG target 2.2, which is to eradicate all types of malnutrition (FAO et al., 2023). However, the extent of the problem varies among poor and unstable countries like Yemen and Somalia, where malnutrition and food insecurity remain significant challenges, and wealthy nations like those in the Gulf Cooperation Council (Ben Hassen & El Bilali, 2022b). Further, in several countries in the MENA, food subsidies, used to ensure that low-income households have access to affordable and nutritious food, have become an integral part of the social contracts. Still, their effectiveness in reducing poverty seems limited (Alfani et al., 2022).

Moreover, food subsidies focus on some stable food items (such as bread, wheat, sugar, tea, cooking oil, milk, and rice), but fresh foods (such as vegetables and fruits) are not included. Consequently, for many low-income households, a healthy diet could be unaffordable. Due to limited food budgets, some consumers are forced to convert from more expensive and nutritious foods (such as fruits and vegetables) to less expensive and subsidized ones (such as rice and bread), reducing the nutritional value and diversity of meals. One of the main advantages of UPA is that it provides an affordable and accessible source of fresh produce, often lacking in low-income urban areas and for marginalized groups. By growing fruits and vegetables in their backyard gardens or community gardens, urban residents can increase their access to healthy food without relying on expensive or distant food sources.

Fourth, in addition to improving food security, UA can provide economic benefits to urban dwellers in the MENA region. Although the economic benefits of UA are relatively small at the community level, it can potentially enhance the resilience of urban communities, particularly in handling economic challenges (Yuan et al., 2022). Urban households can directly benefit from economic advantages and expenses associated with agricultural production, such as self-employment, exchanging products, earning income from sales, and saving money on food and health-related costs (Van Veenhuizen & Danso, 2007). With most of the population living in urban areas, there is a significant market demand for fresh and locally produced food. UA can help meet this demand by creating new opportunities for small-scale farmers and entrepreneurs to grow and sell their products. This, in turn, can boost the local economy by creating jobs, increasing income, and generating tax revenue. Further, UPA can provide employment and economic opportunities, especially for vulnerable groups like women and refugees, who may face barriers to accessing traditional labor markets (Tomkins et al., 2019).

In fact, due to ongoing political instability, conflicts (cf. Libya, Palestine, Iraq, Syria, Somalia, Yemen), and economic hardships (cf. Lebanon), the MENA region has been home to many refugees over the recent years. The number of forcibly displaced and stateless people in the MENA region increased from the 15.8 million recorded in 2020, reaching 16 million by the end of 2021. Most displaced individuals currently reside in non-camp environments, specifically in urban areas. This situation places additional strain on local governments responsible for addressing their daily needs (United Nations, 2020). The situation is particularly challenging given that most displaced people in the region live in extreme poverty and face significant barriers to accessing basic services. For instance, Lebanon, the world's top refugee-hosting country per capita, is facing a dire situation where nine out of every ten refugee households live in extreme poverty (UNHCR, 2021). This situation has increased tensions with host communities due to competition for scarce resources (Daher et al., 2021). In Iraq, many out-of-camp households have accumulated high levels of debt, facing difficulties in accessing adequate housing, purchasing food, ensuring their children attend school, and obtaining healthcare (UNHCR, 2021). Accordingly, the prevalence of acute food insecurity and malnutrition in Jordan and Lebanon is closely related to refugee hosting (Da Silva & Shenggen, 2017). Indeed, UPA has many advantages for refugees (Box 9.1).

BOX 9.1   URBAN AND PERI-URBAN AGRICULTURE ADVANTAGES FOR REFUGEES

- Food security: Refugees resettled in urban areas often struggle to access affordable and healthy food. UPA can provide them with a reliable source of fresh produce, reducing their reliance on expensive, processed food options.
- Economic opportunities: UPA can provide refugees with opportunities to generate income and develop skills valuable for future employment. This can help refugees become self-sufficient and less reliant on aid.
- Psychological benefits: For refugees who have experienced trauma and displacement, UPA can provide a healing and therapeutic space. Engaging in gardening activities and connecting with nature can reduce stress, anxiety, and depression.
- Community building: UPA can also help refugees build social connections and a sense of community in their new environment. Working together in community gardens can foster a sense of belonging and provide a space for cultural exchange.

Source: Authors' elaboration.

Moreover, the air quality in the MENA region is a significant concern, as more than 98% of the population is exposed to particulate matter levels that exceed the limits set by the World Health Organization. The top 10 most polluted countries in the world include five Arab countries, namely Egypt, Kuwait, Libya, Qatar, and Saudi Arabia (United Nations, 2017b). Additionally, among the top 10 countries with the highest number of deaths caused by air pollution, three Arab countries are present: Egypt, Iraq, and Saudi Arabia (United Nations, 2020). Consequently, UPA can play a significant role in environmental sustainability by promoting resource conservation and reducing greenhouse gas emissions. UPA can reduce the heat island effect and improve air quality by transforming vacant lots, rooftops, and other underutilized urban spaces into productive green spaces. UPA can significantly mitigate climate change by implementing sustainable practices such as sequestering carbon in the soil, reducing the need for long-distance transportation of food (cf. food miles), and minimizing the use of fossil fuels in food production. In addition, UPA can promote water conservation by using efficient irrigation systems, harvesting rainwater, and recycling graywater. These practices can help reduce water usage and protect scarce water resources, which are increasingly under pressure in the MENA region (UNESCO, 2023).

CHALLENGES HINDERING THE DEVELOPMENT OF UPA IN THE MENA REGION

Despite its numerous benefits, including increased food security, improved economic opportunities, and enhanced environmental sustainability, UPA also faces multiple challenges in the MENA region.

First, one major challenge of the UPA in the MENA region is the limited availability of land and space. Limited availability of cultivable land, productive soil, and water resources in the MENA region hampers agriculture and public use (Zdruli & Zucca, 2023). According to extensive studies, it has been reported that merely 6.8% of the entire land area, approximately 0.21 hectares per person, is appropriate for agriculture (FAO & ITPS, 2015). In some extreme cases, such as Libya, this percentage falls below 2% (Zdruli, 2014). Further, with a growing population, urban areas are becoming increasingly densely populated. For instance, most Jordanians currently live in flats with minimal access to private or public land for possible agricultural operations (Al-Asad & Zureikat, 2018). Accordingly, finding suitable and affordable agricultural land can be difficult, forcing many urban farmers to cultivate crops in small, cramped spaces or on rooftop gardens, balcony gardens, and even window boxes. While these spaces are ideal for growing herbs and small vegetables, they often lack the necessary space for larger crops, leading to reduced crop yields. This reduction in output makes it difficult for farmers to generate a profit, as they cannot sell enough crops to make their efforts economically viable. In addition, the limited availability of land for UA drives up the cost of land for those who can find it. This can further limit the profitability of UA, making it difficult for urban farmers to grow crops, invest in the necessary equipment, and provide for their families. As new technologies, such as vertical farming and hydroponics, are being developed, efforts are being made to solve this problem. However, the cost of the initial setup remains a significant concern, which could make it unattainable for lower-income populations in developing countries (Yuan et al., 2022).

Further, in the region, there has been a tendency for urban planning to prioritize real estate development, often resulting in changes in land use that negatively impact agriculture. Additionally, laws designed to convert communal lands to private ownership have been applied, further exacerbating the situation. As a result, there has been a reduction in the availability of cultivable land and productive soil. Real estate is an important economic engine in many Middle Eastern and developing countries. It has become a very effective means of collecting surpluses created by nonproductive economic sectors such as remittances. Land's commercialization and transformation into a capital asset have resulted in speculation, with the value of land decided by its prospective return on investment. Agriculture, a generally low-returning industry, was doomed. As a result, speculation and real estate are among the most significant impediments to the creation or execution of zoning and planning restrictions that would elevate UA above a random and transient use of space (Zurayk, 2010). Nevertheless, gentrification has been identified as a development engine in several cities, like Beirut and Cairo, leading to de-territorialization and discordant land use cohabitation (Trovato et al., 2016).

Second, despite being an old activity in the region, UPA continues to suffer from a lack of recognition by planners, agriculturists, policy-makers, researchers, and even its practitioners. For instance, the youth in Egypt are becoming increasingly aware of urban farming, thanks to initiatives such as "Grow Your Own" in Cairo governorate, "Your Roof Is Your Paradise" in Beheira governorate, "Shagara Roofs" in El Abour City, and the "Egyptian Switchers Community." These initiatives aim to promote and encourage urban gardening as a means of sustainable food production and to promote environmental awareness and civic engagement (Groening, 2016). However, despite these efforts, UPA continues to suffer from a lack of recognition and support. Consequently, there is a lack of research, resources, and extension services available to support UA. There is a significant gap in knowledge about the existing urban agricultural lands and other fertile areas in cities, and few strategies and policies exist to promote or support UPA activities (Nasr & Padilla, 2004). To date, the most successful examples of UA have been observed in cities and regions where life has been defined by conflicts and extreme poverty, such as in the Gaza Strip, Palestine, and refugee camps. In these areas, the urgent need for survival has driven communities toward UPA to address food insecurity (Tawk et al., 2014). One of the most successful initiatives in the region is the Gaza Urban and Peri-Urban Agriculture Platform (GUPAP) (Box 9.2).

Notably, there is a scarcity of comprehensive government programs that aim to promote UPA in the MENA region. In many countries, there continues to be a predominant emphasis on traditional, agrarian-centered farming as the primary approach to guaranteeing food security. UPA often encounters policy ambiguity, characterized by the absence of well-defined legislation or support systems that may facilitate its development. As a result, many UPA-related initiatives significantly depend on non-governmental organizations, foreign assistance, or community-driven endeavors rather than being effectively incorporated into a comprehensive governmental framework addressing food security and climate resilience. The lack of governmental policy signifies a wasted chance to develop food systems that are both resilient and sustainable in response to pressing issues such as climate change and rising urbanization. Therefore, while current programs show potential, their effectiveness is constrained without a more comprehensive legislative framework to expand these efforts.

BOX 9.2   GAZA URBAN AND PERI-URBAN AGRICULTURE PLATFORM

- Gaza's economy has been devastated by the lengthy fighting and Israeli-imposed blockade. The movement of people and commodities is severely limited; 90% of factories and workshops have had to shut down; 80% of people need assistance; and exports have lately fallen to less than 2%.
- Gaza is a densely populated area where 90% of all agriculture can be considered urban or semi-urban.
- Gaza Urban and Peri-Urban Agriculture Platform (GUPAP) is a successful initiative created in 2013 to promote UPA in Gaza.
- It comprises a multi-stakeholder, interactive, and participatory space that brings together important players interested in building a resilient Palestinian agriculture industry in Gaza.
- GUPAP currently has around 80 members, which include national and local government institutions, non-governmental civil institutions, women organizations/cooperatives and activists, private sector initiatives/actors/stakeholders, research and educational institutions, micro-finance institutions, urban women agripreneurs' forums, and other relevant actors.
- Throughout the years, GUPAP has developed institutional capabilities and collaboration among its members, resulting in notable results. This includes examining and promoting five policies concerning municipal and national urban and peri-urban agricultural development, as well as the rights of women agripreneurs.

Source: Authors' elaboration based on GUPAP (2023) and Oxfam International (2016).

Third, the inadequate government support for UPA in the MENA region is a significant obstacle for urban farmers seeking to access the training and resources they need to succeed. Consequently, many urban farmers in the region lack the technical know-how and resources necessary to succeed. Without the assistance of the government, many farmers may be unable to acquire the skills and resources they require to optimize their productivity and improve the quality of their produce. This lack of government support can, therefore, impede the growth and development of UPA in the MENA region. For instance, there is relatively no regulation of UPA activity in Cairo (Egypt). Since the state, police, and military have little authority over informal settlements, social capital seems to be the key organizing factor in these places. Yet, from the standpoint of an adaptive process, the absence of structuration of the urban farmer's community underlines the challenge of disseminating new behaviors outside the extended family or neighborhood (Daburon et al., 2017).

Further, many cities in the MENA region have attempted to initiate initiatives related to UA. However, very few of these projects have succeeded in the long term. One of the main reasons for the lack of success is the absence of institutional stability and continuity. When a new administration takes over, existing UA projects are often deprioritized or even discontinued altogether. This lack of continuity means that urban farmers are unable to rely on sustained government support, which is essential for the growth and development of UPA in the region (Box 9.3).

Fourth, UPA in the region also faces regulatory challenges, as many cities lack clear regulations and guidelines for the practice. This can result in urban farms being forced to shut down or move or in farmers facing fines or legal action for violating zoning regulations. In addition, the lack of regulations can result in health and safety concerns for both farmers and consumers, as there are often no standards for food safety and quality in UA. In general, participatory urban governance still needs strengthening across the region. While urbanization is increasingly becoming a priority in the national agendas of Arab states, most national urban policies are in the early stages of development (ESCWA, 2017).

BOX 9.3   URBAN AGRICULTURE IN AMMAN (JORDAN): THE CASE OF THE JABAL AL-QAL'AH DISTRICT PROJECT

- Several UA initiatives were developed by the Greater Amman Municipality (GAM); however, due to a lack of institutional continuity and varying levels of support from different GAM administrations, none of these initiatives have been sustained.
- One notable project, initiated during Omar Maani's mayoralty (2006–2011), was carried out in Amman's Jabal al-Qal'ah district in 2010. Initially, the project aimed to clean up the rooftops in the area of discarded items and decorate them with layers of pumice stone. The district is situated in a vital touristic and cultural heritage area of the city, close to the Amman Citadel.
- But the idea extended and expanded to include the development of rooftop gardens for these residences. As a result, a GAM team led by Hisham al-'Umari, who later led a department within GAM devoted to UA, provided basic training to members of the local community as well as plants, containers, and beds for growing the plants.
- GAM heartily supported the idea then, and plans were to expand it to span 10,000 roofs around the city.
- Unfortunately, as Mr. Maani's term expired, the original project was eventually abandoned. Today, none of the initially planted roofs are still standing.

Source: Authors' elaboration based on Al-Asad and Zureikat (2018).

Despite these challenges, UA can potentially play a significant role in addressing food insecurity and poverty in the MENA region. By supporting the development of UA, governments and international organizations can help to create economic opportunities, increase food security, and improve environmental sustainability.

CONCLUSION

The MENA region faces multiple challenges, such as water scarcity, urbanization, and poverty, significantly impacting food production and security. UPA is a promising solution to food insecurity and climate change in the MENA region. UPA can address these challenges by providing fresh and nutritious food, creating socio-economic opportunities, and reducing the food system's carbon footprint. Some successful UPA initiatives in the MENA region, such as those in Gaza Strip and refugee camps, demonstrate that UPA can be a viable option for addressing food insecurity in challenging environments.

Further, countries in the MENA still face many policy and governance challenges that hamper the effective implementation of the SDGs (Ben Hassen & El Bilali, 2022c), and UPA can contribute to achieving several SDGs. For instance, UPA can help achieve SDG 1 (No Poverty) through job opportunities and affordable food for low-income communities. It can also contribute to SDG 2 (Zero Hunger) by increasing food security and reducing malnutrition. UPA can also help achieve SDG 3 (Good Health and Well-being) by promoting healthy diets and lifestyles, as well as providing healing spaces from trauma. However, despite the potential of UPA, it faces various challenges in the MENA region, which poses significant barriers to its development. These include a lack of recognition, limited government support, limited availability of land and space, and the dominance of the real estate sector.

To overcome these challenges, policy-makers and urban planners need to recognize the potential of UPA in contributing to food security, climate change mitigation, and socio-economic development. This includes developing supportive policies, providing technical and financial assistance, and incorporating UPA into urban planning and zoning. In fact, developing the UPA in the MENA region requires a holistic approach through several actions:

1. Integrate UPA into urban planning: Urban planning should be reoriented to incorporate UPA into land use planning and urban design. This can be achieved by revising zoning regulations and building codes to allow for UPA practices in urban and peri-urban areas.
2. Support small-scale farmers: Governments and international organizations should support small-scale urban farmers. This can be done by providing them access to resources such as land, water, and capital, as well as finance, technical assistance, and training and education programs to improve their productivity and efficiency.
3. Raising awareness: Governments, civil society organizations, and international organizations should raise awareness about the benefits of UPA and its potential to contribute to sustainable development in the MENA region. This can be done through public education campaigns, training programs, and workshops to promote the importance of UPA and how it can be integrated into urban planning and development.
4. Promote sustainable practices: UPA should be promoted as a sustainable and resilient practice that can help mitigate climate change and contribute to food security. This can be achieved by promoting sustainable agricultural practices, such as organic farming and conservation agriculture, and by raising awareness of the benefits of UPA for the environment and public health.

In conclusion, UPA has the potential to contribute to addressing the challenges of food insecurity, social vulnerability, and climate change in the MENA region. However, fully realizing this potential requires a supportive policy and institutional framework, research and extension services, and collaboration among various stakeholders. UPA is not a panacea for the challenges facing the MENA region, but it can be an essential part of a broader strategy to create more resilient and sustainable food systems that promote social, economic, and environmental well-being.

NOTE

1 The region includes Algeria, Bahrain, Egypt, Iraq, Jordan, Kuwait, Lebanon, Libya, Mauritania, Morocco, Oman, Qatar, Saudi Arabia, Sudan, Syria, Tunisia, United Arab Emirates (UAE), and Yemen, as well as West Bank and Gaza Strip.

REFERENCES

Abiad, M. G., & Meho, L. I., (2018), Food loss and food waste research in the Arab world: a systematic review. *Food Security*, *10*(2), 311–322. https://doi.org/10.1007/s12571-018-0782-7

Acharya, G., Cassou, E., Jaffee, S., & Ludher, E. K., (2021), *RICH Food, Smart City: How Building Reliable, Inclusive, Competitive, and Healthy Food Systems is Smart Policy for Urban Asia*. https://openknowledge.worldbank.org/entities/publication/0799614f-c0de-5113-bbd3-35d96822b702

Al-Asad, M., & Zureikat, L., (2018), *Urban Agriculture in Amman A Holistic View*. Friedrich-Ebert-Stiftung. https://library.fes.de/pdf-files/bueros/amman/15779.pdf

Alfani, F., Genoni, M. E., & Hoogeveen, J., (2022), *MENA: Uniform cash transfers are better than subsidies*. World Bank Blog. https://blogs.worldbank.org/arabvoices/mena-uniform-cash-transfers-are-better-subsidies

Al-Saidi, M., (2023), Caught off guard and beaten: The Ukraine war and food security in the Middle East. *Frontiers in Nutrition*, *10*. https://doi.org/10.3389/fnut.2023.983346

Babar, Z., & Kamrava, M., (2014), Food security and food sovereignty in the middle East. In Z. Babar & S. Mirgani (Eds.), *Food Security in the Middle East* (pp. 1–18). Oxford University Press. https://doi.org/10.1093/acprof:oso/9780199361786.003.0001

Bayoumi, M., Luomi, M., Fuller, G., Al-Sarihi, A., Salem, F., & Verheyen, S., (2022), *Arab Region SDG Index and Dashboard Report 2022*. Mohammed Bin Rashid School of Government, Anwar Gargash Diplomatic Academy and UN Sustainable Development Solutions Network. www.ArabSDGIndex.com

Ben Hassen, T., & El Bilali, H., (2022a), Impacts of the Russia-Ukraine War on global food security: Towards more sustainable and resilient food systems? *Foods*, *11*(15), 2301. https://doi.org/10.3390/foods11152301

Ben Hassen, T., & El Bilali, H., (2022b), Impacts of the COVID-19 pandemic on food security and food consumption: Preliminary insights from the Gulf Cooperation Council region. *Cogent Social Sciences*, *8*(1). https://doi.org/10.1080/23311886.2022.2064608

Ben Hassen, T., & El Bilali, H., (2022c), Sustainable Development Goals in the Middle East and North Africa (MENA) Region: Policy and Governance. In W. Leal Filho, I. R. Abubakar, I. da Silva, R. Pretorius, & K. Tarabieh (Eds.), *SDGs in Africa and the Middle East Region. Implementing the UN Sustainable Development Goals - Regional Perspectives* (pp. 1–16). Springer. https://doi.org/10.1007/978-3-030-91260-4_20-1

Ben Hassen, T., El Bilali, H., & Allahyari, M. S., (2022), Food shopping during the COVID-19 pandemic: an exploratory study in four Near Eastern countries. *Journal of Islamic Marketing*. https://doi.org/10.1108/JIMA-12-2021-0404

Daburon, A., Alary, V., Ali, A., El Sorougy, M., & Tourrand, J. F., (2017), Urban Farms Under Pressure: Cairo's Dairy Producers, Egypt. In CT. Soulard, C. Perrin, & E. Valette (Eds.), *Toward Sustainable Relations Between Agriculture and the City. Urban Agriculture* (Urban Agriculture, pp. 73–88). Springer. https://doi.org/10.1007/978-3-319-71037-2_5

Daher, B., Hamie, S., Pappas, K., Nahidul Karim, M., & Thomas, T., (2021), Toward resilient water-energy-food systems under shocks: Understanding the impact of migration, pandemics, and natural disasters. *Sustainability*, *13*(16), 9402. https://doi.org/10.3390/su13169402

Da Silva, J. G., & Shenggen, F., (2017), *Conflict, Migration and Food Security: The Role of Agriculture and Rural Development*. https://www.fao.org/3/i7896e/i7896e.pdf

El Bilali, H., & Ben Hassen, T., (2020), Food waste in the countries of the Gulf cooperation council: A systematic review. *Foods*, *9*(4), 463. https://doi.org/10.3390/foods9040463

El Bilali, H., Strassner, C., & Ben Hassen, T., (2021), Sustainable agri-food systems: Environment, economy, society, and policy. *Sustainability*, *13*(11), 6260. https://doi.org/10.3390/su13116260

ESCWA, (2017), *Arab Climate Change Assessment Report - Main Report*. https://www.riccar.org/index.php/publications/arab-climate-change-assessment-report-main-report

FAO, (2011), *Global Food Losses and Food Waste: extent, causes and prevention*. https://www.fao.org/3/mb060e/mb060e.pdf

FAO, (2013), *First Regional Training Workshop "Launching Workshop of the Regional Water Scarcity Initiative."*

FAO, (2014), *Reducing Food Loss and Waste in the Near East and North Africa*. Fact Sheet - Regional Conference for the Near East (NERC-32), February 2014. https://www.fao.org/3/as212e/as212e.pdf

FAO, (2015), *Regional Overview of Food Insecurity Near East and North Africa Near East and North Africa*. https://www.fao.org/publications/card/fr/c/b715647e-a958-4b88-87fc-9d1c3364b161/

FAO, (2020), *Urban food systems and COVID-19: The role of cities and local governments in responding to the emergency*. https://www.fao.org/3/ca8600en/CA8600EN.pdf

FAO, (2022), *The State of Land and Water Resources for Food and Agriculture in the Near East and North Africa region*. FAO. https://doi.org/10.4060/CC0265EN

FAO, IFAD, UNICEF, WFP, WHO, & UNESCWA, (2023), *Near East and North Africa - Regional Overview of Food Security and Nutrition 2022*. FAO; IFAD; UNICEF; WHO; ESCWA (United Nations Economic and Social Commission for Western Asia); WFP; https://doi.org/10.4060/CC4773EN

FAO, & ITPS, (2015), Status of the World's Soil Resources (SWSR) - Main Report. In *Status of the World's Soil Resouces (SWSR) - Main Report. Food and Agriculture Organization of the United Nations and Intergovernmental Technical Panel on Soils, Rome, Italy*. Food and Agriculture Organization of the United Nations. https://hal.archives-ouvertes.fr/hal-01241064/

FAO, Rikolto, & RUAF, (2022), *Urban and Peri-Urban Agriculture Sourcebook- From Production to Food Systems*. FAO. https://doi.org/10.4060/CB9722EN

Groening, G., (2016), Urban horticulture - gardens as elements of an urbanizing world. *European Journal of Horticultural Science*, *81*(6), 285–296. https://doi.org/10.17660/eJHS.2016/81.6.1

GUPAP, (2023), *About GUPAP*. https://gupap.org/en/who-we-are/

Kamel, I. M., & El Bilali, H., (2022), Urban and peri-urban agriculture in Egypt. *AGROFOR International Journal*, *7*(1), 48–56. https://doi.org/10.7251/AGRENG2201048K

Lal, R., (2020), Home gardening and urban agriculture for advancing food and nutritional security in response to the COVID-19 pandemic. *Food Security*, *12*(4), 871–876. https://doi.org/10.1007/s12571-020-01058-3

Lewis, P., Monem, M. A., & Impiglia, A., (2018), *Impacts of Climate Change on Farming Systems and Livelihoods in the Near East and North Africa*. https://www.fao.org/3/CA1439EN/ca1439en.pdf

Liddle, B., (2017), Urbanization and inequality/poverty. *Urban Science*, *1*(4), 35. https://doi.org/10.3390/urbansci1040035

Martellozzo, F., Landry, J.-S., Plouffe, D., Seufert, V., Rowhani, P., & Ramankutty, N., (2014), Urban agriculture: a global analysis of the space constraint to meet urban vegetable demand. *Environmental Research Letters*, *9*(6), 064025. https://doi.org/10.1088/1748-9326/9/6/064025

Nasr, J., & Padilla, M., (2004), *Interfaces : Agricultures et villes à l'Est et au Sud de la Méditerranée*. Delta.

OECD, & FAO, (2018), *OECD-FAO Agricultural Outlook 2018-2027*. OECD. https://doi.org/https://doi.org/10.1787/agr_outlook-2018-en

Oxfam International, (2016), *Enhancing Market-Oriented Urban Agriculture in the Gaza Strip: Networking for policy change and resilience*. www.oxfam.org

Rabbi, M. F., Ben Hassen, T., El Bilali, H., Raheem, D., & Raposo, A., (2023), Food Security Challenges in Europe in the Context of the Prolonged Russian-Ukrainian Conflict. *Sustainability*, *15*(6), 4745. https://doi.org/10.3390/su15064745

Smit, J., Ratta, A., & Nasr, J., (1996), *Urban Agriculture: Food, Jobs and Sustainable Cities* (UNDP, Habitat II Series, Ed.). https://www.cityfarmer.org/smitbook90.html

Tawk, S. T., Said, M. A., & Hamadeh, S., (2014), Urban Agriculture and Food Security in the Middle Eastern Context. In *Food Security in the Middle East* (pp. 159–184). Oxford University Press. https://doi.org/10.1093/acprof:oso/9780199361786.003.0007

Tomkins, M., Yousef, S., Adam-Bradford, A., Perkins, C., Grosrenaud, E., Mctough, M., & Viljoen, A., (2019), Cultivating Refuge: The Role of Urban Agriculture Amongst Refugees and Forced Migrants in the Kurdistan Region of Iraq. *International Journal of Design & Nature and Ecodynamics*, *14*(2), 103–118. https://doi.org/10.2495/DNE-V14-N2-103-118

Trovato, M. G., Farajalla, N., & Truglio, O., (2016), Gentrification Versus Territorialisation: The Peri-Urban Agriculture Area in Beirut. In B. Maheshwari, B. Thoradeniya, & V. P. Singh (Eds.), *Balanced Urban Development: Options and Strategies for Liveable Cities* (pp. 481–498). Springer. https://doi.org/10.1007/978-3-319-28112-4_29

UN Department of Economic and Social Affairs, (2018), *2018 World Urbanization Prospects*. https://population.un.org/wup/

UNESCO, (2023), *The United Nations World Water Development Report 2023: Partnerships and Cooperation for Water*. https://unesdoc.unesco.org/ark:/48223/pf0000384655

UNHCR, (2021), *Global Report 2021*. https://reporting.unhcr.org/globalreport2021

UNICEF, (2021), *Running Dry: The impact of water scarcity on children in the Middle East and North Africa*. https://www.unicef.org/mena/reports/running-dry-impact-water-scarcity-children

United Nations, (2017a), *World Population Prospects 2017*. https://population.un.org/wpp/publications/files/wpp2017_keyfindings.pdf

United Nations, (2017b), *HABITAT III Regional Report: Arab Region-Towards Inclusive, Safe, Resilient and Sustainable Arab Cities*. https://habitat3.org/wp-content/uploads/Habitat-III-Regional-Report-Arab-Region.pdf

United Nations, (2020), *Arab Sustainable Development Report 2020*. https://asdr.unescwa.org/sdgs/pdf/en/ASDR2020-Final-Online.pdf

Van Veenhuizen, R., & Danso, G., (2007), *Profitability and sustainability of urban and peri-urban agriculture.* FAO. https://ruaf.org/document/profitability-and-sustainability-of-urban-and-peri-urban-agriculture/

WFP, (2022), *Food Security and Diets in Urban Asia: How Resilient are Food Systems in Times of COVID-19?* https://www.wfp.org/publications/food-security-and-diets-urban-asia-how-resilient-are-food-systems-times-covid-19

World Bank, (2013), *Urban Agriculture: Findings from Four City Case Studies.* Urban Development Series Knowledge Papers No. 18. https://documents.worldbank.org/en/publication/documents-reports/documentdetail/434431468331834592/urban-agriculture-findings-from-four-city-case-studies

World Bank, (2014), *Turn Down the Heat: Confronting the New Climate Normal.* Washington, DC: World Bank. https://doi.org/10.1596/978-1-4648-0437-3

World Bank, (2018), *Beyond Scarcity: Water Security in the Middle East and North Africa.* The World Bank. https://doi.org/10.1596/978-1-4648-1144-9

Yuan, G. N., Marquez, G. P. B., Deng, H., Iu, A., Fabella, M., Salonga, R. B., Ashardiono, F., & Cartagena, J. A., (2022), A review on urban agriculture: technology, socio-economy, and policy. *Heliyon*, *8*(11), e11583. https://doi.org/10.1016/j.heliyon.2022.e11583

Zdruli, P., (2014), Land resources of the Mediterranean: status, pressures, trends and impacts on future regional development. *Land Degradation & Development*, *25*(4), 373–384. https://doi.org/10.1002/ldr.2150

Zdruli, P., & Zucca, C., (2023), *Restoring Land and Soil Health to Ensure Sustainable and Resilient Agriculture in the Near East and North Africa region. State of Land and Water Resources for Food and Agriculture thematic paper.* FAO. https://doi.org/10.4060/CC1137EN

Zezza, A., & Tasciotti, L., (2010), Urban agriculture, poverty, and food security: Empirical evidence from a sample of developing countries. *Food Policy*, *35*(4), 265–273. https://doi.org/10.1016/j.foodpol.2010.04.007

Zurayk, R., (2010), From incidental to essential: Urban agriculture in the Middle East. *Journal of Agriculture, Food Systems, and Community Development*, *1*(2), 13–16. https://doi.org/10.5304/jafscd.2010.012.001

10 Urban Agriculture and the Food–Energy–Water Nexus in Qatar and Drylands

From Permaculture Gardens to Urban Living Labs

Anna Grichting Solder

INTRODUCTION: URBAN AGRICULTURE AND THE FOOD–WATER–ENERGY NEXUS

The projects on urban agriculture presented here were undertaken over seven years at Qatar University (QU), beginning in 2011, and were conducted through courses in architecture and urban design, undergraduate and graduate research projects, international research grants, and exhibitions and public events, with participation in local, regional, and international conferences. The decision to incorporate the topic into the curriculum and research at QU in a system approach was motivated by Qatar's reliance on imports for its food supplies, a lack of awareness of, and initiatives on, urban agriculture, and the scarcity of research and projects on growing food locally. This new subject of urbanism, also known as agro-urbanism or food urbanism, has been growing over the past ten years and is no longer seen as a "fringe interest" – and this was highlighted by the upsurge in community gardening during the recent COVID pandemic. Policymakers, planners, research organizations, and financiers from developed and developing nations are all paying attention to the issue. Considering the climate catastrophe in particular, feeding our urbanizing globe has become essential, and municipal actors are rising to the occasion (RUAF, 2023). Cities are incrementally reappropriating their food, and after having been pushed out beyond the city boundaries, agriculture is returning to urban spaces in the form of urban agriculture.

Despite the wide range of sometimes divergent objectives, relocalizing food production inside urban spaces forms part of a broader move by cities to reconquer the food system (Bricas & Conaré, 2019). As an emerging practice, food production within the urban context is performed by various actors, including private individuals, groups, or associations; public administrations; and professional farmers. It is not limited to a closed circle of professionals in the primary sector (Food Urbanism, 2023). Moreover, urban agriculture contributes to food security and safety in two ways: first, it increases the amount of food available to people in cities, and, second, it allows fresh vegetables, fruits, and meat to be made available to urban consumers. While small-scale and localized food production has a long history, including individual allotments, which have been popular in Europe since the late 18th century, the integration of such farming practices within the economic and ecological system of towns and cities is a newer development (Viljoen et al, 2005). This means that urban resources such as compost from food waste and wastewater from urban drainage are made use of, while urban problems such as the pressure on land and development also have to be negotiated.

Productive landscapes have been reintroduced into city design in these ways: urban and peri-urban agriculture, rural–urban linkages and landscape development, urban food systems, and city region

DOI: 10.1201/9781003359425-10

food systems, and it is also increasingly integrated into buildings, whether repurposed on roofs or in industrial buildings, or as new forms of vertical agriculture. Food-growing areas are being added to schools, institutions, towns, and public places, becoming a standard element of landscape design. Vertical farming or using roofs for farming and integrating this into buildings has multiple positive effects, like producing food, increasing a building's thermal insulation, and reducing the heat island effect. As an alternative to decorative green lawns and barren trees, it would also be much more efficient to sow crop seeds and plant fruit trees. Regenerative and sustainable urbanism is about increasing the quality of life by bringing more resources within a short distance and increasing the quality of products that are offered. This is achieved by integrating both productive and non-productive landscapes to enhance the capability of a community or a city through food-producing parcels, contributing to the ecological value of the land and tying in environmental, social, and economic issues.

QATAR: FOOD, WATER, AND ENERGY

Qatar is one of the world's largest exporters of liquid natural gas (Reuters, 2023), with the third largest gas reserves after Russia and Iran, and plans to expand production. The country is considered one of the most highly water-stressed countries in the world, with little in the way of natural water resources. It is almost entirely reliant on the desalination of seawater to provide for municipal and industrial needs, while the growing agricultural sector has historically been dependent upon the desalination of brackish groundwater (Lawler et al. 2023). Desalination – which provides 99% of the drinking water – is a process that has adverse effects on the marine environment with increased salination of the Gulf waters accompanied by an increase in temperature (Water Action Hub, 2023). Since 2017, local food production in Qatar has intensified due to the 2017 blockade by the neighboring countries of Saudi Arabia, the United Arab Emirates, Bahrain, and Egypt (Wellesley, 2023). This crisis forced the country to find alternative sources for its food imports and act as a catalyst to accelerate research and investment to increase local production in Qatar. While research was being carried out on food, water, and energy in Qatar, it was primarily based on technology and science, not including society and ecology, and very little was based on systems and nexus thinking.

The nexus between food, water, and energy underpins security issues in many parts of the world. Qatar has large non-renewable energy reserves but needs to secure all three supplies to ensure its security. The recent blockade highlighted the country's lack of domestic food production and reliance on its neighbors for food imports. Emerging architecture and urban and landscape design models are addressing the new challenges of climate change, resource scarcity, insecurity, and the loss of biodiversity and applying more integrative and systems-thinking approaches. Urban agriculture can build a regenerative relationship between natural systems and human communities, dramatically improving the generative capacity of buildings, landscapes, infrastructure, and cities. We need to think about soil, water, terrain, and climate – how nature works and how we can enhance the urban ecosystems – making them productive rather than consumptive urban systems. This new way of thinking about food in the urban ecosystem needs to be integrated at the academic, research, policy, and business levels with an ecosystem of knowledge and expertise in urban agriculture. For example, current legislation in Qatar and the region tends to encourage rather than limit food waste, which is not often recycled for composting. Therefore, a valuable resource for food growing is being wasted. At the same time, it is necessary to think about new types of food and protein sources that use less land, water, and energy, such as microalgae and other superfoods.

INTEGRATING URBAN AGRICULTURE INTO RESEARCH AND TEACHING: PROJECTS, PROCESSES, AND STAKEHOLDERS

This initiative aspires to develop new approaches in the emerging cross-disciplinary fields of urban agriculture, highlighting the different disciplines and integrative processes needed to understand urban ecosystems and food infrastructures and design and implement innovative ways of transforming the territory to become more productive. It was based on deepening traditional knowledge and local ecosystems and inviting internationally well-known researchers and practitioners to share expertise and practical knowledge on emerging practices. The sustained projects and initiatives over the years allowed to build up a body of knowledge and experience, as well as a network of researchers, partners, and practitioners, and resulted in the receiving of a highly competitive international research grant attributed to the Belmont Forum for the Sustainable Urbanization Global Initiative as well as the establishment of the QU Living Laboratory.

To increase food security in Qatar and to develop better urban systems and landscapes related to the food, water, and energy nexus, it is necessary to integrate the production of food into the design of urban and architectural systems to increase the efficient use of non-renewable resource use and to increase local production of food. This should not only be approached from a technical and engineering approach, as has been the case previously but should be integrated into a holistic strategy that includes interdisciplinary approaches and is integrated into the educational and administrative systems. Solutions are needed not only at the governmental and industrial scales but also at smaller community and business scales. A multi-scale approach is necessary to address infrastructural requirements, and challenges such as water resources and land use for food production at larger scales, combined with small-scale approaches through individual, residential, and community food-growing projects that are accompanied by training and incentives, and that, when multiplied, can also significantly increase the food production in Qatar.

Combined with a multi-scale approach, it is also important to involve multiple stakeholders. By bringing together international experts with local academics, researchers, students, architects, landscape architects, planners, agricultural businesses, local NGOs, local administrations, and government institutions, there is an opportunity to develop creative responses to the questions of food supply and food security, from local residential initiatives to larger urban and rural farms, including the better integration of food production into public buildings and landscapes. The projects and research are aligned with the Qatar National Research Strategies on Food, Water, and Energy Security and with developing and strengthening interdisciplinary research partnerships between Qatar's academic, private, and public components and internationally to enable more sustainable urbanization and a healthier environment.

Urban agriculture was integrated into several architecture and urbanism courses that were taught at QU for seven years (2011–2018). Students designed or developed/regenerated architectural and urban landscapes to make them productive, addressing a wide range of urban agriculture typologies, including vertical farming, community gardening, edible boulevards, urban farms, and rooftop farming. These projects and experiments were conducted at various scales in selected locations, primarily in Qatar (Grichting, 2023).

The timeline (see Table 10.1) illustrates and highlights the different pedagogical and research components that integrate urban agriculture, which include lecture and design courses; student and faculty research; university and public events; local, regional, and international conferences; and continued stakeholder engagement. Creating the network of stakeholders was an essential component of the work, engaging students, researchers, experts, and partners in the projects from the academic, professional, government, and civil society fields. The pedagogy is predicated by a critical design issue – the integration of food production into architectural, urban, and landscape projects – and includes alternative learning/teaching methods with activities outside the classroom. As well as addressing food, soil, and biodiversity, the pedagogical approach integrates questions of socio-spatial justice, migrant communities, gender sensitivities, community-based design, and social engagement and is based on experiential learning, inquiry-based learning, and outcome-based learning.

The lectures on food urbanism that were part of the package covered the basics of food systems, the relationship between food, water, energy, and waste, the idea of footprints, or the carbon footprint associated with growing and moving food from farm to table. At the graduate level, these locations featured expansive landscapes, industrial infrastructures, shopping malls, and residential compounds. Students were assigned specific sites in Qatar to work on. The focus of the work for the undergraduate students was the QU campus, with an edible garden serving as an experimental landscape. Students conducted urban agriculture case studies on a global scale in relation to the projects they were working on. The undergraduate course's objectives were to motivate students to incorporate food production into their studio ideas, whether for structures, surrounding landscapes, rooftops, or urban plans. The teaching approach includes one specific element where students are given projects from the previous class and are expected to expand on those projects and situations to create a body of knowledge, designs, and scenarios.

At the graduate level, students worked on implementing urban food strategies, systems, and designs in different architectural typologies and urban contexts (Grichting et al., 2016). These included dense urban development, gated residential communities, migrant workers housing, educational institutions, the transformation of offshore oil fields, a suburban treated wastewater pond, abandoned quarries, reactivated wadis, rural desert landscapes, and the university campus. In a course on urban planning legislation, graduate students studied policy documents at the global and local levels to identify gaps and propose how the design systems and strategies for urban agriculture and the food and water–energy nexus could be translated into urban policies. This aspect of policy will be elaborated on further in the chapter.

TABLE 10.1

Timeline of Projects, Initiatives, and Events on Urban Agriculture, 2012–2020

Date	Name	Type	Actors–Stakeholders–Partners	Outputs–Results
30 November 2012	Qatar Sustainability Expo – CoP18. Designing Urban Food Systems in an Emerging Drylands Metropolis. Qatar Foundation	Conference Presentations	Undergraduate and Graduate Students with Faculty	Newspaper article
24–25 November 2013	Qatar Foundation Annual Research Forum. Designing Productive Landscapes in an Emerging Desert Metropolis: The Case of Doha. Qatar National Convention Center	Conference Presentation	Faculty and Students	Conference proceeding
2013–2014	Designing for Food Security. Investigating Productive Landscapes in Doha	UREP Research Project. Funded by QNRF. 30'000 USD	Undergraduate Students, Permaculture specialists, Faculty Advisor	http://blogs.qu.edu.qa/foodurbanismdoha/
July 2013	Gulf Research Meeting Cambridge. Gulf Cities as Interfaces. Designing Productive Landscapes in an Emerging Desert Metropolis. The Case of Doha	Speaker. Workshop and publication.	Researchers and Academics Specialized in the Gulf Region Masters' Students from Qatar University	Publication of proceedings Workshop
20 May 2014	"Healthy Habits & Juicing Festival". Launch of the Edible Garden at Qatar University. UREP Research Project. Designing for Food Security. Investigating Productive Landscapes in Doha	Event with students, faculty, and guests as part of Architecture Day	Student Association, Faculty, Landscape Engineers QU, Capital Projects Engineers, NGO Friends of the Environment, NGO Sprout Middle East, SAIC Agriculture Company	http://blogs.qu.edu.qa/foodurbanismdoha/?p=569

(Continued)

TABLE 10.1 (*Continued*)

Timeline of Projects, Initiatives, and Events on Urban Agriculture, 2012–2020

Date	Name	Type	Actors–Stakeholders–Partners	Outputs–Results
27–28 October 2014	Future Landscape and Public Realm. Case Study: Integrating Plant Life into Architecture and Building Design – Hanging Gardens of Doha. Intercontinental Hotel, Doha	Conference Speaker	Planning Authorities, Public Space Specialists, Landscape Architects, Landscape Firms, Students, and Researchers	Stakeholder engagement, knowledge sharing
10–11 March 2015	Qatar Projects. Landscaping Projects for Increased Health, Food Security and Biodiversity. MEED International Conference, Grand Hyatt Hotel, Doha. 10–11 March 2015	Conference Speaker	Public Administrations, Real Estate Developers, Professional Architects, Engineers, and Planners, Academics, and Students	Stakeholder engagement, knowledge sharing
25 October 2015	Future Landscape and Public Realm. Qatar. Food Urbanism. A Focus on Horticulture and Student Life, Intercontinental the City, Qatar. 25 October 2015	Conference Speaker	Planning Authorities, Public Space Specialists, Landscape Architects, Landscape Firms, Students, and Researchers	Stakeholder engagement, knowledge sharing
7–9 October 2015	7th International Aesop Sustainable Food Planning Conference, Politecnico di Torino. A Productive Permaculture Campus in the Desert. Visions for Qatar University	Speaker. Paper Presentation and Publication	Academics, Researchers, Students, Public Administration, Keynote Speaker Carlo Petrini, Founder of Slow Food	Publication of proceedings
1 April 2015	Landscapes as Infrastructure for Biodiversity and Food Security. Perspectives from Switzerland and Qatar. Lectures and Workshop on Green Roofs at Qatar University. Exhibition on Swiss Landscapes and food Urbanism Projects in Qatar	Exhibition and Workshop	Experts on Green Roofs, Students and Faculty, SAIC Agriculture Company, Nakheel Landscape Company, Embassy of Switzerland, Zurich University of Applied Sciences	Qatar TV television documentary
6 May 2016	Qatar Green Building Council. The Hanging Gardens of Doha. Seminar. Lecture. Symbiotic Systems and Productive Landscapes	Invited Lecture with Edouard Francois Architect.	Qatar Green Building Council Professional Architects, Landscape, Water Engineers Researchers Faculty and Students	QGBC website
19 March 2016	Permaculture Workshop. Soul to Soil from Waste to Fertility, Building Soil Fertility with Compost. Workshop and Building of a Compost Facility at Qatar University International Workshop: Sustainable Urbanism: New Directions. Qatar Foundation Annual Research Conference. ARC16	CWSP Grant for International Workshop. Funded by QNRF. Sponsored by SAIC, AlNakheel, and the Embassy of Switzerland 65'000 USD	Ecological Systems Designer, Permaculture Experts, Faculty and Students, Landscape Designers and Gardeners, Migrant Workers, Qatar University, Schools, Professional Architects, International Architects, and Academics	Webzine publication composting. Webzine publication permaculture workshop Webzine on Urbanista

(*Continued*)

TABLE 10.1 (*Continued*)

Timeline of Projects, Initiatives, and Events on Urban Agriculture, 2012–2020

Date	Name	Type	Actors–Stakeholders–Partners	Outputs–Results
24–25 October 2016	3rd Annual Future Landscape and Public Realm Qatar. Blue Urbanism: Examining the Relationship between the Urban Landscape and Coastal Cities. InterContinental Hotel, Doha	Member of the Advisory Board, Invited Speaker, and Panel Moderator	Planning Authorities Public Space Specialists Landscape Architects Landscape Firms Students and Researchers	Stakeholder engagement, knowledge sharing
5–6 December 2016	2nd Annual Future Drainage and Stormwater Networks Qatar Sustainable Drainage: Designing Landscapes with TSE and Stormwater. 5th City Center Rotana, Doha, Qatar	Professional Conference. Invited as Chairperson and Speaker	Public Works Authorities Water Engineers Water Companies Students and Researchers	Stakeholder engagement, knowledge sharing
1–2 May 2017	Future Drainage and Stormwater Networks Abu Dhabi. Constructed and Coastal Wetlands in Urban Water Systems in the Gulf Region. Anantara Eastern Mangroves Hotel and Spa, Abu Dhabi, UAE	Professional Conference. Invited as Chairperson and Speaker	Public Works Authorities Water Engineers Water Companies Students and Researchers	Stakeholder engagement, knowledge sharing
30 and 31 October 2017	4th Annual Future Landscape and Public Realm Qatar Conference Landscape Urbanism for Climate Change and Food Security. The Doha Corniche. InterContinental Hotel Qatar	Professional Conference. Member of the Advisory Board, ChairPerson and speaker	Planning Authorities Public Space Specialists Landscape Architects Landscape Firms Students and Researchers	Stakeholder engagement, knowledge sharing
5–6 December 2017	3rd Annual Future Drainage and Stormwater Networks, Qatar Conference. Sustainable Drainage: Designing Landscapes with TSE and Stormwater. City Center Doha, Qatar	Professional Conference. Invited as Chairperson and speaker	Public Works Authorities Water Engineers Water Companies Students and Researchers	Stakeholder engagement, knowledge sharing
6 December 2017	Seminar on Food Security. Qatar Foundation Research Outcomes Designing for Food Security. Productive Landscapes in Doha. Qatar Science and Technology Park	Research Seminar. Invited Poster Presentation	Faculty Researchers Students Funding Agency QNRF	Stakeholder engagement with government, academia, and professionals
28 November 2017	The United Nations' Role to Prevent and Solve the Conflicts in the Agricultural Context. Urban Agriculture. Faculty of Organic Agricultural Sciences at the University of Kassel Germany	Invited Lecture and discussion Via Skype	Faculty Students International Researchers in Organic Agriculture	Publication in *Future of Food Journal*
5–6 March 2018	Agritech, Aquaculture, and Food Safety. UK and Gulf Countries Science Collaboration Symposium Invited Speaker. Urban Agriculture in Drylands. The Case of Qatar. Sultan Qaboos University, Muscat, Oman	Invited speaker. Organized and funded by the British Council Oman	International Researchers Working on the Gulf Region	Knowledge sharing with specialists in food and agriculture
20–22 March 2018	Agriteq Qatar Conference Integrating Food Production into Architecture and Urban Planning in Qatar and Drylands. DECC Doha Exhibition and Convention Center	Invited as Chairperson and speaker	Ministry of Agriculture Agriculture Companies Landscaping Companies International Experts	Networking with experts. Engaging stakeholders and partners

(*Continued*)

TABLE 10.1 *(Continued)*

Timeline of Projects, Initiatives, and Events on Urban Agriculture, 2012–2020

Date	Name	Type	Actors–Stakeholders–Partners	Outputs–Results
2018–2020	The Moveable Nexus: Design-Led Urban Food, Water and Energy Management Innovation in New Boundary Conditions of Change. Research Teams Qatar University, Keio University, Japan, University of Michigan, Delft University of Technology, Queen's University Belfast, and University of Technology Sydney. http://m-nex.net	International Research Grant. Belmont Forum SUGI. Funded by QNRF and International Research Agencies. QNRF 600'000 Euros	Qatar University International Universities Government Administrations NGOs Faculty, Researchers, Students University Capital Projects Agricultural Firms and Farms Permaculture Specialists Architecture firms	International Workshop and Conference in Qatar Qatar University Living Laboratory Workshop Sydney Workshop Detroit
2019	CWSP Grant from Qatar National Research Fund. Grant for Conference Workshop Sponsorship for Productive Urbanism for Food Security in Qatar and Drylands	Research Grant International. QNRF 25'000 USD	International Experts and Researchers on Food Urbanism Local Partners and Administrations in Qatar	
7–8 March 2018	Middle East Landscape Conference. Urban Agriculture in Drylands. Designing Landscapes for Productive Cities and Communities, Tehran, Iran	Keynote Speaker	Municipality Representatives International Landscape Architects and Researchers Students and Faculty	Knowledge sharing with specialists
24–28 February 2019	MNEX Doha. International Conference and Workshop Curator, Coordinator, and Lead Research Consultant. Qatar University, Doha, Qatar	International Conference and Workshop. Curator, Lead Researcher	Researchers, Government Administration, NGOs, Faculty, Students, University Administration, Agricultural Firms and Farms, Permaculturists Architects	Website report
27 March 2019	Kazan State University of Architecture and Engineering. Future Cities and the Food, Water, Energy Nexus. BFFT Space, Kazan, Tatarstan, Russia, 27th March 2019	Invited Lecture	Faculty, Students, and Researchers	Knowledge sharing with students and faculty
11–15 November 2019	UCLG World Summit of Local And Regional Leaders Durban 2019. MNEX Doha. Qatar University Living Lab. In Urban Living Labs: Experimenting for Sustainable Urban Development. ICC Congress Center, Durban, South Africa	Invited speaker for a panel and workshop. Hosted and funded by JPI Urban Europe	European and International Researchers Working on Living Laboratories Municipal Leaders and Mayors NGOs and Activists	Report UCLG website
28 April 2020	JPI Urban Europe Lunch Talks #12 The Role of Urban Design in the Food–Water–Energy Nexus. "Taking Action on the Urban Land-Use and Infrastructure Dilemma"	Invited Speaker. Webinar Presentation. EU Research	Researchers Working on European-Funded Projects	Website of webinar

FROM PERMACULTURE GARDEN AND EDIBLE CAMPUS TO THE UNIVERSITY LIVING LAB

The research, courses, and studio projects resulted in a number of urban agriculture projects at different scales. The graduate student projects covered a wide range of different urban sites and landscapes in Qatar, and some were applied to other territories such as Cyprus. With regard to the undergraduate students, the urban agriculture projects and research were focused on the QU campus or were integrated into the student's design projects. The permaculture garden designed in the College of Engineering was a collective undertaking implemented over several years through different courses and research grants, with the support of the College, the University Building Services, the Environmental Science Center, the Embassy of Switzerland, and several private partners – Al Sulaiteen Agricultural Company and Al Nakheel Landscapes (Grichting, 2017). This small-scale project was a practical application of permaculture (Mollison et al., 1981) for food production on the university campus (see Figure 10.1).

At the scale of campus, several proposals and designs to create productive landscapes at QU were produced by the students (both graduate and undergraduate) over the years. The continuity over the years allowed the students to build on the knowledge and designs of the previous students. The plans and designs served as a tool and a vision to launch an interdisciplinary and multi-stakeholder involvement in a strategy and plan to improve food production and biodiversity on the campus. The projects consider the analysis of the food cycle process, the optimal use of scarce resources such as water and energy, and the recycling of waste (organic waste, in particular). Case studies were conducted on international university campuses that were implementing urban agriculture, for example, McGill University's School of Architecture, to learn about and see what approaches and strategies they have used to create an edible campus (McGill, 2023).

The results were different versions of a master plan for the campus of QU that showed the land uses and functions, backed by various types of charts that showed the current state and the suggested future design, with ideas for the unused green fields and new buildings, as well as projects to redesign the existing landscapes and buildings, including productive green roofs. The final master plan includes all types of buildings, rooftops, productive and edible landscapes, and the central park with water features and biodiversity hotspots.

The culmination of the process was the MNEX international joint research project "The Moveable Nexus: Design-Led Urban Food, Energy and Water Management Innovation in New Boundary Conditions of Change" granted by Belmont Forum/JPI Urban Europe led by Keio University, Japan, under the international consortium with seven partners from six countries (MNEX, 2023). Bringing together a consortium of international researchers and universities – Keio University Japan, University of Michigan, Delft University of Technology, Queen's University Belfast, University of Technology in Sydney, and QU – the project resulted in the creation of the Living Laboratory at QU, engaging stakeholders from within the university and from the government, professional, and civil society sectors (Figure 10.2).

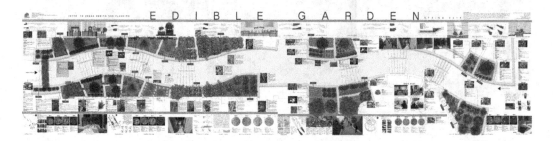

FIGURE 10.1 The Permaculture Food Garden at the Female College of Engineering, Qatar University.

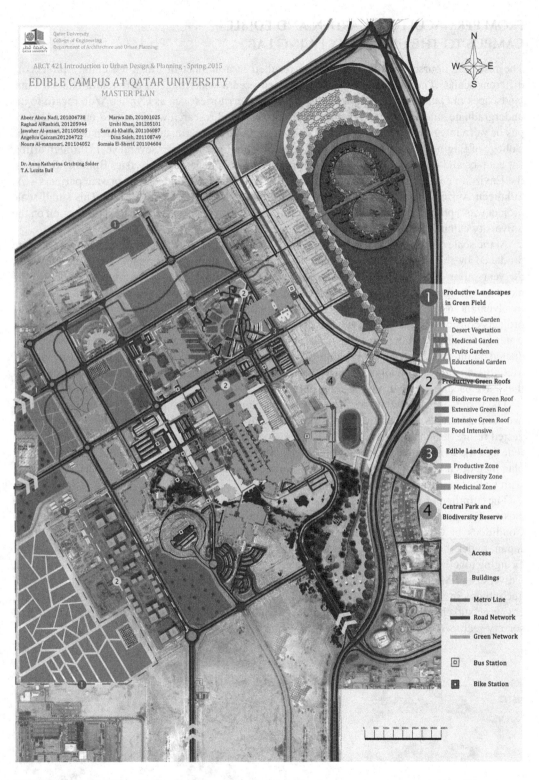

FIGURE 10.2   The Qatar University Master Plan for food urbanism.

"Living laboratories comprise spaces that are owned and managed by key stakeholders in the living laboratory, and, more often than not, universities are obvious partners with the resources to fulfil their institutional and scientific requirements". QU campus in Doha covers approximately 800 hectares and also has a University Farm for research located in rural areas. The Living Laboratory for the food–water–energy nexus builds on the previous campus projects presented in this chapter. It also seeks to integrate ongoing research and networks at QU related to food, water, and energy such as new food crops, halophytes, and microalgae; reuse of water (treated sewage effluent (TSE) and produced water from industry and air conditioning); and recycling of organic waste and soil production and renewable energies. Considered a microcosm of the city, the Living Lab engages all the university communities including researchers, faculty, administration staff, students, technical staff, and workers and brings in stakeholders from the public and private sectors (König, 2013). See Figure 10.3. The Facilities and Services Department and Capital Projects Department at QU assisted with data, maps, and campus information during the project. Arab Engineering Bureau, the oldest and largest Qatari architectural firm, is a partner to participate in the design of charrettes and the design of the master plans and food–energy–water (FEW) systems. Global Farm and Turba Farm were key partners in developing the two contrasting living lab scenarios on the QU campus, the permaculture, and technology-based farming. The aim of the research project was to combine the nature-based systems and the technological systems of food–water–energy nexus design to increase food production on the campus, while minimizing the use of water and energy and increasing biodiversity and quality of soil.

The Doha Living Lab at QU focused on a campus master plan as an "urban and living water machine" that produces food with a minimum consumption of energy and water. It addressed the quality of life, through an increase in the quality of food and the quality of air, through reduced CO_2 and carbon capture, as well as fostering bio-cultural diversity in the species and types of food that are grown and the communities that are involved in the production. The stakeholders and university communities are an important part in the elaboration and implementation of a living laboratory, based on the premise that it is not only technologies but also the active involvement of communities and change in behaviors and lifestyle that will build a more resilient society and sustainably affect the food–water–energy nexus, both quantitatively and qualitatively.

The Living Lab placed an important focus on the soil and how the desert soil can become fertile with efficient composting and recycling systems and effective watering and nutrients. With that, it looked at different forms of land use, including the use of roofs as an additional layer of land use, as well as complementary systems of food production using natural (permaculture and traditional) and technological food systems – that is, soil-based and water-based vertical farming. It proposed to implement this through a Food–Water–Energy Nexus Master Plan, which integrated new nexus-based architecture buildings and retrofitted existing structures. Through the modeling of a dynamic master plan, the different scenarios elaborated could be quantified and qualified as to how they best achieve the objectives of the research.

Dynamic Adaptive Master Plan and Urban Interventions

The Doha Living Lab combines technological FEW systems and interventions with natural FEW systems (permaculture) to create a resilient campus and foster the biodiversity of fauna and flora on the campus. The design charrettes held during the International Workshop in February 2019 and the local Design Workshop held in April 2019, as well as some previous projects on the edible campus, resulted in the following design scenarios that include:

- an Edible Permaculture Boulevard at the garden scale – College of Engineering
- an urban Edible Boulevard – technology based – connecting the metro station to the campus and leading to the student housing
- an urban Edible Boulevard – permaculture and soil based – connecting the food forest and biodiversity Wadi to the campus buildings

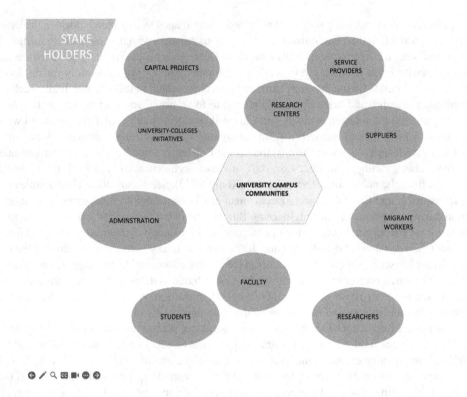

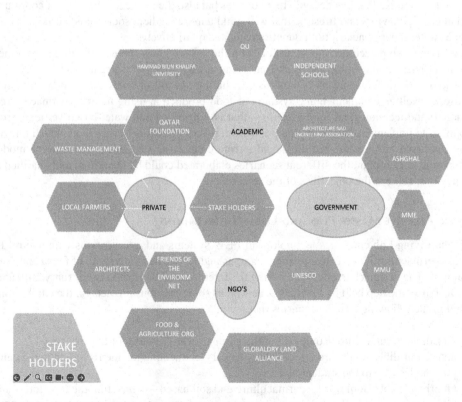

FIGURE 10.3   Stakeholders of the Qatar University Living Lab.

- a FEW Cube at the intersection of the two urban boulevards, which would house the new Agricultural College of QU and the FEW Research Center
- a water machine tower for the algae systems as a food–water–energy microorganism
- the restoration of the historic campus building to house new FEW-based activities
- the building of a biochar facility to produce good soil on the campus
- the creation of a permaculture food forest in the Lower Wadi area
- the transformations of surfaces into permeable surfaces and soil
- the transformation of decorative landscapes into edible landscapes
- the use of roofs for water–energy–food production
- walkability, electric trams, electric personal vehicles, shaded connections between buildings, and green roofs

THE QU LIVING LABORATORY: FROM RESEARCH TO DESIGN TO POLICY

The QU Living Lab followed the basic concept of the MNEX research project, based on mobilizing social and natural resources, moving stakeholders to action, and bridging local and global knowledge. An important outcome of the project was the policy recommendations that are necessary to transform the social and built environment and protect precious resources. Building on the research developed with master's students in the urban planning legislation course, and as part of the MNEX research project, we questioned how the MNEX design and urban interventions could be implemented at the policy level. We reflected on the three following themes.

MOBILIZING SOCIAL AND NATURAL RESOURCES TO CREATE MORE WITH LESS THROUGH DESIGN SOLUTIONS

Expanding the effectiveness of food production in the city with minimal water availability, exploring new food crops and protein sources including halophytes and microalgae, reuse of water (TSE and produced water from industry and air conditioning), recycling of organic waste and soil production, use of renewable energies, and decentralized and local on-site water treatment and energy production – these systems are proposed to be integrated into the buildings and landscapes of the campus, and they are expressed in the form of the Master Plan and Nexus Flow diagrams. At the policy level, these interventions require incentives or changes in legislation – for example, for the production of food using TSE or grey water. Bringing in stakeholders from the public sector, private sector, research, and academia on a practical project can jump-start or accelerate the discussion and communication regarding novel approaches to nexus planning. Visualizations of the master plan, flow diagrams, and data modeling will contribute to enhancing and forwarding the discussion toward policy changes in the use of different types of water, incentives for on-site treatment and waste processing, the value of local species and biodiversity, etc. The presence of professional farmers with the decision makers and engineers from public works can produce a policy that is sound and implementable.

MOVING STAKEHOLDERS TO ACTION THROUGH CROSS-SECTORAL DIALOGUE

The QU Living Lab engages the university communities and brings in stakeholders from the public and private sectors. Engagement tools – such as phone applications and stakeholder platforms – can contribute to informing communities and gathering data and model responses. Stakeholder action can result in the implementation of policy at the university level – relating to campus management and design and students' responsibilities and engagement – and also at the government level, especially with regard to the involvement of migrant workers in the creation of community gardens and the production of their own food. It can also result in incentives for the university communities to reduce their FEW prints through personal and collective action and behavior. While Qatar does not have a strong history of public participation, the traditional Majless space and system is a forum of discussion and exchange, which is confined to the Qatari nationals and separated between men and women. The proposal to use phone applications for stakeholder engagement is a means to overcome social and language barriers and, coupled with more traditional and physical meetings and workshops, can assist in creating an inclusive process that can both inform communities and gather knowledge and data from them. The community gardens on the campus – for children, Nepali workers, Filipino maids, students, faculty, and researchers – are also spaces of engagement and exchange, and the types of crops favored by the different communities, the biodiversity that is generated, the social fabric that is created – all these can be measured and transformed into scientific data. From this, policies can be developed with the communities that can contribute to creating productive landscapes and community gardens that improve the resilience of the FEW nexus in the city of Doha. A phone application is proposed as a tool that also serves to encourage communities and stakeholders to engage in policy discussions and to propose policy recommendations that could enhance their FEW print performances. The model of how QU engages its community in the FEW nexus design can be a model for the city and the country on how to engage the larger community.

MOVING AROUND LOCAL AND GLOBAL TO WHAT IS LOCALLY NEEDED WITH THE SUPPORT OF INTERNATIONALLY DEVELOPED GUIDING PRINCIPLES

Working with the international MNEX teams, the MNEX at QU aims to support Qatar's emerging knowledge economy by increasing knowledge, awareness, and participation in the food–water–energy nexus using art, science, and culture. This can take the form of public participation workshops, outreach events, artistic installations, and international events such as the MNEX International Workshops (MNEX Doha, 2019). As an example, a permaculture and composting workshop to build a composting pit at QU was held in 2016 as part of an international workshop conducted by an American soil specialist (see Figures 10.4 and 10.5). This workshop resulted in the implementation of a composting system by the landscape engineers and gardeners at QU, and international expertise was integrated into the campus (Grichting & Bullivant, 2016). The international scope of the project can generate and produce policies at a higher level, and the recommendations of an international team can assist in pushing forward policy recommendations at the national level.

In conclusion, we can summarize the following actions:

- Assessing the current policies in Qatar regarding food security, urban farming, and sustainable development
- Analyzing how the Qatar University Living Lab would affect or be affected by the different policies available in Qatar
- Analyzing the gaps within the different policy documents and how the QU Lab can bridge these gaps

The recommendations are summarized in Table 10.2.

CONCLUSIONS AND REFLECTIONS

This chapter shares the author's experience in bringing urban agriculture into the Architectural and Urban Design Curriculum of the university, while communicating and exchanging knowledge, visions, and projects with multiple local, regional, and international experts and stakeholders. From a fringe topic, not necessarily recognized as belonging to architecture or urbanism, the importance of food systems and urban agriculture became top priorities of government policies and research agendas following the blockade of Qatar in 2017. Students at QU have already been working on such scenarios since 2012, and student projects increasingly integrated food production into their designs, but the importance of this approach was truly recognized after the crisis.

FIGURE 10.4 Creating a composting system at Qatar University.

FIGURE 10.5 Planting the Edible Permaculture Garden.

Building incrementally on design and research projects at the university, while building a network of stakeholders, partners, and experts, the method demonstrates a multipronged approach for teaching, research, and knowledge building within an academic institution. With the support of research funding, as well as the financial and in-kind support of private actors – farms, agricultural companies, etc. – and partnerships with government entities and the Swiss Embassy, it was possible

TABLE 10.2

Summary of Policy Recommendations for the Qatar University Living Laboratory

Title	Problem	Approach
WATER. A Framework for diverse water sources and integrated water systems (The Urban Water Machine)	Due to the lack of freshwater in Qatar and the environmental costs of desalination, it is necessary to consider multiple sources of water to produce food, including treated sewage effluent (TSE), saltwater, produced water, and grey water. This will increase resilience and water security for Qatar.	Policies and incentives related to the following need to be developed. Downscaling infrastructure and creating decentralized water treatment plants integrated into buildings and neighborhoods. Addressing policy on TSE to allow the use in food production (not only animal feed). Addressing cultural barriers against eating food produced with TSE by introducing labeling systems (as in organic food labeling).
ENERGY. Incentives for renewable energy alternatives	Energy is abundant and very cheap, and there is no incentive to produce alternative and renewable energies. The environmental costs of energy production and consumption – CO_2 emissions and poor air quality – must be mitigated to create resilient energy systems and energy security for Qatar.	Policies and incentives related to the following need to be developed: Downscaling energy infrastructure and creating on-site energy production systems – solar, biogas, microalgae, etc. Reducing energy consumption in buildings and landscapes – including the production of food. Create policy incentives to implement alternative energy sources by taxing non-renewable fuels.
FOOD. Developing urban farming for better resource use and for productive ecosystems.	The recent embargo on Qatar has demonstrated the fragility of relying only on external food sources. Until recently, local Qatari farms were not very productive. A lot of land in the city is vacant, and there is great potential for the use of rooftops.	Policies and incentives related to the following need to be developed. Land use planning and master planning that integrates urban farming in Qatar needs to be developed. The fifth façade, or rooftop, should be explored as a space for growing food locally, reducing food miles, and reusing grey water. Land use should include spaces for recycling waste, including food waste to create compost for soil-based agriculture. New farming technologies – aquaponics, hydroponics – associated with nature-based and soil-based food-growing systems (permaculture approach) should be combined to create diverse, symbiotic, and resilient food-growing systems. New types of crops, superfoods, or traditional crops and species that are high in nutrients and protein should also be cultivated. Adapt animal fodder to high-quality and local species that increase the dairy yield and quality of the milk, thereby reducing energy and water to produce the same amount of milk.
HEALTH. Health issues due to poor diet for rich and poor. Lack of knowledge on the interrelation between healthy soil, healthy food, and healthy gut	Health issues due to poor diet for both wealthy communities and migrant workers. There are obesity and diabetes, due to the high content of fats, free sugars, and salt, and illnesses due to overconsumption of animal proteins.	Policies and incentives related to the following need to be developed: Create policies and practices that include diet and awareness of nutrition at all levels of education and in all communities. Provide access to nutritious and fresh foods to all residents through community gardens and greenhouses.

(Continued)

TABLE 10.2 (*Continued*)

Summary of Policy Recommendations for the Qatar University Living Laboratory

Title	Problem	Approach
	Alternatively, there are populations such as migrant construction workers, who suffer from high blood pressure, blood sugar issues, and dehydration.	Improve soil quality to increase food quality and gut health. Use public art and actions to communicate the soil–food–gut health.
COMMUNITIES. Participative engagement of stakeholders to responsible action and sustainable solution	Loss of biodiversity through urbanization, salination of groundwater and seawater, desertification, etc. Social and cultural segregation and high-income inequalities in a future knowledge economy. Social issues include the imbalance of natives to foreigners: 313,000 Qataris reside with 2.3 million others.	Policies and incentives related to the following need to be developed: Stakeholder participation as a process for designing and implementing resilient food water and energy systems for biological and cultural diversity. Bio-cultural diversity refers to the continuing co-evolution and adaptation between biological and cultural diversities and reflects people's ways of living with nature, generating local ecological knowledge and practices across generations that allow societies across the world to manage their resources sustainably while also maintaining cultural identity and social structures (Ramsar, 2023). Build on the knowledge of migrant communities on food growing – Nepali farmers, Filipinos, etc. and at the same time increase their knowledge on food–water–energy through public engagement and community gardens.

to create and deliver a number of projects, events, and publications related to food urbanism and the food–water–energy nexus. The experiential learning and design-build aspect of the project (Edible Boulevard, Composting pit, etc.), as well as collaborations with permaculture experts, international landscape architects and green roof specialists, and scientists from the Environmental Science Centre at QU, gave students tools and knowledge in both science-based approaches and practical implantation experience and design methods. It also introduced them to working with different disciplines and experts, finding a common language, and translating knowledge and findings into design. One obvious collaboration for students and researchers working on urban agriculture would have been with the Agriculture Faculty or College. There is no such educational facility in Qatar, and in 2020 QU decided to open an Agricultural Research Station located on the QU Farm, with the idea of establishing a specialized center for agricultural research. One of our partners, Al Sulaiteen, established an agricultural research and training center in 2011 (SAIC, 2023)

The Gulf region also does not have any Landscape Architecture programs, as the subject of urban agriculture is often integrated into the landscape curriculum. Leading educational institutions today are working with the emerging discipline of landscape urbanism, in which the topics of productive landscapes, urban agriculture, systems design, and food–water–energy nexus thinking can be introduced. This cross-disciplinary and integrative way of thinking was at the foundation of this work and process.

The key reasons for conducting this research and putting a prototype into practice at QU were the lack of productive urban land, society awareness of food production and preparation, difficulties with food insecurity, and unchecked urban growth. Universities play a clear role in spreading information and developing highly qualified employees to meet perceived economic needs as they are crucial institutions in processes of social change and development. To encourage students to grow their own food on campus, universities can address the problem of food and water security and integrate it into the curriculum in different fields.

The university campus offers unmatched opportunities to teach, do research, demonstrate, and learn about all facets of sustainability and environmental challenges, and it has been referred to as a microcosm of the greater community (König, 2013). As the ones who will significantly alter their communities, the students, who will play this role, are encouraged to support new social and cultural values.

REFERENCES

Bricas, Nicolas & Conaré, Damien. (2019). "Historical perspectives on the ties between cities and food" *Field Actions Science Reports*, Special Issue 20 | 2019. https://journals.openedition.org/factsreports/5594

Food Urbanism Blog. (2023). www.foodurbanism.org. Accessed 10th June 2023.

Grichting, Anna & Bullivant, Lucy. (2016). *Sustainable Urbanism: New Directions*. Proceedings of the International Workshop. Qatar Foundation 21st March 2016. Published by Qatar University. Published as a Webzine with Urbanista. https://www.urbanista.org/issues/sustainable-urbanism

Grichting, Anna. (2017). A productive permaculture campus in the desert: visions for Qatar University. *Future of Food: Journal on Food, Agriculture and Society*, 5(1), 21–33.

Grichting, Anna. (2023) "Educational Frameworks for designing Regenerative Food Systems in the Arabian Gulf. The Case of Qatar." In H. Harriss, A. M. Salama & A. G. Lara (Eds.), *The Routledge Companion to Architectural Pedagogies of the Global South*. Taylor & Francis.

Grichting, Anna; Alamawi, Rana & Asadi Rama. 2016. "Designing Productive Landscapes in an Emerging Desert Metropolis: Food Systems and Urban Interfaces in Doha." In G. Katodrytis & S. Syed (Eds.), Gulf *Cities as Interfaces*. Gulf Research Center.

König, Ariane. (2013). *Regenerative Sustainable Development of Universities and Cities. The Role of Living Laboratories*. Edited volume. Edward Elgar Publishing Ltd, Cheltenham, UK.

Lawler, J., Mazzoni, A., Shannak, S. (2023). *The Domestic Water Sector in Qatar*. In: *Sustainable Qatar. Gulf Studies* Cochrane, L., Al-Hababi, R. (eds), vol 9. Springer, Singapore. https://doi.org/10.1007/978-981-19-7398-7_11

McGill University. (2023). *Making the Edible Campus*. McGill Sustainability. Accessed 10th Aug. 2023. https://www.mcgill.ca/mchg/files/mchg/MakingtheEdibleCampus.pdf.

MNEX. (2019) Doha Design Workshop: *Building a Food Infrastructure - The Qatar Experience*. https://m-nex.net/m-nex-doha-design-workshop-february-2019building-a-food-infrastructure-the-qatar-experience/

MNEX. (2023) *Moveable Nexus: Design-led Urban Food, Energy and Water Management Innovation in New Boundary Conditions of Change*. https://m-nex.net/about/. Accessed 10 Aug. 2023.

Mollison, B., Holmgren, D., Barnhart, E. (1981) *Permaculture 1: A Perennial Agriculture for human civilizations*. International Tree Crop Institute, USA

Ramsar. (2023). https://www.ramsar.org/activities/bio-cultural-diversity. Accessed 1th Aug. 2023.

Reuters. (2023) https://www.reuters.com/business/energy/qatarenergy-signs-27-year-lng-deal-with-chinas-sinopec-2022-11-21/. Accessed 10th Aug. 2023.

RUAF. (2023) *Global Partnership on Sustainable Urban Agriculture and Food Systems*. www.ruaf.org. Accessed 8th Aug. 2023.

SAIC: Al Sulaiteen. (2023). https://www.alsulaiteengroup.com/company/al-sulaiteen-agricultural-research-study-and-training-center.html. Accessed 10th Aug. 2023. Viljoen, André; Bohn, Katrin & Howe, Joe. (2005). *Continuous Productive Urban Landscapes: Designing Urban Agriculture for Sustainable Cities*. Elsevier, 2005.

Water Action Hub. (2023). https://wateractionhub.org/geos/country/179/d/qatar/. Accessed 8th Aug. 2023.

Wellesley, Laura. (2023). *How Qatar's Food System Has Adapted to the Blockade*. 14th November 2019. Chatham House. https://www.chathamhouse.org/2019/11/how-qatars-food-system-has-adapted-blockade. Accessed 9th Aug. 2023.

11 Cultivating Change in Cities
Exploring and Classifying the Determinants of Urban Agriculture in India

Maitreyi Koduganti and Sheetal Patil

INTRODUCTION

In an era of unprecedented urban expansion, marked by the rapid migration of populations to urban centers, the sustainable coexistence of nature and concrete has emerged as a formidable challenge. With estimates of the addition of nearly 273 million people by 2050, India will not only be the world's most populous country (World Population Prospects, 2019) but also be almost half urbanized. The burgeoning urban centers are rapidly expanding, giving rise to various socio-economic challenges and opportunities. On the one hand, India's urbanization is nothing short of a modern marvel, as millions are drawn from rural areas to partake in the allure of urban life. However, on the other hand, this migration and urban sprawl have sparked a growing demand for resources, housing, and sustenance, leading to complex socio-economic dynamics that shape the very landscape of urban agriculture (UA) practices. It is within this labyrinth of challenges that UA has emerged as a promising solution, offering a unique bridge between urbanization, sustainability, and well-being (Maxwell et al., 1998; Ackerman et al., 2014; Padgham et al., 2015; Mntambo 2017; Soga et al., 2017; Azunre et al., 2019).

The essence of UA in India transcends the mere cultivation of crops within city limits. It signifies the fusion of traditional agricultural knowledge with modern practices, adapting to local contexts while embracing innovation (Rao et al., 2022). Moreover, it is driven by the socio-economic forces that underscore India's urban dynamics, making it a subject of profound interest and importance.

CURRENT STATE OF EXISTING LITERATURE: IDENTIFYING GAPS

In this chapter, we aim to explore the shared elements of urbanization and agricultural practices, which revolve around the interconnected dynamics of human society and the natural environment. As urbanization draws more individuals into cities, a noticeable gap emerges between human and environmental systems. Nevertheless, urbanization exerts a considerable influence on agricultural infrastructure, farm production costs, and overall farm income, potentially offering more opportunities than challenges in agriculture (Wu et al., 2011). Notably, agriculture is not confined to remote rural areas but also extends into urban settings, offering a plethora of prospects and solutions for sustainability and resilience (Azunre et al., 2019; Knaus & Haase, 2020; Pungas, 2020)

Existing literature on UA predominantly emphasizes its advantages, challenges, and outcomes (Malberg et al., 2020; Lwasa et al., 2014; Padgham et al., 2015). Only a few studies venture beyond mere descriptive analyses of these outcomes, delving into their distinct effects across various dimensions, including social, economic, cultural, and spatial aspects. It is important to note that the variability in outcomes may stem from differences in practices within these intersecting areas. This represents a notable void in the current body of research, which we intend to address in this chapter.

DOI: 10.1201/9781003359425-11

Our inquiry begins with the understanding that socio-economic factors that play a pivotal role in shaping the landscape of UA practices. At the core of this interplay are the individuals and communities involved in UA, each propelled by a distinct set of influences that amalgamate elements of economics, culture, and environment. These determinants collectively serve as the foundation for the varied spectrum of urban agricultural practices observed throughout India, encompassing rooftop gardening, community farming, vertical agriculture, and more. Consequently, the primary objective of this chapter is to delve into and thoroughly explore the intricate network of socio-economic factors that shape UA practices.

The second section of this chapter elaborates on the methodology, tools, and analytical approaches employed. In the third section, we delve into the outcomes regarding the determinants of UA, categorizing them into three main groups: demographic, contextual, and inherent. Finally, we provide a succinct discussion of the findings and outline the path forward.

METHODOLOGY

RESEARCH APPROACH

A substantial portion of the data utilized in this chapter stems from broader research on the 'Urban and Peri-urban Agriculture as Green Infrastructure (UPAGrI)' study (of which the authors were a core part, which was designed to investigate the impact of UA on sustainability and well-being (Patil et al., submitted). Focusing on Bengaluru and Pune as crucial research sites, the present study attempts to provide an in-depth understanding of how socio-economic factors like space, time, gardening resources, economic resources, knowledge, skills, and individual choices determine the influence of UA practices within these two Indian cities. To unpack this, we used a mixed method approach, which encompassed a comprehensive online survey that was administered across 29 Indian cities from 2021 to 2022, as well as conducting online in-depth interviews with 50 key informants in Bengaluru and Pune (Singh et al., 2021). A detailed literature and policy review was also conducted to establish a foundational understanding of UA practices, outcomes, and the underlying factors governing them.

COLLECTING DATA

Scoping visits: To gain insights into various UA practices and establish deeper connections with local stakeholders, we conducted scoping visits with NGOs, government officials, community-based organizations, fellow researchers, and policymakers.

Online survey: Following our initial scoping visits, we conducted an extensive online survey using the Open Data Kit (ODK) platform. This survey, divided into seven distinct sections, delved into various aspects of UA – farming practices, the extent of UA activities, resource utilization, training participation, and experiences related to UA during the COVID-19 pandemic. The survey, targeting households, garnered 441 responses from both formal and informal[1] settlements spanning 29 Indian cities (Patil et al., submitted).

Key-informant interviews: Within the broader context of the ODK survey, we selected 50 participants from diverse demographic backgrounds, encompassing various age groups, income levels, genders, and socio-cultural backgrounds, drawn from two distinct Indian cities. This selective sampling aimed to investigate the multifaceted factors influencing UA outcomes. Moreover, as an additional dimension to our inquiry, participants willingly shared visual materials such as photographs and videos showcasing their gardens and agricultural produce. This voluntary sharing of visual content gave us a deeper insight into their emotions, perceptions, and the intangible outcomes of their UA endeavors.

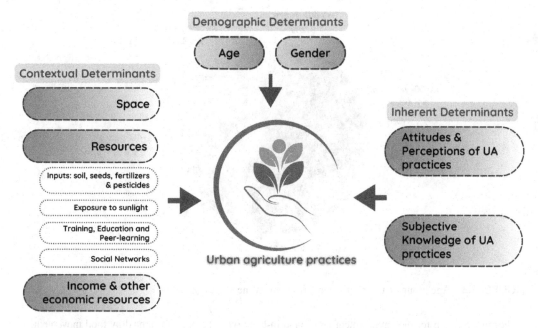

FIGURE 11.1 Depicting socio-economic determinants of UA practices. (Source: Authors' depiction (also see Alemu et al., 2020).)

ANALYZING THE DATA:

During our study, we took precautions to safeguard the privacy and confidentiality of the participants by anonymizing[2] both qualitative and quantitative data. The quantitative data from the online survey was visually presented using frequency tables, graphs, and crosstabs. These graphical representations were employed to analyze the influence of various factors such as age, gender, income, occupation, and available space on the outcomes of UA. Conversely, we adopted a thematic analysis approach (Braun & Clarke, 2012), to systematically encode the qualitative data.

OBSERVATIONS

We found that several determinants shaped UA practices, and we divide them broadly into (a) demographic determinants (age, gender), (b) contextual determinants (space, resources, income, and other economic resources), and (c) inherent determinants (attitudes and perceptions of UA, knowledge of UA) (see Figure 11.1). In the following section, we delve into a comprehensive exploration of each of these determinants.

DEMOGRAPHIC DETERMINANTS

AGE

Our analysis noted that a significant proportion of those involved in UA practices fell within the age range of 41 to 60 years (as depicted in Figure 11.2). Furthermore, the primary incentive driving these participants to participate in UA was predominantly recreational. Approximately, 97% of respondents residing in informal settlements and 65% of respondents in formal settlements indicated that their engagement in UA was motivated by leisure and as a personal hobby.

We also observed that younger individuals involved in UA frequently approach this practice with an inclination toward innovation and a keen aspiration for sustainability. A 28-year-old female entrepreneur and founder of iKheti, Pune, added,

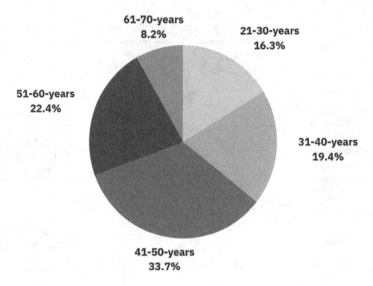

FIGURE 11.2   Age group of UA respondents from an online survey (*n*=361).

> For me, concern for the environment goes hand-in-hand with the need for a healthy food movement.
> Hence, I wanted to create a platform for both individuals & communities to grow healthy consumable
> crops within their premises & promote sustainable urban farming.
>
> *(P_SI_17)*

We also noticed that UA practitioners from diverse age groups collaborate by combining technology
and innovative approaches while concurrently embracing traditional and organic farming methods,
which the older age group more commonly favors. For instance, the (male)founder of an aquaponics
unit in Bengaluru mentioned,

> We have members across all age groups working on different aspects of aquaponics. Younger engi-
> neers and agricultural specialists work on technology that improves the efficiency of the system, and
> middle-aged and older folks look at how we can grow native and traditional herbs and plants using these
> techniques. We rely on collective knowledge.
>
> *(B_SI_17)*

Different age groups seemed to bring unique motivations, physical capacities, access and approaches
to technology use, and resource allocation strategies to their urban farming endeavors.

GENDER

Based on the findings of our online survey, it was apparent that women constituted the majority,
taking on significant roles in ensuring food and nutritional security and contributing to the health
and well-being of households (see Figure 11.3). Notably, a substantial portion of these women fell
within the age range of 41 to 60.

We noted distinctions in the perspectives of men and women regarding UA practices, and these
disparities influenced their respective roles, responsibilities, and the outcomes of these practices.
More than two-thirds of women involved in UA did so out of passion or as a leisure activity, while
over half of the men cultivated their food, primarily for health-related benefits. Similarly, there
appeared to be variations in the types of plants cultivated by men and women. Women primarily
focused on growing ornamental plants, leafy greens, herbs, and medicinal species, while men were

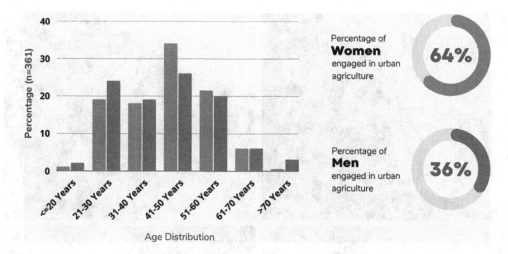

FIGURE 11.3   Distribution of men and women practicing UA across different age groups in both settlements. (Source: Authors' depiction from online survey.)

more inclined to cultivate a broader array of fruits and vegetables. A 40-year-old female urban gardener in Bengaluru affirmed these gendered roles (traditional view of women as caregivers and men as providers) and stated,

> Women have more affinity towards nature and aesthetics, which includes beautifying their homes and terraces. Apart from growing fresh and safe food, we want to teach our children how to grow their food and nurture them. However, men see gardens as more functional spaces, more like a project, and focus on how much they can grow and reap from their gardens.

(B_SI_6)

Most participants in formal and informal settlements indicated that women were primarily responsible for the daily upkeep of gardens. Meanwhile, tasks involving heavy lifting and the transportation of bulky garden materials and tools were typically managed by men. Although women take the lead in UA practices, it is noteworthy that men also make substantial contributions to cultivation efforts.

CONTEXTUAL DETERMINANTS

SPACE

We observed that the extent to which UA can be carried out was influenced by available space, including backyards, rooftops, terraces, vacant plots, footpaths, or community gardens (see figure 11.4). The size and type of available space also impacted the choice of plants grown. For example, respondents residing in independent houses often had access to larger areas for cultivation, allowing them to grow more extensive beds of crops such as cereals and fruit trees, in addition to vegetables and herbs. A female home gardener from Bengaluru noted,

> During COVID-19 lockdowns, we expanded our gardens – we planted more tomatoes, turnips, drumsticks, potatoes, onions, cluster beans, strawberries, turmeric, and even mangoes on our house's first and second floor. If not for this space, we would not have been able to yield so much fresh and nutritious produce.

(B_FI_3)

FIGURE 11.4   Different types of UA spaces in formal and informal settlements.

Conversely, individuals living in large housing societies and complexes within formal settlements often have limited space, typically confined to small balconies. As noted by a mid-forties male UA practitioner in Pune, *'in a typical 8 × 10ft balcony, one can only keep 10–20 pots, where they can grow few ornamentals, herbs, and creepers'* (P_SI_20). Moreover, many of these societies impose restrictions on residents, preventing them from cultivating larger plants or gardening in communal areas. Hence, these individuals collaborated with their neighbors and fellow gardeners to maximize their cultivation within the available space.

It's important to note that the significance of space varies among respondents. In formal settlements, the availability and access to *sufficient* space play a crucial role in determining the extent of UA practices and their outcomes. In contrast, for those in informal settlements, the mere ability to *access* space, whether through ownership, leasing, or community arrangements, emerged as a critical factor in their ability to engage in UA activities. A 56-year-old male construction worker in Bengaluru notes, *'This land is not ours, so how can we grow here?'* Furthermore, local regulations and zoning laws seemed to influence where and how UA can be practiced. Indicating this challenge, a (male)wage laborer from Bengaluru noted, *'(this) land is empty, so I am growing here. If the municipality decided to build a house here, there is a risk that they would come and clear my gardens. I am not allowed to farm here officially'* (B_LIS_12). Evidently, many individuals in informal settlements lack ownership of the land beneath their homes, limiting their ability to use it for cultivation. A 43-year-old female daily wage laborer in Bengaluru notes, *'I wish to grow so many more plants, but where is the space?*

I know how to grow potatoes and all such vegetables, but I don't have enough space' (B_LIS_6). This uncertainty of land access discouraged them from participating in UA activities despite their motivation and knowledge of growing.

RESOURCE AVAILABILITY

Availability and access to inputs (soil, seeds, fertilizers, and pesticides), training and education, and social networks emerged as a second determinant that seemed to have a profound impact on the extent of UA activities and outcomes.

a. **Inputs:**

Soil: At the outset, it's crucial to emphasize that access to nourishing and fertile soil is a fundamental requirement for promoting robust plant growth. In formal settlements, more than 25% of respondents across both settlements pointed out that the soil quality in urban regions exhibits substantial variations, particularly regarding nutrient composition, pH levels, and the presence of contaminants. Due to the limited availability of better soil in urban areas, several respondents ventured into alternative gardening methods. A female home gardener from Pune in her early forties mentioned, *'We ventured into soil-less farming technique- hydroponics. Even a small area like a windowsill is enough for growing a few greens and vegetables' (P_SI_9).* Similarly, a female resident (in her mid-fifties) from informal settlements in Bengaluru notes, *'wherever new construction or demolition happens, we get the soil from there. The subsurface soil (soil below a few centimeters of the topsoil) is very fertile. It doesn't need too many inputs and is moist too' (B_LIS_14).* Though access and availability of good quality soil emerged as a significant barrier in both formal and informal settlements, it allowed respondents to venture into alternative methods like hydroponic, soil-less farming, and composting that were prominent within formal settlements.

Seeds: More than a third of participants noted that the availability of high-quality seeds impacts the selection of crops, their suitability for urban environments, cost factors, and the overall diversity of UA practices. Many respondents encountered challenges in sourcing readily available seeds in urban areas and procuring smaller quantities of seeds proved problematic. A male vertical farming specialist from Pune notes, *'Mostly seeds come in bulk, and not many at household level find that useful. Growing ready-made saplings can overcome that' (P_SI_20).* Several enterprises like UrbanMali, iKheti, and My Dream Garden provide ready-made saplings and smaller seed quantities for setting up smaller rooftop and backyard gardens. A home gardener in her mid-sixties from formal settlements in Pune mentioned, *'We exchange seeds from friends, sometimes I collect my seeds after planting. One can create your resources, like seed banks' (P_FI_1).* In contrast, most respondents in informal settlements obtain seeds from their ancestral lands. Additionally, initiatives like Hariyalee Seeds rejuvenate traditional and heirloom seed varieties, facilitating their conservation and exchange within UA communities.

Fertilizers & Pesticides: Approximately 40% of respondents emphasized the critical role of accessible fertilizers and pesticides, encompassing organic choices like compost and manure, as well as synthetic alternatives, in maintaining soil fertility and plant well-being. In urban environments where high-quality soil is often scarce, the demand for consistent fertilizers and pesticides to improve soil health becomes more pronounced. While synthetic fertilizers are readily available in the market, many urban households prefer organic inputs, particularly in formal settlements. Notably, over a third of formal settlement households prepared homemade natural pesticides such as neem oil and *Jeevamrutha*, employing them on their plants. An urban farmer from Pune in her early thirties indicated, '*Sometimes if there is a fungus on our plants, we use Mitrkeeda – a white powder that needs to be thrown around the garden. They are microbes for pest control that are non-chemical. We don't want to spray chemicals*' (P_FI_5).

We noted that more than 70% of formal settlement residents and around 20% of informal settlement residents practiced composting, using their kitchen waste as a soil amendment. Home garden waste composting is a longstanding tradition in Indian households (Pandey et al., 2018). Echoing the practice of most households in informal settlements, a resident in her early forties from Pune noted, '*I dump the leftover cooked food, kitchen waste, and used flowers at the topmost layer of the soil of these plants. I have been doing this all my life*' (P_LIS_2).

b. **Exposure to sunlight:** We gathered that sunlight levels determine the plant selection in UA spaces. Seasonal variations in sunlight patterns impact UA's layout, planting schedules, and harvest times. An urban farmer in her late thirties from Pune mentioned,

My apartment does not receive much sunlight. I have tried growing chilies and tomatoes (sometimes they are successful, and sometimes they are not). Right now, I cultivate more herbs, lemongrass, curry leaves, and other plants that require less sunlight.

(P_SI_7)

In informal settlements with closely spaced homes, limited sunlight exposure hindered plant growth. Respondents in these areas adopted various tactics to maximize and regulate sunlight (see Figure 11.5). A 46-year-old female tailor in Pune's informal settlement explained:

We do not have any space on the ground to grow plants. Hence, we used a plank as a base for growing plants on our roof. This way, the space does not get blocked for other people living in the wadi to walk. Second, the plants get enough sunlight as they are located on an elevation.

(P_LIS_1)

c. **Training, education, and peer learning:** Education and training regarding plant selection, seed choice, soil management, pest control, and farm design significantly influence UA practices. Several independent UA practitioners, including B_FI_9, P_FI_4, and B_FI_7, provided in-person and online training, primarily in formal settlements. They offered guidance on establishing home gardens, initiating plant cultivation, and maintaining them. During the COVID-19 lockdown, at least two-thirds of respondents from formal settlements increasingly turned to online platforms, such as Facebook and Zoom, for conducting these educational sessions (see Figure 11.6).

Enterprises like UrbanMali and My Dream Garden and NGOs like INORA and Garden City Farmers organized workshops that united UA practitioners from various age groups. These workshops served as platforms for knowledge exchange and mutual learning as practitioners shared their insights and experiences. A gardening group in Pune expressed their involvement, stating,

We organize talks on gardening, birds, biodiversity, and benefits of UA, where we invite people, especially schoolchildren, and take them around our garden and give them information about the trees, and climbers. This will motivate the younger generation to farm and invest in nature.

(P_SI_15)

However, more than 35% of all respondents stated that they had not participated in any formal training or workshops related to UA. Instead, at least 40% of respondents across

FIGURE 11.5 Plants grown along the roof and sides of houses in informal settlements.

formal and informal settlements depended on peer learning through friends and neighbors (see Figure 11.6). A male informal settlement construction worker in his late forties in Pune pointed out,

If we face any challenges in growing, we ask our neighbors or relatives from our native cities. Farming is a part of our life; we learn from our experiences.

We observed that respondents within informal settlements barely attended online training or workshops, partly because of limited mobile phone data and knowledge about these trainings. In sum, we saw that peer learning, training, and education played a significant role in providing the knowledge, skills, and awareness necessary for urban farmers to navigate the unique challenges and opportunities of urban environments.

d. **Social networks:**[3] Our observations indicated that social networks considerably determine UA practices. As hubs for knowledge exchange, they facilitated the sharing of gardening insights, problem-solving, and the exchange of resources and harvests. Emphasizing the advantages of resource sharing, a formal settlement home gardener in Bengaluru stated, *'We are a group of gardeners – we save and share seeds among ourselves and try to keep*

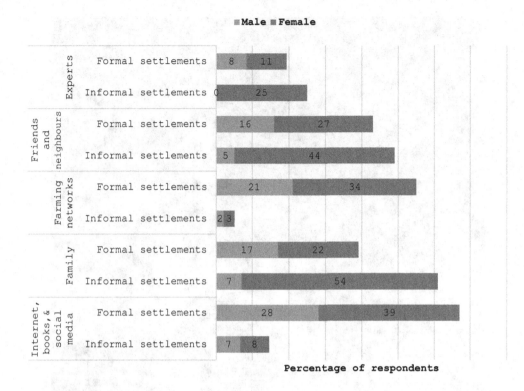

FIGURE 11.6 Source of information for UA practices in formal (*n*=300) and informal settlements (*n*=61). (Source: Author's depiction from online survey.)

them alive in our garden. Now, we rarely buy seeds' (B_SI_3). As previously mentioned, numerous respondents in formal settlements turned to online platforms like Zoom and Facebook to expand their network of like-minded UA practitioners. Additionally, they formed smaller WhatsApp groups based on their residential areas.

However, social networks took on a distinctive form within informal settlements. Respondents in these areas often participated in networks primarily shaped by their shared living conditions, common challenges, and the need for mutual support. We noticed that these networks typically displayed a less formal structure than those in formal settlements, leaning toward a more community-driven model. Residents relied on their neighbors for UA assistance and troubleshooting information (see Figure 11.6). A 50-year-old female construction laborer offered insight, stating when we go to our village, my neighbor *'waters and takes care of plants. I do the same when she is away' (B_LIS_5).* Many residents could not participate in online networks or WhatsApp groups due to their limited familiarity with online platforms and restricted mobile data. However, their connections with relatives and farmers in their places of origin played a crucial role in providing support and information and fostering community cohesion.

INCOME AND OTHER ECONOMIC RESOURCES

As an economic determinant, income significantly influenced various aspects of UA practices, such as their types, scale, inputs, training, and resource availability. Our online survey revealed that more than 30% of respondents earned up to 0.5 million INR annually, with fewer than 6% earning over 3 million INR annually.

In our examination of UA practices in formal and informal settlements, we observed that respondents with higher incomes could invest in advanced farming techniques, including soil-less farming, hydroponics, and vertical farming. They could also afford higher-quality seeds, fertilizers, and other inputs. Constrained by lower incomes and smaller plot sizes, respondents in informal settlements favored low-cost and often traditional gardening methods, such as container gardening, rooftop gardening, or community gardening. Highlighting the financial constraints that limited their input purchases, a middle-aged man from an informal settlement in Bengaluru emphasized, '*Wherever I go to work, I collect (these) pots and bring them here. Many people throw these slightly damaged pots outside, and I bring them and plant my seeds in them*' *(B_LIS_9)*. Furthermore, respondents with a higher income were able to expand and diversify their gardens, especially in the case of a home gardener in Bengaluru, who added, '*During the COVID-19 lockdown, I started gardening even on the second floor of my terrace, where I grew onions, herbs, beans, turmeric, beetroot, carrots, and other fresh vegetables*' *(B_FI_3)*. This was not the case for many informal settlement residents. A male grocery store owner within an informal settlement in Pune reflected,

> Everyone living in this area leaves for work early and returns late evening. Hence, no one has the time to grow plants or vegetables. More than that, space is a huge constraint. Hence, the thought of expanding or even having an urban garden does not cross people's minds.

(P_LIS_1)

We noted limited subsidies specifically aimed at supporting UA activities. However, certain municipalities, notably the Pune Municipal Corporation (PMC), initiated economic incentives to encourage adopting sustainable practices. PMC, for instance, introduced two tax rebate schemes in 2008 to promote rainwater harvesting and organic waste management for household-level sustainability. These initiatives not only incentivize such practices but also serve as models illustrating their manifold benefits. Such recognition has the potential to raise awareness and stimulate increased citizen engagement in urban greening. Nevertheless, it's important to note that most of these benefits are primarily for individual homeowners and apartment complexes. Extending these incentives and providing access to more space and resources would likely further motivate their participation in UA practices.

INHERENT DETERMINANTS

In the context of UA, inherent determinants refer to the individual characteristics, perceptions, and conditions that influence individual choices and preferences. In the following sections, we specifically identified attitudes, perceptions, and subjective knowledge of UA practices as inherent determinants of UA practices.

ATTITUDES AND PERCEPTIONS OF UA PRACTICES

We noted that individuals' personal attitudes and perceptions played a pivotal role in shaping their UA practices. Specifically, these factors profoundly impacted their motivations, the choices they made regarding methods and resources, and the extent to which they were actively involved in UA. These attitudes and perceptions served as either catalysts or impediments to adopting UA practices. For instance, individuals with favorable attitudes toward cultivation, such as those who valued sustainability, were environmentally conscious and held a strong passion for nature preservation, were more inclined to embrace UA. As expressed by a 37-year-old female IT professional from Pune, '*My garden is my getaway from the city's chaos. The greenery gives me peace and a pause to think*' *(P_SI_13)* (also see Koduganti & Singh, 2022). Our survey found that more than 70%

of respondents residing in formal urban settlements and approximately 40% of those in informal settlements believed that UA practices contributed to an enhancement in nutritional intake, as illustrated in Figure 11.7. Moreover, nearly all the respondents believed that engaging in UA activities positively influenced their mental well-being, fostering a sense of tranquility and happiness. For instance, a 30-year-old homeowner who practiced gardening emphasized how her garden served as a source of rejuvenation for her and noted, '*Whenever I am exhausted and suspect getting a migraine, I go to my farm and weed-out or water my plants and I calm down. Garden helps us keep healthy mentally, not just physically*' (B_SI_3).

On the contrary, several respondents harbored doubts regarding UA practices. In this regard, we observed that merely 40% of respondents in formal urban settlements and less than 20% of those in informal settlements believed that UA practices led to cost savings on vegetables. This skepticism was evident in the comments of a male home gardener from Bengaluru, who remarked, '*I find urban farming unsustainable because I can harvest only 2 chilies or one tomato over two months. I still must go to the market to get vegetables. Hence, I see no benefit in continuing or investing in the practice*' (B_SI_16). Most survey participants believed that UA was economically unsustainable or imposed financial strain, leading to deterrence in their participation. Factors contributing to this perception included the substantial initial expenses and the perceived low returns, particularly regarding crop yield and income generation, which discouraged their active involvement (see Figure 11.6). Additionally, it is worth noting that various residential complexes and housing societies in Pune and Bengaluru prohibited residents from farming or composting activities in communal areas. These restrictions were typically justified by concerns related to unpleasant odors and pest infestations.

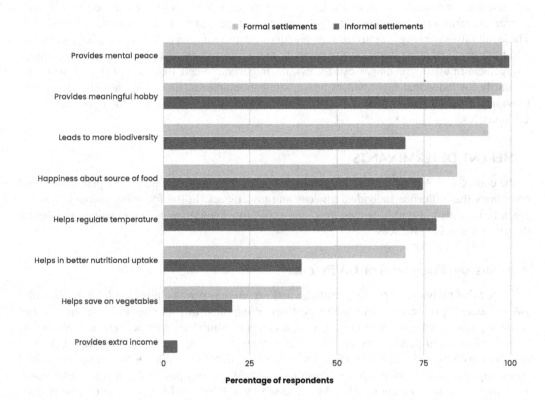

FIGURE 11.7 Respondents' perception regarding the benefits of UA practices. Formal settlements (*n*=300) and informal settlements (*n*=61). (Source: Authors' depiction from an online survey.)

Therefore, people's attitudes and perceptions about UA profoundly impact whether they engage in these practices and the extent to which they embrace them. A positive attitude can lead to tremendous enthusiasm and commitment, while negative or indifferent attitudes can act as barriers to adoption by UA.

SUBJECTIVE KNOWLEDGE OF UA PRACTICES

In the context of UA, subjective knowledge, defined as how a person thinks, knows, or learns (Han, 2019), plays a pivotal role in shaping and guiding agricultural practices. Influenced by experiential thinking, narrative processing, and making sense of local wisdom significantly influenced how individuals and communities engage in UA, ultimately impacting the choices, methods, and overall success of their UA pursuits. Gardening, whether it involves cultivating a few pots of basil or ornamental plants, is a deeply ingrained tradition within most households. This knowledge, which encompasses the art of cultivation, was typically transmitted through generations and is deeply entwined with local cultural practices.

It is worth noting that the extent of subjective knowledge regarding farming techniques differed between formal and informal settlements. Such a difference can be partly attributed to the fact that a significant proportion of residents in informal settlements come from a farming background, endowing them with a distinctive advantage over others in terms of agricultural expertise. For instance, a middle-aged female UA practitioner from an informal settlement in Bengaluru indicated experiences,

> I am from a village in Tamil Nadu. I grew up there and farmed all my life. So, I felt the need to grow plants and gardens here in the city, for example, you see (this) mango tree. I just planted the seed, and it was so small before ... and look at it now, it's so big! And it gives fruits too.

> *(B_LIS_6)*

However, within formal settlements, a home gardener and a teacher in Pune reflected, '*There is very limited knowledge about how to grow plants. A few households try to grow banyan in small pots, while some don't compost because they feel such sites can become breeding grounds for mosquitoes.*' In such cases, training, education, workshops, and awareness programs supplemented subjective knowledge regarding the best times for planting, soil management, seeds to be used, pest management, and so on amid residents in formal settlements. Such programs enhance subjective knowledge while fostering community collaboration and the exchange of ideas and best practices among urban farmers.

In sum, we observed that subjective knowledge about UA practices blends traditional wisdom, local context, and cultural factors to ensure the sustainability and success of urban farming initiatives, while also helping communities adapt to changing environmental and economic conditions.

DISCUSSION AND WAY FORWARD

While UA practices exhibit a wide range of diversity, the individuals engaging in these practices can be categorized based on their social, cultural, economic, and spatial characteristics, significantly influencing the nature of their UA activities. Demographic determinants, such as age, have introduced technology into UA, while gender has played a defining role in the choice of plants and division of labor. Our findings echo those of various other studies like Olivier (2019), which underscore that gender is primarily referenced concerning the demographic profiles of respondents. For example, in studies conducted in various regions such as Cape Town (Olivier & Heinecken, 2017), Italy (Mancini et al., 2021), North America (Grebitus et al., 2017), and Durban (Khumalo &

Sibanda, 2019), women have been found to play significant roles in UA, with younger women often displaying a more positive attitude toward urban farming. Although gender's influence on UA is evident, the exact gender relations regarding the division of labor remain unclear due to the prevailing stereotype associating women with nature and nurturing. By often perceiving UA as a hobby rather than unpaid work or an extension of domestic duties, it has been labeled as a predominantly feminine activity.

Regarding contextual determinants, the availability of space and resources determined the scale of UA, influencing the choice between small and large plants and trees. The tenure of the available space also defined the type of UA, with rented or leased spaces and informal settlements leading to practices involving pots and recycled materials. These contextual determinants also impacted practitioners' capacity to incorporate allied activities such as composting, water recycling, and waste material usage. These contextual determinants align with environmental sustainability objectives, positioning UA as a crucial nature-based solution. Training, education, peer learning, and social networks serve as determinants that spread awareness about 'best UA practices' and environmental consciousness and sustainability.

Inherent determinants provide a more personal and profound insight into the diverse UA practices. Attitudes shape specific goals and, consequently, perceptions of the benefits of UA. Subjective knowledge empowers practitioners to adapt to local contexts and blend traditional wisdom with modern practices.

Demographic, contextual, and inherent determinants result in varying outcomes for UA practitioners, often leading to unequal adaptation and unjust scaling. To address these challenges, we emphasize the importance of proactive urban planning and engagement with relevant institutional and policy frameworks. Leveraging urban and peri-urban agriculture can address the interconnected food–energy–water–health–livelihoods nexus. For example, by using UA to create *green jobs* in cities, including UA in the curriculum of primary and secondary schools to foster environmental awareness, and deliberating urban plans and building designs that facilitate and incentivize sustainable practices like waste recycling and food cultivation (Roy & Rao, 2021; IIHS, 2023).

ACKNOWLEDGMENTS

We wish to express our gratitude for the financial assistance provided by the British Academy's 'Urban Infrastructures of Well-Being' program, funded through the UK Government's Global Challenges Research Fund, for our research project titled 'Urban and Peri-urban Agriculture as Green Infrastructure (UPAGrI): Implication on Well-being and Sustainability in the Global South.' We sincerely thank all our respondents for their precious time and valuable input. We thank Chandni Singh for her guidance and feedback in structuring the chapter. Additionally, we extend our gratitude to the colleagues who contributed to the larger UPAGrI project – Parama Roy, Prathijna Poonacha, Nitya Rao, Amruth Kiran, Swarnika Sharma, Ashwin Mahalingam, Gayatri Naik, Pragati Khabiya, Kruthika Nagananda, Bala Panchanathan, Ombeni Swai, and Betty Mntambo – for their valuable intellectual support and camaraderie during various phases of this research.

NOTES

1 In this study, the term "formal settlements" pertains to legally sanctioned residential areas known for their well-established property rights and provision of fundamental infrastructure and services. Conversely, "informal settlements" encompass regions lacking clear legal recognition, characterized by substandard housing conditions and restricted access to essential services and infrastructure. It is crucial to recognize that the demarcation between formal and informal settlements may not always be straightforward in real-world scenarios, as urban areas often exhibit a spectrum of formality and informality (Patil et al., submitted).

2 We have provided evidence through a coding system that includes designations for study locations (B for Bengaluru, P for Pune). Following these location codes are identifiers for various respondent categories (SV for Scoping visit, SI for Stakeholder interview excluding farmers, FI for Farmer interview with individual farmers), and finally, a numerical serial code for each specific respondent within their respective category. For instance, the fifth stakeholder interview conducted in Bengaluru would be denoted as B_SI_5.

3 In the realm of urban agriculture, social networks encompass the relationships and connections among individuals, organizations, and communities engaged in or supporting urban agriculture initiatives (Diehl, 2020).

REFERENCES

Ackerman, K., Conard, M., Culligan, P., Plunz, R., Sutto, M. P., & Whittinghill, L. (2014). Sustainable food systems for future cities: The potential of urban agriculture. *The Economic and Social Review*, *45*(2), 189–206.

Alemu, M. H., & Grebitus, C. (2020). Towards sustainable urban food systems: Analyzing contextual and intra-psychic drivers of growing food in small-scale urban agriculture. *PloS One*, *15*(12), e0243949. https://doi.org/10.1371/journal.pone.0243949

Azunre, G. A., Amponsah, O., Peprah, C., Takyi, S. A., & Braimah, I. (2019). A review of the role of urban agriculture in the sustainable city discourse. *Cities*, *93*, 104–119. https://doi.org/10.1016/j.cities.2019.04.006

Braun, V., & Clarke, V. (2012). Thematic analysis. American Psychological Association.

Diehl, J. A. (2020). Growing for Sydney: Exploring the urban food system through farmers' social networks. *Sustainability*, *12*(8), 3346. https://doi.org/10.3390/su12083346

Grebitus, C., Printezis, I., & Printezis, A. (2017). Relationship between consumer behavior and success of urban agriculture. *Ecological Economics*, *136*, 189–200. https://doi.org/10.1016/j.ecolecon.2017.02.010

Han, T. I. (2019). Objective knowledge, subjective knowledge, and prior experience of organic cotton apparel. *Fashion and Textiles*, *6*(1), 4. https://doi.org/10.1186/s40691-018-0168-7

Khumalo, N. Z., & Sibanda, M. (2019). Does urban and peri-urban agriculture contribute to household food security? An assessment of the food security status of households in Tongaat, eThekwini Municipality. *Sustainability*, *11*(4), 1082. https://doi.org/10.3390/su11041082

IIHS. (2023). *An Action Agenda for Urban Agriculture in India*. Indian Institute for Human Settlements. Retrieved September 19, 2023, from https://iihs.co.in/knowledge-gateway/wp-content/uploads/2023/05/An-Action-Agenda-for-Urban-Agriculture-in-India.pdf

Knaus, M., & Haase, D. (2020). Green roof effects on daytime heat in a prefabricated residential neighbourhood in Berlin, Germany. *Urban Forestry & Urban Greening*, *53*, 126738. https://doi.org/10.1016/j.ufug.2020.126738

Koduganti, M., & Singh, C. (2022, October 20). Green Shoots of Health, Happiness, and Wellbeing. Growing Greener Cities. Retrieved September 6, 2023, from https://exhibition.upagri.net/exhibition-rooms/healthier-happier-citizens/green-shoots-of-health-happiness-and-wellbeing/

Lwasa, S., Mugagga, F., Wahab, B., Simon, D., Connors, J., & Griffith, C. (2014). Urban and peri-urban agriculture and forestry: Transcending poverty alleviation to climate change mitigation and adaptation. *Urban climate*, *7*, 92–106. https://doi.org/10.1016/j.uclim.2013.10.007

Malberg Dyg, P., Christensen, S., & Peterson, C. J. (2020). Community gardens and wellbeing amongst vulnerable populations: A thematic review. *Health Promotion International*, *35*(4), 790–803. https://doi.org/10.1093/heapro/daz067

Mancini, M. C., Arfini, F., Antonioli, F., & Guareschi, M. (2021). Alternative agri-food systems under a market agencements approach: The case of multifunctional farming activity in a peri-urban area. *Environments*, *8*(7), 61. https://doi.org/10.3390/environments8070061

Maxwell, D., Larbi, W. O., Lamptey, G. M., Zakariah, S., & Armar-Klemesu, M. (1998). Farming in the shadow of the city: Changes in land rights and livelihoods in peri-urban Accra. *Cities Feeding People series;* rept. 23.

Mntambo, B. (2017). *Intra-household gender relations and urban agriculture: The case of vegetable cultivation in Morogoro Municipality, Tanzania* (Doctoral dissertation, University of East Anglia).

Olivier, D. W. (2019). Urban agriculture promotes sustainable livelihoods in Cape Town. *Development Southern Africa*, *36*(1), 17–32. https://doi.org/10.1080/0376835X.2018.1456907

Olivier, D. W., & Heinecken, L. (2017). The personal and social benefits of urban agriculture experienced by cultivators on the Cape Flats. *Development Southern Africa*, *34*(2), 168–181. https://doi.org/10.1080/03 76835X.2016.1259988

Padgham, J., Jabbour, J., & Dietrich, K. (2015). Managing change and building resilience: A multi-stressor analysis of urban and peri-urban agriculture in Africa and Asia. *Urban Climate*, *12*, 183–204. https://doi. org/10.1016/j.uclim.2015.04.003

Pandey, R. U., Surjan, A., & Kapshe, M. (2018). Exploring linkages between sustainable consumption and prevailing green practices in reuse and recycling of household waste: Case of Bhopal city in India. *Journal of Cleaner Production*, *173*, 49–59. https://doi.org/10.1016/j.jclepro.2017.03.227

Patil, S., Singh, C., Koduganti, M., Poonacha, P., & Sharma, S. (submitted). Urban Agriculture in India: Mapping its landscape and sustainability and well-being outcomes. *Local Environment*.

Pungas, L. (2020). Food self-provisioning as an answer to the metabolic rift: The case of 'Dacha Resilience'in Estonia. *Journal of Rural Studies*, *68*, 75–86. https://doi.org/10.1016/j.jrurstud.2019.02.010

Rao, N., Patil, S., Singh, C., Roy, P., Pryor, C., Poonacha, P., & Genes, M. (2022). Cultivating sustainable and healthy cities: A systematic literature review of the outcomes of urban and peri-urban agriculture. *Sustainable Cities and Society*, *85*, 104063. https://doi.org/10.1016/j.scs.2022.104063

Roy, P., & Rao, N. (2021). *The Role of Urban Agriculture in India's Green Building Policy*. Urban and Peri-urban Agriculture as Green Infrastructure (UPAGrI). Retrieved January 8, 2023, from https://8a6ac 3e7-e78e-4937-baf4-bbb9f028d464.filesusr.com/ugd/d6ecfb_903c17f542cf423d888597823909fd2c.pdf

Singh, C., Patil, S., Poonacha, P., Koduganti, M., & Sharma, S. (2021). When the "field" moves online: reflections on virtual data collection during COVID-19. *Ecology, Economy and Society-the INSEE Journal*, *4*(2), 149–157. https://doi.org/10.37773/ees.v4i2.414

Soga, M., Cox, D. T., Yamaura, Y., Gaston, K. J., Kurisu, K., & Hanaki, K. (2017). Health benefits of urban allotment gardening: Improved physical and psychological well-being and social integration. *International Journal of Environmental Research and Public Health*, *14*(1), 71. https://doi.org/10.3390/ijerph14010071

World Population Prospects. (2019). *United Nations*. https://www.un.org/development/desa/pd/news/world-population-prospects-2019-0

Wu, J., Fisher, M., & Pascual, U. (2011). Urbanization and the viability of local agricultural economies. *Land Economics*, *87*(1), 109–125. https://doi.org/10.3368/le.87.1.109

12 Impact of Urban Farming on the Built Environment

Mohd Khalid Hassan and Sadaf Faridi

INTRODUCTION

The built environment of the cities is under tremendous pressure from the growing population and increasing density. Industrialization is one of the major concerns leading to the advancement and development of the urban process; it has also caused contradictions between human settlements and natural resources. To sustainably accommodate infrastructural developments, it has become imperative to integrate natural ecology with heavily constructed and developed cities. Urban farming is one such emerging field that has the potential to reduce the environmental burden of the urban built environment along with providing a solution for the food crisis and climate change.

Urban farming is supplemental food production in cities practiced in different parts of the world during wars to overcome food shortages and as a community developmental program. "Allotment gardens" in Germany, "Pingree Potato Patches" in Detroit, "Victory gardens" in the United States, United Kingdom, Canada, Australia, and Germany during World War I and World War II, "Community gardens" in the United Kingdom, "P-Patch" in Seattle, and "The Severn Project" in Bristol are all examples of how urban agriculture (UA) has been used time and again, to deal with food crises and for building communities.

Since the Food and Agriculture Organization (FAO) Committee on Agriculture (15th Session) advocated the creation of an interdepartmental initiative on UA in 1999, the FAO has been working on UA. Later, under the banner "Food for the Cities," this effort was included in FAO's Priority Areas for Interdisciplinary Action. However, it is increasingly understood that urban farming has benefits beyond those listed above. It is currently becoming a tool for improving the built environment of cities.

Building-based farming is rapidly being recognized as an innovative technique for expanding agricultural practices in urban environments, as ground-based farming has become a significant limitation due to conflicting demands for space.

TYPOLOGY OF GROWING SPACES FOR URBAN FARMING

Several types of spaces in a city can be utilized as growing spaces for UA. These could be broadly classified into the following categories (Skara, 2019):

1. Soil-bound spaces: This typology encompasses urban derelict land, family gardens, allotment gardens, communal gardens, squatter gardens, parks, and other open areas, as well as guerrilla gardening. Gardening on one's own is one example of this trend (private level and community level).
2. Movable and soil-independent systems: This typology includes mobile containers and growing boxes. An example of this typology includes the plant and fish farming box invented in Belgium.
3. Building-bound space: This typology includes rooftops, mainly open rooftops, flat roofs, covered rooftops, and roofs with inclination. Facades are also included, primarily open

DOI: 10.1201/9781003359425-12

and covered facades. The confined area includes building additions like balconies and windowsills.

4. Water-bound spaces: This typology includes urban streams, stagnant waters, and amphibia systems. Some examples include ponds, lakes, and floating islands.

These different growing spaces contribute to affecting the built environment in different ways.

IMPACT OF URBAN FARMING ON THE BUILT ENVIRONMENT

Urban farming significantly impacts improving the environmental quality of urban areas and changing the stagnant and static condition of the city into a healthy, vibrant, and productive lifestyle. Urban farming affects the microclimate in the following ways.

MODERATION OF INTERNAL BUILDING TEMPERATURE

Plants that grow on building facades offer protection from the sun by obstructing sunlight with their leaves; due to this shade, less sunlight hits the facades, which results in less radiation impacting the building and lowering the inside temperature. On the exteriors of buildings, dense vegetation offers extra wind protection. Plants minimize the impacts of wind, improving wintertime insulation. It is considered that every 0.5°C drop in interior temperature results in an 8% reduction in power used for air conditioning. Local air temperatures will be cooled in two ways, i.e., green walls (GWs) and roofs. At first, the sun's heat is less readily absorbed by walls behind the greenery. Second, green roofs and walls will evaporate, or chill, heated air by evaporating water. Bituminous materials, masonry, and concrete absorb the bulk of the sun's energy, which is then reradiated as perceptible warmth. About 15% to 50% of the radiation by concrete, asphalt, or masonry is subjected to and reflected by them (Villanova, 2013). The heating of these surfaces can be decreased, especially in densely populated metropolitan areas, by greening pavement surfaces with plants to filter radiation before it hits hard surfaces.

Even at night, the temperature in an urban warming island environment is more significant owing to the surface's ability to absorb heat and radiation after dusk (Villanova, 2013). In 1982, Krusche et al. investigated how solar energy is converted on a GW building facade. A GW facade gets 100% solar energy but utilizes only 20–40% for evapotranspiration, 5–20% for plant photosynthesis, 5–30% for leaf reflection, and 10–50% for heat transfer.

It has been demonstrated that plants on a GW may evapo-transpire significantly and lower the head transmission of hard surfaces. UV rays shorten the life of materials, so reducing the amount of sunshine the building's components are exposed to can increase their durability. Buildings benefit from this, and doing so saves money on maintenance expenses.

The benefits increase as the planted surfaces get denser and thicker. Insulation material reduces the impact of the temperature differential between inside and outside. In cold weather, the insulating layer lowers the heat transmission rate to the outside.

In summer, the opposite condition occurs; the rate of heat movement from the exterior to the interior is slowed down. Many approaches may be used to increase the insulation level of vertically greened surfaces by surrounding the structure with vegetation. Wind reduces a structure's energy efficiency by 50%; a vegetation layer acts as a cushion to stop the wind from flowing along a building surface, and through vegetation, thermal endurance can be reduced from 23 to 12 W/m^2 K (Villanova, 2013).

Apart from vertical greened surfaces, rooftop farms are a growing trend in UA. The effectiveness of rooftop gardens has been studied in light of the regional climate and environmental conditions through research conducted in Saudi Arabia. The size of the effects is determined in the said experiment using a one-sample t-test. The experiment employed model structures of materials comparable to those used to construct actual buildings, both with and without rooftop gardens. For some time, the inside temperatures of each building were monitored, and the findings were compared. The outcomes of the study indicate that there is a temperature differential between the two structures, particularly at the peaks. Installing rooftop gardens on a wide scale is anticipated to lower energy usage and, eventually, energy costs.

Research conducted in Saudi Arabia also showed that the mean temperature difference is considerably different from zero, which strongly shows that the difference was not the result of chance but was most likely caused by the rooftop garden's ability to lower the interior temperatures of the buildings. The difference in interior temperature was seen to grow during peak hours. This demonstrates that there is an inverse relationship between the inside temperature of a building with a rooftop garden and the exterior temperature (Gashgari Raneem, 2018).

Reducing Urban Heat Island Effect

An urban heat island (UHI) is a metropolitan region with a substantially higher temperature than the nearby rural area, especially during the winter. This phenomenon is known as the "UHI effect" in the cities to avoid any misunderstanding about global warming.

UHI effect: Cities' daytime surface temperature can be increased up to 10° higher than nearby suburban and rural locations (Gashgari Raneem, 2018). Cross-regional disparities in energy gains and losses cause this phenomenon. There are various factors responsible for cities becoming UHIs, such as the following:

1. During the day, solar radiation is absorbed by solid surfaces like buildings and pavement. Still, in rural areas, soil and vegetation quickly lose the energy that has been absorbed near the ground. Evaporative cooling is the term used for this.
2. Increased runoff caused by the abundance of impervious surfaces in cities further reduces evaporative cooling. Latent heat is used simultaneously with the evaporation of water, which converts it from a liquid to a gas and cools the surroundings.
3. The heat produced by buildings and automobiles may add to a city's solar heat.
4. In metropolitan settings, hard surfaces such as asphalt roadways, black roofs on buildings, and other structures better transfer heat than vegetation, leading to greater air temperatures.
5. Solar energy is trapped in downtown regions due to building reflections and infrared heat absorption brought on by the canyon-like surface created by the tall structures. At night, this heat is reflected over metropolitan areas where it is trapped throughout the day, forming a bubble of air significantly warmer than the surroundings.

Hot gas emissions from cities increase daily, as there isn't enough flora to absorb them.

The reduced level of evaporation is another reason temperatures in cities are often higher than those in the surrounding areas.

The UHI effect is becoming increasingly problematic in overpopulated cities due to asphalt and concrete of roads and buildings functioning as enormous heaters as well as industries, autos, and air conditioners emitting hot gases, which increases the air temperature of the urban areas. Living walls and green facades create unique microclimates distinct from the surrounding climate and help tremendously in evaporative cooling.

The GWs have an impact on both the outside and interior climates. This is because green facades absorb less heat than non-greened facades, which helps to reduce night-time heat radiation. As a result, a green facade helps to reduce the effect of UHIs.

It is a fact that, due to water evaporation, water surfaces like lakes allow temperatures to rise more slowly. This trait is also present in vegetation and woodlands that contain plants and water.

Further, rainwater that runs off onto hard surfaces such as glass and concrete directly goes to the sewage system and does not contribute to evaporation. Although transpiration and evaporation processes can contribute to the amount of water in the air, plants also buffer water on their leaf surfaces for a more extended period. Reducing concrete and asphalt surfaces and increasing plants may help make the metropolitan environment more sustainable.

Improving Air Quality

Most cities have bad air quality because of the numerous industrial and automobile emissions. Alarming amounts of carbon monoxide, nitrogen oxides, sulfur oxides, particulates, ozone, and volatile organic compounds (VOCs) are frequently detected. Many cities also lead to the growth of dust domes that contain a variety of hazardous airborne pollutants. In North America, nitrogen dioxide concentrations in metropolitan areas are frequently 10–100 times greater than in non-urban regions, primarily due to automobile exhaust. Compared to rural areas, condensation nuclei, particles, and gaseous admixtures are detected at ten times greater levels. Approximately 2% of yearly deaths in the United States are attributed to air pollution, which is a significant source of numerous health problems for city dwellers, particularly kids (Sam, 2011).

The UHI effect also causes an increase in air pollution and its concentration. A significant percentage of this rise is attributed to energy-related emissions resulting from the high temperature. In addition, warmer air immediately raises ground-level ozone concentrations. When VOCs and nitrogen oxides combine in heat and sunshine, surface ozone is created.

Ozone exposure can increase bronchitis, emphysema, and asthma symptoms and induce a range of respiratory conditions, including throat and lung irritation, inflammation, wheezing, and breathing difficulties.

In addition to providing cooling benefits, green spaces, including rooftop gardens, can help reduce air pollution levels through their effects on atmospheric deposition and microclimate. Vegetation filters the air during dry deposition by removing airborne debris. These green spaces' leaf surfaces serve as natural drains for pollutants. Ozone levels might be decreased due to this drop in temperature because of the slowing of photochemical processes. Urban CO_2 is reduced in part through plant photosynthesis. Since rooftop gardens frequently include taller plants with more giant leaves than typical green roofs, they may provide significant ozone and air pollution mitigation benefits beyond regular green roofs.

UA must be compared to other urban green spaces to gauge the scope of its air pollution reduction advantages. Various studies have demonstrated that trees are particularly good at reducing urban air pollution. Urban trees are responsible for the annual removal of 711,000 metric tons of air pollution (Wang, 2021.). However, growing enough trees in heavily populated metropolitan areas is not always feasible to reduce air pollution significantly. It is considered that typical green roofs may also remove contaminants. Acidic gaseous compound and particulate matter levels decreased over a 4,000 m² green rooftop built in Singapore by 37% and 6%, respectively (Wang, 2021). Similarly, Toronto research estimated that 7.87 metric tons of airborne pollutants might be removed by every 109 hectares of rooftop gardens. While researching to evaluate Chicago's intense green roofs' capacity to remove pollutants, Yang et al. discovered that installing intensive green roofs over Chicago's remaining green roofs might eliminate up to 2,046.89 metric tons of air pollutants. It is found that rooftop gardens can eliminate air pollutants like those found in non-agricultural rooftop gardens.

Numerous cities are undertaking greening programs to improve local air quality and reduce the UHI effect due to increased awareness of the advantages of green space on air quality. To deal with the problems of excessive urban heat and poor air quality, UA in rooftops and community gardens/urban farms offers an alternative type of green space that may be included in urban planning. Additionally, since community gardens are somewhat less expensive than more expensive greening initiatives and rooftop gardens, using UA to enhance air quality may be a better option.

As discussed earlier, plants may filter airborne particles to lower their concentration. Small dust particles stick to the leaves and plant stem surfaces, lowering the dust in the air (particulate matter 2.5, particulate matter 10). Maintaining clean air is greatly helped by plants' capacity to capture airborne dust, which is tidied up and reintroduced to the soil after rainstorms. Even after rain, the smallest particles stay on the leaves, but, after the leaves fall, a natural mechanism will cause this dust to return to the soil.

In their studies conducted in 1994, Bussotti et al. demonstrated that plant barriers directly beside a roadside (daily traffic level 20,000–50,000 cars) are more helpful in capturing lead (Pb) and cadmium (Cd) particles than plants tested in the rural region. Additionally, dust and other airborne particles are deposited on surfaces constructed of concrete, asphalt, stone, brick, and glass and distribute the dust particles throughout the atmosphere when these materials are overheated in the summer. GWs will help reduce the total number of airborne particles contributing to environmental contamination and significantly lessen the overall volume of particles that cause this phenomenon. As natural air filters, vertical gardens assist in removing harmful contaminants from the air while releasing oxygen. Toxins that are hazardous to human health, such as formaldehyde, benzene, and VOCs, are eliminated or reduced by plants in vertical gardens.

It is not a novel concept to think that plants can influence the local air quality. It is a well-known remedy for air pollution levels that has encouraged many metropolitan regions to increase their number of trees and green areas. Adding grass, climbing ivy, and other green infrastructure to urban areas may significantly improve the air quality at the street level by reducing particulate matter and nitrogen oxide levels by up to 60% and 40%, respectively (Pugh, 2016). Vertical gardens may maximize this benefit indoors and outside since they fit several plants into a small vertical space. GWs are installed indoors in many office buildings and other high-traffic areas to enhance indoor air quality. One can apply the same idea on a smaller scale by including a vertical garden in a house or workplace. Moreover, cities are leveraging the outer walls of some of their more prominent buildings to construct outdoor garden spaces to counteract increased pollution levels. One outstanding illustration is the Easyhome Huan gang Vertical Forest City Complex in Huanggang, China. Many of the structures in this complex, including homes, hotels, and sizable commercial spaces, were constructed using vertical forests to bring the advantages of nature into a densely populated metropolitan area.

Many plants in these gardens function to remove airborne impurities and lower carbon dioxide levels. They contribute this way to a larger initiative in many cities and metropolitan regions to lessen their total carbon footprint.

Reduced Noise Pollution Levels

Plants naturally muffle traffic noise when placed along the walls of more heavily used roads and highways. They may significantly lower the noise levels in neighboring buildings by absorbing and reflecting the noise emitted by these busy roads, which makes up a sizable portion of the noise produced in a metropolitan environment.

The absorption of external noise and the isolation of internal spaces from outside noise are two independent aspects of the acoustic performance of green living systems. Street width, building height, and the acoustic properties of the materials used in the building envelope affect how sound penetrates through the urban fabric in street canyons and other urban settings from loud areas into quiet zones. Building envelopes are frequently composed of hard materials, so there is a significant potential to reduce acoustic waves diffracting across them. This potential can be increased by employing vegetation. There are three ways in which plants might lower sound levels (Dimitrijević, 2017):

Plant components can diffract (scatter) and reflect sound. Leaves, branches, twigs, and trunks all have various effects.

Plant components are capable of absorbing sound. Sound waves produce mechanical vibrations in plant components that dissipate energy by turning the sound energy into heat. Attenuation is further aided by the influence of thermo-viscous boundary layers at vegetation surfaces.

The destructive interference of sound waves can be used to lower sound levels. The direct input seen from the source to the receiver and a contribution from ground reflection may conflict negatively when dirt is present. An acoustically soft soil is a result of vegetation. The acoustical ground effect (soft soil), sometimes known as the ground dip, is a common name for this phenomenon.

According to Dunnet and Kingsbury's study on urban noise reduction by vegetation, urban areas' hard surfaces frequently reflect sound rather than absorb it (Dunnet & Kingsbury, 2008). On the other side, green roofs may reduce noise due to the substrate and the plants that grow there. Plants block higher frequencies, whereas the ground tends to block lower frequencies.

In analyzing the effectiveness of green living walls in reducing and controlling noise pollution, two standard laboratory experiments were carried out in a study , where the modular GW's ability to transmit sound directly was the sole factor considered. The main results indicated that the GW performed effectively at low and high frequencies. The R coefficient, which measures the GW's ability to suppress airborne noise, was lower than that of the other practical options. To demonstrate the possible efficacy of a GW in lowering noise pollution in its surroundings, an acoustic measuring experiment was conducted at a site in Cergy, in the Val d'Oise region, close to Paris, France. As a result of installing the GW on the site, measurements revealed a reduction in the total sound pressure levels (dBA) produced by traffic. Depending on the measurement day, acoustic gains remained low and ranged from 0.6 to 2.5 dBA (Ecotiere & Gauvreau, 2016). The intriguing research discussed installing a vertical indoor greenery system on a wall's north side and evaluated the noise reduction brought on by using a wall face covered with vegetation. The average reduction in dBAs ranged from 2% to 3% depending on the frequency weightings used, except for extreme frequencies, which correspond to the sound frequencies heard by humans. There was a sonic mitigation of 6–8% for the chosen frequency using those two weightings, respectively, for noise that lasted longer than 1 s, which can be considered the most problematic for human sensitivity (Ecotiere, & Gauvreau, 2016).

LOWER ENERGY COSTS

Anyone who resides in a region with a warmer temperature may probably experience higher energy costs during the sweltering summer months due to the high energy demand for air conditioners. The same may be stated for how heat affects energy costs for people who live in cooler climates. One might be surprised to learn that adding vertical plants to the internal or external walls can assist in lowering these energy expenditures.

There are two ways through which adding a garden to the living area can be advantageous. First, the plants and the soil and air around them can serve as insulation. Second, through a process known as evapotranspiration, the plants in the vertical gardens will strive to lower the temperatures both on the surface of the walls and in the surrounding air.

A study conducted in Sweden found that integration of the vertical farm with the building energy system can reduce greenhouse gas (GHG) emissions of the vertical farm by up to 40%. The synergistic development of a vertical farm with its host building and surrounding buildings for more holistic urban farming systems was found to have significant potential benefits for the vertical farm. Additionally, GHG emissions decreased by around 20% due to the usage of brewery leftovers like carbon dioxide and discarded grains. Also, several trade-offs were found in these scenarios where more outstanding infrastructure might accelerate material and water resource depletion. The urban vertical farms' energy and material efficiency may be increased through synergies with neighboring buildings and those with the host building. The study also sheds light on the advantages of using unused space for urban food supply and reducing energy demand for residential building owners.

ENHANCING AESTHETICS

People in urban regions have mostly lost their connection to nature. Thus, they value the periodic presence of greenery in parks, trees, and other scattered areas. GWs have gained notoriety worldwide and, in some cases, have even become tourist attractions. The visual benefit of GWs is well known, and research has shown that being among plants makes people feel happy, which is excellent for their physical health. Moreover, these emotions increase productivity in work settings.

Furthermore, it might be helpful to conceal some barriers. There is a case in Madrid where a living wall system was used to cover a portion of the wall close to a museum; this improved the view for visitors, and it even rose to fame as a major tourist destination.

However, GWs might potentially reduce the amount of graffiti and ugly aesthetics on walls. In this manner, GWs protect structures against vandalism. Urban gardening can enhance people's perceptions of urban environments as being appealing. However, not all strategies are seen favorably. Flowerbeds or floral meadows and well-maintained vegetable plots enhance the aesthetic attractiveness of urban green space. Nevertheless, container gardening practices do not have an awe-inspiring impact and are commonly characterized as chaotic. Although flower locations were preferred above vegetable scenarios, participants in one experiment generally favored the idea of having additional vegetable plots nearby. People favored flowerbeds and flower meadows over vegetable plots regardless of socio-demographic characteristics; hence, both may be advocated in urban gardening sites. Additionally, this would broaden the total diversity, which is advantageous from an ecological and aesthetic standpoint. It can be observed that although not all public land use practices are seen favorably, some of them do add to how attractive metropolitan areas are regarded to be, if vegetable plots, flowerbeds, or flower meadows can also be used to improve the visual appeal of urban green space as compared to traditional lawns and container gardening techniques. The permanent urban gardening areas could benefit from a combination of flower meadows, flowerbeds, and vegetable plots. Moreover, this would improve the total variety, which is advantageous from both an ecological and an aesthetic point of view. Well-maintained flowerbeds and well-managed vegetable plots could enhance the visual value of urban green space against disorganized plant arrangements or ugly containers. Urban gardens will be more appreciated and less likely to be vandalized or neglected if they receive a certain amount of upkeep. People won't be interested in landscapes that they don't like. Regional variations suggest that not all areas of a city may value the development of vegetable gardens or the establishment of flowerbeds similarly. Planners for the city, managers of the parks, and other interested parties are encouraged to conduct a comprehensive analysis of potential urban gardening areas to reduce future disappointments. It is also observed that individuals are more interested in cultivating vegetables than flowers and are generally willing to participate in urban gardening. It is expected that, in the future, more people will dwell in residences without private gardens as urbanization rises, although urban gardens may offer similar enjoyable experiences to those that millions of people enjoy in their private gardens.

INCREASING GHG EMISSION

UA, while touted as a sustainable solution for food production in cities, may inadvertently increase GHG emissions (Raychel Santo, 2016). This seemingly contradictory outcome arises from various factors associated with urban farming practices.

First, transporting inputs such as fertilizers, soil, and seeds to urban farms contributes to carbon emissions. The reliance on fossil fuel-powered vehicles to deliver these resources from distant locations adds to the carbon footprint of UA. Moreover, synthetic fertilizers' energy-intensive production and distribution further exacerbate GHG emissions.

Second, the energy demands of urban farms play a significant role in GHG emissions. Indoor vertical farming systems, popular in urban settings, require artificial lighting and climate control, which consume substantial electricity. Unless renewable energy sources are employed, the electricity required for these operations often comes from fossil fuel-powered grids, releasing more carbon dioxide into the atmosphere.

Third, waste management in UA poses a challenge. If not appropriately managed, organic waste disposal from urban farms can lead to methane emissions—a potent GHG—during decomposition in landfills.

To mitigate these negative impacts, efforts should focus on implementing sustainable practices in UA. This includes minimizing transportation distances through local sourcing, adopting renewable energy solutions, such as solar panels, for powering indoor farms, and establishing efficient waste management systems that promote composting and anaerobic digestion.

Causing Water Pollution

Despite its many benefits, UA can contribute to water pollution due to various factors inherent in its practices. These factors include using fertilizers, pesticides, and inadequate runoff management. One of the main contributors to water pollution in UA is the excessive use of fertilizers. When these fertilizers are applied in excess or during inappropriate weather conditions, they can be washed away by rain or irrigation water, entering nearby water bodies. The high levels of nitrogen and phosphorus in the runoff can lead to eutrophication, causing algal blooms and oxygen depletion, negatively impacting aquatic ecosystems.

Similarly, the use of pesticides and herbicides in urban farming can also result in water pollution. If not applied carefully, these chemicals can leach into the soil or be carried away by rainfall, contaminating nearby streams, rivers, or groundwater sources. This contamination poses risks to aquatic organisms and potentially affects drinking water supplies.

Inadequate runoff management is another factor contributing to water pollution from UA. Urban farms often lack proper drainage systems or measures to capture and treat runoff water. As a result, excess water-carrying pollutants, such as sediment, nutrients, and chemicals, can flow directly into storm drains, which eventually discharge into water bodies.

Implementing best management practices is crucial to mitigate water pollution from UA. These practices include proper nutrient management to minimize fertilizer runoff, integrated pest management strategies to reduce the use of pesticides, and the implementation of green infrastructure such as rain gardens, bioswales, or constructed wetlands to capture and treat runoff water before it enters water bodies.

BUILDING CONSTRUCTION AND DESIGN

Converting vacant space of walls and roofs into something visually beautiful and artistically engaging, vertical and rooftop gardens are an excellent method to maximize restricted space and recover neglected space. They can be compared to a building's natural tapestry. Building construction and design must be considered to accommodate agricultural practices on rooftops and vertical surfaces.

ROOFTOPS

Up to 32% of cities and built-up regions are covered by roofs (Proksch, 2011), representing a significant amount of currently unutilized space in metropolitan areas. In many places, utilizing additional green roof technology to convert these urban rooftops into ecological and environmental resources is becoming standard practice. Urban dwellers have become more interested in a sustainable, nutritious diet in addition to the investment in environmentally conscious infrastructure. The outcome of this rapidly growing interest is that UA schemes are now being formed in many locations to cultivate organic and local vegetables. Combining these two sustainable methods has led to a new kind of green roof called the rooftop farm.

The essential elements of a green roof are a vegetative layer and a base (or growth medium) layer, wherein water is retained, and the vegetation is attached, with a reservoir board or drainage layer to release or store excess water. A root barrier and waterproofing membrane, which is composed comprising a structural support or roof slab and an insulating layer, separates these water-carrying layers from the actual roof structure. The substrate's depth influences the roof's environmental properties, weight, and plant types that are allowed there, which in turn affects the structural needs of the roof. The two main varieties of green roofs are defined by the thickness of their substrate layer: extensive, which has a substrate layer that is less than 6 inches (15 cm), and intensive, which includes a substrate layer larger than 6 inches (15 cm) (Proksch, 2011). While comparing productive green roofs and conventional rooftop farms, construction must be thoroughly examined. Many rooftop farms analyzed in the study by Proksch have continuous substrate layers and surface growth beds. As they fall under the umbrella of intensive green roofs, they may be contrasted in terms of the benefits and performance of buildings. However, raised bed rooftop farms utilize only a fraction of the roof's surface for growing.

The most crucial aspect of a green roof's or rooftop farm's effectiveness is the substrate, which is governed by its composition, depth, and weight. For long-term sustainability, substrates are often composed of 80% (or even more) of lightweight mineral aggregate with 20% (or even less) of organic material (Proksch, 2011). To keep the growth medium's volume intact, the mineral elements with pores minimize weight, retain water, and degrade very slowly. The organic materials break down quickly and become available to plants as nutrition.

CONSTRUCTABILITY AND SUBSTRATE WEIGHT

The increased weight of vegetated roofs, primarily determined by the mass and depth of the substrate, has a significant impact on their constructability. Cases of a few of the educational rooftop farms that had sizable budgets were studied to realize the ideal depth of the growth media and structural support. For instance, the rooftop garden at Trent University can support the weight of saturated soil, equal to around 100 pounds per cubic foot (wet). About 180 pounds per square foot of dead weight is carried by the roof (wet). The load-bearing capability of the roof is a limiting element for retrofitting existing rooftops; hence, lighter substrate mixtures must be devised for these applications. In New York City, the renovated rooftop farms atop pre-war warehouse-style buildings have been built successfully. Typically, their roofs can hold the same weight as each ceiling. For example, the roof of the Brooklyn Grange can support around 130 pounds per square foot in weight. The overall weight of the farm's parts is only between 40 and 50 pounds for every square foot, which is significantly less than the maximum structural load of the roof, even when the soil is entirely saturated with water.

VERTICAL FARMING

Plants planted in planters: A facade might have plants growing out of pots placed all over or at different altitudes. Numerous examples of these green facades are easily found. Planters are occasionally positioned at the base of a facade to allow plants to simulate the illusion of falling leaves. They can also be arranged randomly along the wall. The main characteristic of this type of GW is the need for a continuous watering system. Because plants can only grow in the space supplied by the planters, the size of the planters and the placement of the plants decide how the green facade will develop. It appears that potted plants require more upkeep than those that are rooted in the ground. Additionally, because of how slowly this system grows, it may take several months to completely encircle the facade. Plants buried in the earth are an alternative to plants that climb themselves or need support to flourish.

Greenwave is developing this kind of prefabricated living wall technology. It is constructed from box modules composed of reinforced plastic and fiberglass. Various plants may be grown within these boxes because they are filled with earth. Plants with large roots may also thrive using this strategy if the boxes have enough area. The pipelines for irrigation are fixed beneath the boxes to provide the living wall with water and nutrients.

RELATIONSHIP BETWEEN URBAN FARMING AND SUSTAINABLE CITIES

The relationship between urban farming and building sustainable cities can be understood through the concept of continuous productive urban landscapes (CPULs), which describes the future of the cities as unified urban landscapes designed and deliberated with the inclusion of urban farming within the existing urban built environment.

Bohn and Viljoen Architects initially invented the specific terminology for this, and this innovative concept expresses a different means of looking at cities to the designers of urban spaces. As seen nowadays, under-utilized landscapes can be substituted with productive vegetation, leading cities to the future where local food is produced abundantly in varied landscapes.

With the help of quantifiable and qualitative arguments, CPUL follows a holistic method to suggest that urban farming plays a crucial role in building more resilient and sustainable food structures along with adding several positive layers to the spatial and social value of the cities. The multi-faceted approach of a CPUL focuses on interrelating urban food-growing settings within the urban realm with the people and linking these landscapes to the countryside to enable activities transversely in every part of the urban food system.

The main component of CPULs is an intentionally deliberated and well-designed fusion of productive and continuous urban landscapes. They are built to include natural and living features and are positioned inside an urban-scale landscape concept. The purpose of CPULs is to re-establish a connection between lives and the processes needed to support it by encouraging and enabling urban residents to witness activities and processes often associated with the rural.

CPULs have the potential to promote more sustainable and livable cities by addressing social, spatial, environmental, and economic challenges facing urban areas. CPULs have the potential to create a sense of community and promote social interaction among urban residents who participate in food production and other activities. They can also help to address social inequalities by providing access to fresh, healthy food to low-income communities who may not have easy access to it otherwise. CPULs can help to mitigate the effects of urban sprawl and improve the overall quality of the built environment by providing green spaces and natural habitats for wildlife. They can also help to reduce the UHI effect by increasing green cover and providing shading.

CPULs can contribute to biodiversity conservation by providing habitats for various species and promoting ecological connectivity between urban and peri-urban areas. They can also help reduce cities' carbon footprint by promoting local food production and reducing transportation emissions associated with food transportation. CPULs can create employment opportunities in UA, horticulture, and other industries. They can also contribute to the local economy by providing fresh, healthy food to local markets and restaurants, promoting local food systems, and reducing dependence on imported food.

CONCLUSION

Cities offer various spaces for the accommodation of UA. Spaces such as soil-bound, building-bound, independent containers, and water-bound have a vast potential to be utilized for UA ranging from residential to district and city to regional scale. It is highly relevant to the social, spatial, environmental, and economic aspects of urban life and the city's built environment. Based on research and experiments conducted across the globe, some of the prominent effects have been discussed in this chapter, and inferences have been drawn that UA is effective in moderating internal and external building temperatures, leading to the reduction of energy requirements inside and sinking of heat island effect outside the buildings. In addition to the benefit of lowering the temperature, UA helps to minimize the amount of air pollution through its effects on dry deposition and the microclimate. Vertical gardens and on-ground vegetation also help lower noise pollution levels in the urban built environment. UA has a solid potential to stand out and add to the aesthetic value of our cities against the monotony of concrete, glass, and stone. UA also has negative impacts, such as increased GHG emissions in energy or resource-intensive locations. Despite its many benefits, UA can contribute to water pollution due to various factors inherent in its practices. It also impacts the evolution of construction techniques to accommodate vegetation on rooftops and vertical gardens and open under-utilized land for increasing yield without burdening the structure, along with providing flexibility of arrangement wherever required at that place. Hence, in the popular spotlight, the implications of urban farming on the built environment are extensive and be promoted through practical measures and schemes.

BIBLIOGRAPHY

Azkorra, Z. (2015). Evaluation of green walls as a passive acoustic insulation system for buildings. *Applied Acoustics*, vol. 89, 46–56.

Cong Gonga, C. H. (2017). The research of gray space design of architecture based on green stormwater infrastructure application. *International Conference - Alternative and Renewable Energy Quest, AREQ 2017*, 1-3 February (pp. 219–228). Spain.

Dimitrijević, D. A. P. (2017). Noise pollution reduction and control provided by green living systems in urban areas. *Scientific Proceedings III International Scientific-Technical Conference "Innovations"* (pp. 124–127). Sofia.

Dunnet, N., & N. Kingsbury. (2008). *Planting Green Roofs and Living Walls*. Portland: Timber Press.

Ecotiere, D., & B. Gauvreau. (2016). In-situ evaluation of the acoustic efficiency of a green wall in urban areas. *Hamburg Conference: Internoise 2016*. Hamburg.

Friederike Well, F. L. ((2020)). Blueegreen architecture: A case study analysis considering the synergetic effects of water and vegetation. *Frontiers of Architectural Research*, vol. 9, 191–202.

Gashgari Raneem, L. A. (2018). The Effect of Rooftop Garden on Reducing the Internal Temperature of the Rooms in Buildings. Madrid. *The 4th World Congress on Mechanical, Chemical, and Material Engineering*.

Gbadegesin, A. (1991). Farming in the urban environment of a developing nation - A case study from Ibadan Metropolis in Nigeria. *The Environmentalist*, vol. 11, 105–111.

Gong, P., & D. Wang. (2021.). Design and application of agricultural ecological. Acta Agriculturae Scandinavica, Section B - Soil & Plant Science *Section B — Soil & Plant Science*, vol. 71, issue 9, 783–793.

Heather, K. L. (2012). The environmental benefits of urban agriculture on unused, impermeable and semi-permeable spaces in major cities with a focus on Philadelphia, PA. https://www.researchgate.net/publication/304048480_The_Environmental_Benefits_of_Urban_Agriculture_on_Unused_Impermeable_and_Semi-Permeable_Spaces_in_Major_Cities_With_a_Focus_on_Philadelphia_PA

Hendrik Brieger, P. L.-M. (2016). Does urban gardening increase aesthetic quality of urban areas? A case study from Germany.Urban Forestry & Urban Greening 17

Joan Munoz-Liesa, S. T.-C.-D. (2021). Building-integrated agriculture: Are we shifting environmental impacts? An environmental assessment and structural improvement of urban greenhouses. Resources Conservation and Recycling. 169.

Michael Martin, T. (2022). Estimating the Potential of Building Integration and Regional Synergies to Improve the Environmental Performance of Urban Vertical Farming. Frontiers in Sustainable Food Systems, Volume 6.

Monika Szopinska-Mularz, S. L. (2019). Urban farming in inner-city multi-storey car-parking structures - Adaptive reuse potential. *Future Cities and Environment*, vol. 5, 1–13.

Katia Perini, M. O. (2011). *Greening the building envelope, facade greening and living wall systems. Open Journal of Ecology*, vol. 1, 1–8.

Proksch, G. (2011). *Urban Rooftops as Productive Resources: Rooftop Farming versus Conventional Green Roofs*. Architectural Research Center Consortium: Considering Research. Detroit.

Pugh, T. A. M., et al. (2016). Effectiveness of green infrastructure for improvement of air quality in Urban street canyons. *Environmental Science & Technology*, vol. 46, 7692–7699.

Raychel Santo, A. P. (2016). Vacant Lots to Vibrant Plots. Baltimore, MD: *Johns Hopkins Centre for a Livable Future*.

Skara, S. L. G. et al. (2019). Urban agriculture as a keystone contribution towards securing sustainable and healthy development for cities in the future. *Blue-Green Systems*, vol. 2, no. 1, 1–27.

Sam, S. C. (2011). Green roof urban farming for buildings in high-density urban cities. *Green Roof Conference* (pp. 18–21). Hainan.

Sam, C. M., & Hui, D. S. (2011). Green roof urban farming for buildings in high-density urban cities. Hainan *China World Green Roof Conference* (pp. 18–21). Hainan.

Villanova, M. P. (2013). *First approach to the energetic savings when applied to the Seagram Building in* New York. Barcelona. Thesis Barcelona School of Civil EngineeringConstruction Engineering Department.

Vralsted, R. (2011). *Planning for building-integrated agriculture in Las Vegas*. Las Vegas: Graduate College, University of Nevada.

de Zeeuw, H., van Veenhuizen, R., & Dubbeling, M. (2011). The role of urban agriculture in building resilient cities in developing countries. *Journal of Agricultural Science*, vol. 149, 153–163.

13 Challenges of Urban Agricultural Production from the Aspects of Plant Protection and Food Safety
A Case Study of the Republic of Serbia

Dragana Šunjka, Sanja Lazić, Slavica Vuković,
Siniša Berjan, and Hamid El Bilali

INTRODUCTION

For contemporary agriculture, it is difficult to meet the requirements for food production and, at the same time, the increasing concern for the environment. This is even harder considering that, by 2050, an additional 2 billion people will need nutritious food, while over 80% of food will be consumed in cities (UN, 2018). Developed countries have reached the ultimate threshold of balance between the use of resources (water and soil), ecological sustainability, and biodiversity conservation. Thus, the question of producing enough food for a constantly growing population is raised.

Hence, cities worldwide are launching initiatives designed to relocate their food systems. The reduction of arable land and the increasing demand for food led to the development of urban agriculture as a new concept. According to the FAO (2022),

> *Urban and peri-urban agriculture (UPA) can be defined as practices that yield food and other outputs through agricultural production and related processes (transformation, distribution, marketing, recycling …), taking place on land and other spaces within cities and surrounding regions. It involves urban and peri-urban actors, communities, methods, places, policies, institutions, systems, ecologies and economies, largely using and regenerating local resources to meet changing needs of local populations while serving multiple goals and functions.*

Urban agriculture includes agricultural activities in urban areas, with the production of fresh food for own needs as a main goal. In contrast, the larger-scale production is concentrated on the outskirts of cities. The best examples of urban agriculture are gardens, backyards, community gardens, schools, and rooftops, where food is produced in a small space. However, growing plants in an urban environment depends on different conditions compared to traditional agriculture production; in cities, gardens are arranged according to available space, often near marginal areas, railways, roads, and industry.

It is well known that locally grown food from small producers has a better reputation with consumers compared to imported food available in large supply chains. The placement of urban farming products is mostly done in markets, while, in recent years, doorstep-selling or delivering to stores in their municipalities has become very attractive for producers. Moreover, farmers have formed networks for delivering products directly to food processing companies, hotels, and restaurants.

DOI: 10.1201/9781003359425-13

Nowadays, people are more aware of the value of local consumption, the adverse effects of pesticides, the role of farmers in food systems, and the importance of a diversified, balanced diet. However, although well received by consumers, locally produced food is often less controlled than food from longer supply chains. Food produced in urban farming systems should not meet lower safety standards, which is especially important considering the presence of short supply chains.

In urban agriculture and other agricultural production, food safety hazards mainly originate from plant protection and nutrition, i.e., applying pesticides and fertilizers. Pest control requires intensive plant protection, mainly based on chemical pesticides. However, their intensive and/ or unconscionable use and non-compliance with the pre-harvest interval (PHI) lead to the accumulation of residues in agricultural products and the environment. The additional sources of contamination of agricultural products from urban production are polluted soil, water, and air. Pesticides application in non-agricultural purposes, such as communal hygiene, indoor and outdoor, where they are used for the control of human disease vectors, e.g., mosquitoes, ticks, and fleas (Cooper and Dobson, 2007), nonselective application in parks, lawns, and home gardens, is a common practice, resulting in the occurrence and accumulation of these pollutants in the environment. Herbicides, insecticides, and rodenticides used for those purposes enter agricultural products and food chains through contaminated soil and water. While pesticides have been intensively used in urban environments, there is still a lack of information on the fate and behavior of these toxicants.

Besides this, anthropogenic pollutants in the urban environment originate from traffic, industries, atmospheric deposition, and incinerators (Chen et al., 2005). Consequently, urban farming environments may contain toxic substances such as heavy metals, including lead, zinc, copper, tin, mercury, and arsenic. Plants absorb these elements in the soil and water, which, in excessive amounts, may lead to reduced plant growth, phytotoxicity, or health problems for consumers. Moreover, the highly carcinogenic polycyclic aromatic hydrocarbons from wood stoves and road traffic have been found in vegetables from urban gardens, such as those in urban gardens of Sao Paulo, Brazil (Amato-Lourenco et al., 2017), and Veszprém County, Hungary (Kováts et al., 2021).

The source of pesticide residues, heavy metals, persistent organic pollutants, nitrates, nitrites, and mycotoxins originate from agriculture. However, the level of agrochemicals used by urban farmers has rarely been researched or recorded. Thus, adequate plant protection is crucial to produce healthy and safe food. Using synthetic pesticides and other agrochemicals to control a wide range of pests in urban agriculture must be strictly controlled and reduced.

It is significant to emphasize the importance of the green economy, which aims to achieve sustainable development without environmental degradation. Sustainable food systems are some of the main Green Deal goals,[1] based on the inclusive growth of the European Union (EU) through a set of policy initiatives to be realized by 2050. The most important part is the farm-to-fork strategy,[2] which aims to improve the sustainability of agricultural production through a comprehensive approach. The European Commission suggests a reduction in the use of chemical pesticides by 50% by 2030. The strategy proposes a ban on the use of all pesticides in sensitive areas, such as public parks and gardens. This does not mean the complete prohibition of chemical pesticides, given that, in agricultural production, the control of some pests can be achieved only with this measure, while as it is set in integrated pest management (IPM), chemical pesticides must be the last solution in plant protection. As a result of pesticide reduction, human health, the environment, pollinators, and biodiversity will be protected, and the main goal of the farm-to-fork strategy, sustainable production of safe food, will be achieved.

In this context, the chapter focuses on the safety of the food produced in urban agriculture systems. Through selected case studies in the Republic of Serbia (RS), it provides an overview of the status of urban agriculture from the aspects of food safety to contribute to the understanding of the importance of appropriate plant protection and reduction of synthetic pesticides in urban areas. The results of the case studies are given in "Case Studies".

URBAN AGRICULTURE IN THE RS

The concept of urban agriculture in the Balkans is known, but due to new challenges, global urbanization, the need for environmental protection, and other current issues in food production, it has gained even greater importance. Serbia is a Southeast European country with a long rural tradition. Thus, urban gardening is relatively new. Some forms of urban gardens, private and family-run, existed during the 20th century (Lazić, 2008); however, they intensively developed with the urbanization process.

Depending on the location, there are two main types of urban agriculture (FAO, 2022):

- Intra-urban agriculture in limited areas near the city center, with the production of vegetables, flowers, and seedlings for own needs
- Peri-urban agriculture to produce meat, milk, eggs, fruits, and vegetables, for the market, mostly in the city surroundings

In the RS territory, urban agriculture is organized as garden communities, i.e., open areas where individuals or groups cultivate plants on public or private land, and their size can vary from 50 m² to larger areas of several hectares. They can be located on the territory of the city or the outskirts. It is estimated that, in Belgrade's capital city, gardens cover at least 100 ha of city land. In the capital, but also Novi Sad, Šabac, Pirot, and other cities, many associations and initiatives for permissions enable the expansion of city gardens on land suitable for organic production. One such oasis is Baštalište, the first garden community in Serbia that sprung up at the entrance to Slance, a surrounding area 10 km from Belgrade city center. The concept was formalized in 2013 when a group of citizen associations and individuals established the first formal garden community on private land. Here, on 18.40 ha, vegetables are grown according to the organic production principles.

In most European cities, urban gardens are granted to citizens by the authorities. In the RS, the city of Šabac is the first municipality that has granted its citizens land for urban farming. Even though different forms of agriculture are represented in urban settlements throughout Serbia, authentic urban agriculture is present only in the largest metropolitan areas, which include the cities of Belgrade and Novi Sad. These metropolitan areas enclose 5,032 km² with a population of around 2.1 million in 2021.[3]

Urban agriculture has a vital role in supplying the urban areas of Belgrade and Novi Sad with fresh food for city markets (Živanović-Miljković et al., 2022). It is estimated that around 30,000 farms in over a hundred Belgrade villages (Green Ring) could feed the city. Intensive agricultural production, modern orchards, livestock farms, and the food industry continue to form the sector's backbone. However, a growing number of minor- and medium-sized farmers have adopted different urban agriculture models and leveraged the greater purchasing power of urban consumers to directly market and sell value-added food, often in combination with on-farm services (Filipović et al., 2013).

The main characteristic of urban agriculture is the local production of food for local use, i.e., a closed supply system within the local community that enables the recultivation of city areas, employment of the population, more food and healthier nutrition, increasing the economic power of the city and its inhabitants, and stronger connections between residents (Gasperi et al., 2016; Sanyé-Mengual et al., 2019; Orsini et al., 2020).

Nowadays, urban agriculture considers the ecological development mode of producing pollution-free, green, organic vegetables using fewer pesticides and chemical fertilizers to reduce pesticide residues in vegetables (Bao, 2020). However, it has not entirely got rid of the use of pesticides in practice (Montiel-Leon et al., 2019). In addition to many advantages, urban agriculture has some health and environmental risks, such as the potential use of contaminated land and water, the application of pesticides, and the possibility of organic agrochemicals runoff into water sources (Meftaul et al., 2020; Aubry and Manouchehri, 2021; Ping et al., 2022). For such reasons, urban agriculture requires proper attention.

PLANT PROTECTION AND PESTICIDE RESIDUES

Crop protection with the aim to achieve high yields and appropriate food quality is the most important task of contemporary agricultural production. Nowadays, plant protection relies almost wholly on using chemical compounds to control disease-causing agents, damaging insects, and weeds (Mandal et al., 2020). Their long-term application and successful control of the mentioned organisms have led to a situation where agricultural producers almost completely excluded traditional methods of plant protection and replaced them with pesticides. According to official data from the Food and Agriculture Organization (FAO) of the United Nations, if pesticides were not used for just one year, pests would reduce world food production by 25–30% (FAO, 2023). However, the intensive and particularly inadequate application of chemical pesticides causes growing concerns about food safety. Pesticide residues, resulting from applying plant protection products (PPPs) on crops used for food, can potentially pose a risk to human health (EFSA, 2022; Handford et al., 2015). Due to this, a comprehensive legislative framework has been established in the EU. It defines rules for approving active substances in PPPs, their use, and their residues in food. To ensure a high level of consumer protection, legal limits, or so-called maximum residue levels (MRLs), in the EU are established in Regulation (EC) No. 396/2005.[4] MRLs are set in the EU for more than 1,300 pesticides covering 378 food products/food groups. Moreover, a default MRL of 0.01 mg/kg applies to nearly 690 pesticides not explicitly mentioned in the MRL legislation. Through EU and national control programs, Regulation (EC) 396/2005 obligates Member States to carry out controls to ensure that food on the market complies with the legal limits. Although MRLs set by the European Commission, Environmental Protection Agency, and Codex Alimentarius are recognized worldwide, many countries define MRLs. Thus, MRLs in agricultural products in the RS[5] are defined by the regulation on maximum permitted quantities of residues of PPPs in food and feed; however, it is completely harmonized with the EU legislation. Regarding pesticide residues, it is essential to emphasize the significance of awareness of the quality of agricultural products intended for infants and children; the maximum allowable amounts of pesticide residues for food are ten times less than those for adults (EFSA, 2018). This means that pesticide residues in agricultural products must not be detected, and, for these sensitive groups, it is best to use organically produced fruits, vegetables, and cereals. Besides the MRL, the basic standards for pesticides include the determination of the acceptable daily intake, i.e., the total amount of a pesticide that can be ingested daily over a lifetime without significant health risk (Keikotlhaile and Spanoghe, 2011; Ambrus and Yang, 2016).

The presence of pesticides in the urban environment represents a threat to human health. Plants can easily absorb pesticides from the soil, water, and air and transfer them to the vegetative and reproductive parts. Even though several studies have focused on pesticide residues in agricultural products in developed and developing countries (Prodhan et al., 2018; Lehmann et al., 2017), there is still a lack of data on pesticide residues in fruits and vegetables grown in urban areas.

In the past decade, food containing pesticide residues has increased enormously (Pesticide Action Network (PAN), 2014); thus, it has become a significant safety concern for consumers worldwide. Therefore, it is essential to provide insight into food pesticide residues before consumption (Hasan et al., 2017). For safe and healthy food, continuous monitoring programs of chemical pesticide residues and adequate handling during food preparation, particularly for fruits and vegetables mostly consumed fresh, are extremely important. From the surface, pesticide residues can easily be removed by washing under running water, immersion, and removing the outer leaves and flowers, while cuticular pesticide residues are eliminated by peeling. Furthermore, residues inside the fruit can be removed or reduced by heat treatment if they are thermolabile compounds, while exposure to low temperatures (freezing) does not affect pesticides.

Considering the number of treatments required to protect against harmful agents, especially in apples, peaches, strawberries, grapes, tomatoes, peppers, spinach, and lettuce are at risk of increased pesticide residues. Many diseases and pests occur before ripening or in the technological maturity phase, and applying pesticides in this period represents an additional risk.

CASE STUDIES

SIGNIFICANCE OF THE STUDY

The high quality of fruits and vegetables matters to both consumers and producers; thus, the determination and quantification of pesticide residues in food are of the most importance. That is the reason why research on pesticide residues in fruits and vegetables grown in urban agriculture is highly justified. Besides direct use, i.e., application to control harmful organisms, pesticides can enter crops from contaminated soil and water. Thus, the quality of soil and water used in urban agriculture requires monitoring.

Besides pesticides, different environmental pollutants endanger plants due to their harmfulness and potential for interaction. Lead (Pb), rather than cadmium (Cd), is less mobile and could be easily transferred to plants in acidic soils poor in organic matter. Therefore, heavy metals in the urban environment represent pollutants of particular concern. Thus, in the past few years, the Government of the City of Novi Sad financed monitoring programs for the presence of these pollutants in agricultural products and the environment in urban areas.

METHODOLOGY

The monitoring programs of pesticide residues and heavy metals are based on projects which aim to reduce the potential risks to human health and the environment associated with the use of crop protection products. All sampling sites were in Novi Sad, a city situated in the northern part of Serbia. Fruit and vegetable samples, originating from intra- and peri-urban agriculture, were collected at the green markets that primarily offer these products. The quality of the environment in the urban area of Novi Sad was evaluated by analyzing soil and water. The pesticide residues and heavy metals content were analyzed using validated methods. Detailed information is provided in subsections "Monitoring Program of Pesticide Residues and Heavy Metals in Soil, Monitoring Program of Pesticide Residues and Heavy Metals in Fruits, and Monitoring Program of Pesticide Residues in Vegetables from the Ecological Agriculture".

STUDY AREA

Novi Sad is the second-largest city in the RS. It is a part of the Panonian Basin, which is a major agricultural area, and as is often said, "this plain can feed Europe". Citizens of Novi Sad are supplied with fresh fruits and vegetables mainly from urban agricultural production. Namely, besides city gardens as a form of urban agriculture, in the surrounding Novi Sad, intensive peri-urban agricultural production, as a primary source of food supply for the urban population, was developed. However, due to the application of PPPs and other pollutants in the environment, both in urban and in the surrounding areas, the question arises of the quality of agricultural products and the soil and water used for their growing.

Within different projects, the research was carried out in the Laboratory for Biological Research and Pesticides at the Faculty of Agriculture, University of Novi Sad. The following subsections discuss the outcomes of this research.

MONITORING PROGRAM OF PESTICIDE RESIDUES AND HEAVY METALS IN SOIL

The most crucial prerequisite to producing healthy food is soil quality, which requires the determination of pesticide residues and heavy metals. Contaminated soil and water could be sources of pollutants in urban agriculture. The analysis of these compounds in soils of Novi Sad covered non-agricultural and arable land. The samples from the gardens were taken in the suburbs where small producers, following conventional production and use of PPPs, grow fruits and vegetables for their own needs and the markets. Conversely, non-agricultural soil samples were taken along city roads and parks. During 2022, soil sampling was performed with a probe at a 0 to 30 cm depth, with 80 soil samples (40 from non-agricultural areas and 40 from gardens). To evaluate the presence of pesticides in soil samples, 48 active substances were analyzed. A modified solid phase performed the extraction of pesticides. A modified solid-phase extraction performed the extraction of pesticides, while for its simultaneous determination in soil, GC/MS and HPLC/DAD were used (Lazić et al., 2015).

The results of pesticide residue analysis are shown in Figure 13.1. Pesticides were found in 80% of arable soils. Regarding insecticides, organochlorines, abamectin, α-cypermethrin, bifenthrin, and chlorpyrifos were detected. Organochlorine insecticides have not been used for many years, but since they are highly persistent compounds, their metabolites remain in the environment. Other insecticides are still used to control pests in many crops, intensively grown in home gardens (Source: Ministry of Agriculture, Forestry and Water Management, Republic of Serbia, 2023).

The presence of two herbicides (terbuthylazine and pendimethalin) was detected in the analyzed agricultural soil samples. Fortunately, herbicide residues were not found significantly, considering weeds in non-agricultural areas and gardens are mostly removed mechanically.

Analysis of fungicide residues in agricultural soils revealed the presence of 14 active substances, mostly triazoles.

It is also important to highlight that the presence of pesticides in the soil can also impact the number of microorganisms. Their reduced number influences the degradation process of various pollutants, and, as a result, these substances remain longer and can enter the food chain.

Residues of the analyzed pesticides in non-agricultural soils were not detected, or their amounts were below the limit of detection (LOD). It is expected, considering that the monitoring program focused on active substances used in agricultural production. In urban areas, pesticides are mostly used to control weeds, ticks, and mosquitoes or for communal hygiene purposes (van de Merwe et al., 2018).

Further, using the atomic absorption spectrometry method, the content of copper, cadmium, and lead in the soil samples was evaluated (Table 13.1). Heavy metals detected in agricultural and urban soils originate from industrial pollution and traffic. In different amounts, they were also found in soil samples of some gardens near major roads, and their increased content can also be expected in agricultural products. Results showed a higher amount of copper in arable than in non-agricultural soils. This high value results from the long-term use of copper-based PPP to control a wide range of plant disease-causing agents. In this study, the average lead content was much higher in non-agricultural soils, while arsenic content was slightly higher in agricultural than non-agricultural soils. Mercury content in non-agricultural soil samples was

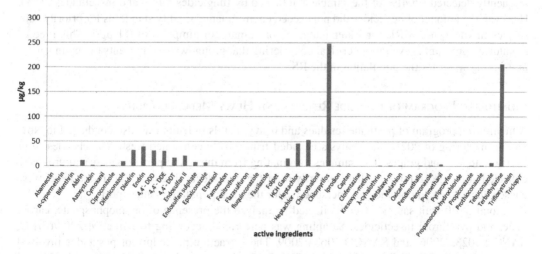

FIGURE 13.1 Pesticide residues in agricultural soil in the territory of Novi Sad.

TABLE 13.1

Heavy Metals in Agricultural and Non-agricultural Soil Samples

Heavy Metal	Agricultural Soil (mg/kg)	Non-agricultural Soil
Copper (Cu)	40.10	32.00
Lead (Pb)	18.82	68.87
Cadmium (Cd)	0.23	0.27
Arsenic (As)	8.92	6.95
Mercury (Hg)	0.05	0.10

twice as high as the average content in agricultural soils. Heavy metals adsorbed in the soil can pass into the soil solution, become available for plants, and reach the food chain.

The monitoring program results provided insight into the application of PPP by small producers and the ecological status of the soil in the city of Novi Sad's territory. According to the results, it could be expected that the heavy metals found in the soil could also be found in agricultural products due to their bioavailability to plants.

MONITORING PROGRAM OF PESTICIDE RESIDUES IN WATER

In the peri-urban area of Novi Sad, with intensive production of fruits and vegetables for urban markets, monitoring was carried out to evaluate the content of pesticides in drainage water, widely used for irrigation. Besides the occurrence of pesticide residues in drinking water, control of their presence in water used for irrigation in agriculture, particularly in urban areas, is very important (Lazić et al., 2012; Lazić et al., 2013). This primarily refers to the rivers and groundwater, considering that the water and land quality in conventional production, especially in organic agriculture, is crucial. The presence of pesticide residues in these matrices may cause yield reduction and a decrease in quality due to their uptake.

The monitoring program covered 32 sampling sites during the summer of 2021. Analysis was performed with solid-phase disc extraction, followed by GC/MS and HPLC. In analyzed samples, 22 pesticides were detected – 8 fungicides, 9 herbicides, and 5 insecticides. Herbicides were most frequently detected (in 41% of the samples), followed by fungicides (36%) and insecticides (23%). However, the amount of pesticides did not exceed environmental quality standards for priority substances in water, nor MRL for water intended for human consumption of 0.1 µg/l. This type of monitoring program is even more critical considering that groundwater represents the main source of drinking water for the population in the RS.

MONITORING PROGRAM OF PESTICIDE RESIDUES AND HEAVY METALS IN FRUITS

A monitoring program of pesticide residues and heavy metals in fruits was also conducted in Novi Sad in the spring of 2021. The analysis included fruit mostly consumed fresh, viz. cherries, sour cherries, apples, and grapes. The samples, originating from intra- and peri-urban agriculture, were collected at the green markets that primarily offer these products (Figure 13.2). These fruits were chosen due to their importance and perspective for the local population and export.

A total of 40 fruit samples were collected to analyze the presence of organophosphate, carbamate, and pyrethroid insecticides. Sampling was performed according to Directive 2002/63/EC, SANCO 10232/2006, and SANCO 10684/2009. The extraction procedure of pesticides involved a widely used QuEChERS-based method (EN 15662) (Anastassiades et al., 2003), while GC and

FIGURE 13.2   Green market in Novi Sad.

HPLC analyzed residues. Methods were validated according to SANTE[6] criteria and applied to actual fruit samples.

Considering that fruits grow mostly within conventional farming, with intensive plant protection with several pesticide treatments (some in the ripening stage), the occurrence of pesticide residues could be expected. As a result, in apples, dimethoate was found in 40% of the analyzed samples, while its content exceeded MRL of 0.01 mg/kg in 30%. Organophosphate residues were not found in grapes, cherries, and sour cherries or were below LOD of 0.005 mg/kg. Of the analyzed 14 compounds from the carbamate group, the presence of carbaryl and pirimicarb was determined in the fruit samples; although the presence of pirimicarb was confirmed in 40% of grape and 20% of cherry samples, its amount did not exceed the MRL. The monitoring program covered analysis of several pyrethroid insecticides (bifenthrin, lambda-cyhalothrin, beta-cyfluthrin, cypermethrin, alpha-cypermethrin, zeta-cypermethrin, fenvalerate, and deltamethrin). However, their presence in the analyzed fruit samples was not detected, or the content was below the LOD (0.001 mg/kg).

Moreover, within this monitoring program, heavy metal analysis was carried out in fruit samples, emphasizing the presence of lead and cadmium. The maximum allowed concentrations of cadmium and lead in fruit are 0.05 and 1 mg/kg, respectively. In fresh material, cadmium was not detected in the analyzed samples of sour cherries, while the level of lead was significantly below the MRL of 1 mg/kg. The cadmium and lead content in cherries, apples, and grapes did not exceed the prescribed values.

Considering the new list of active substances of PPPs allowed for use, many pesticides found in our analyzed samples, due to their persistence and toxicological characteristics, are prohibited these days, and the presence of their residues is not allowed (source Ministry of Agriculture, Forestry and Water Management, Republic of Serbia, 2023).

Although it is in Europe, Serbia is still not a part of the EU. However, most of the obligations and legislation related to food safety and environmental protection prescribed in the EU are under establishment. In the latest report on the overall pesticide monitoring programs conducted in the EU in 2020, 88,141 conventional and integrated agriculture samples were analyzed in the presence of 659 pesticides, with an average of 264 pesticides per sample. Of all analyzed samples, 94.9% fell within legal limits; of these, 54.6% did not contain quantifiable residues (results below the limit of quantification (LOQ) for each pesticide analyzed), while 40.3% of the samples contained quantified residues not exceeding the legal limits. MRLs were exceeded in 5.1% of the samples (4,475), an increase (3.9%) compared to previous monitoring in 2019. Based on the results, 3.6% of all the samples triggered legal sanctions or enforcement actions, higher than in 2019 (2.3%) (EFSA, 2022).

Monitoring Program of Pesticide Residues in Vegetables from Ecological Agriculture

Nowadays, contemporary agriculture focuses on producing quality food without harmful agro-chemicals rather than on yields. This goal could be achieved by ecological agricultural production (Figure 13.3). In the RS, this type of agriculture is under intensive development, with constantly increasing areas, with a similar trend in neighboring countries Montenegro and Croatia, while, in Bosnia and Herzegovina, it is still in the initial phase. However, the average share of areas under certified organic agricultural production in the Balkan region is only 0.39% (Znaor and Landau, 2014).

In recent years, the demand for organic products has significantly grown due to the number of associations dealing with healthy nutrition, consumer protection, environment protection, and promoting the benefits of organic food.

Increased interest in organic food consumption coincides with increasing awareness of the importance of nutrition, health, food safety, and environmental protection. Consumers' perception of the importance of personal health is usually ahead of the perception of the need to protect the environment. In other words, organic food is consumed and bought more for personal health than for caring for the environment (Hamzaoui-Essoussi and Zahaf, 2012). The increased consumption of organic food is also the result of the awareness that it presents a lower risk than conventionally produced food.

The fact that most agrochemicals are prohibited in certified organic and unacceptable in ecological forms of production is the basis for the assumption that they would not be found in agricultural products. However, due to the persistent pesticides accumulated in the environment, their residues can still be found in these products.

Evaluation of the presence of pesticide residues in vegetables from ecological production of small agricultural producers from the area of Novi Sad, covered a total of 23 samples of vegetables (chard, broccoli, pumpkin, celery, beetroot, cauliflower, kohlrabi, cucumber, potato, cabbage,

FIGURE 13.3 Organic gardening with raised beds. (Source: Courtesy of Dragan Močević.)

pepper, eggplant, parsley, carrot, and zucchini), taken at green markets. The presence of 50 pesticides from the group of organochlorines and their metabolites, organophosphorus, pyrethroids, and carbamates, were analyzed (Pucarević et al., 2015). The extraction of pesticide residues was performed on the solid phase using acetonitrile (Fillion et al., 2000). The extracts were analyzed using gas chromatography with mass detection (GC/MS).

The results indicate that pesticide residues in 72% of the vegetable samples were not detected or below the LOD (0.001 mg/kg). Of the residues in a few samples, organochlorine, pyrethroids, and organophosphorus insecticides were found, while carbamate-based pesticides were not detected in the tested vegetable samples. The residues of persistent pesticides, banned for a long time, indicate that they were found in vegetables because of migration from the environment. Namely, because of their presence in soil and/or water, pesticides quickly could enter plants, and root vegetables are especially threatened (Saldahna et al., 2012).

It is important to emphasize that no specific MRLs are established for organic agricultural products. The set MRLs apply to both organic and conventional food products.

In the EU in 2020, monitoring of the pesticide residues present in organic agricultural products was monitored on 5,783 samples (EFSA, 2022). Overall, 80.1% of those samples did not contain residues, 18.4% of samples residues were below or at the MRL value, and 87 samples were detected with residue levels above the MRLs (1.5%). Sporadically, some strictly prohibited pesticides such as chlorpyrifos, anthraquinone, and lambda-cyhalothrin were found.

The presence of pesticide residues in organic agricultural products could be the consequence of spray drift, contaminations origin from the environment (soil, water), contaminations during the handling, packaging, storage, or processing, or even incorrect labeling of conventionally produced food as organic food (EFSA, 2022).

PLANT PROTECTION IN URBAN AGRICULTURE

In intra-urban areas, it is difficult to achieve organic agriculture due to pesticide application in communal hygiene and pollution related to traffic and industry (Antisari et al., 2015). However, the contamination of agricultural products grown in such an environment and the potential food safety risks are rarely studied (Petit et al., 2011). In the research conducted in France, concerns that around 10% of agricultural areas, including market gardens and "green belts", were located close to major roads have been raised. To overcome this and minimize the potential risk, some guidelines, such as the use of "isolation distances" between roads and fields, are proposed (Rémy and Aubry, 2008). Thus, the presence of pesticides and proximity to traffic and industry can limit the cultivation of fruit and vegetables in urban areas following the principles of organic agriculture.

Results obtained in monitoring programs have shown the presence of pesticides and heavy metals in agricultural products and the environment in urban areas. Because of intensive pesticide application and/or their persistence in the environment, analyzed fruits and vegetables from conventional and ecological agriculture accumulated some pollutants. Conventional farming uses chemical pesticides as the primary measure for pest control thus facing a variety of serious challenges, from food safety to environmental threats. Moreover, applying active substances with the same mode of action for a long time can create selection pressure and lead to the development of resistant pest populations and, therefore, increase the use of pesticides.

A significant challenge of plant protection in urban agriculture represents high crop diversity. The first evidence of some new, invasive pests and diseases, such as spotted wing drosophila (*Drosophila suzukii* Matsumura) (Diptera: Drosophilidae) and the cereal aphid *Sipha maydis* Passerini (Hemiptera: Aphididae), is recorded in urban farms. They are often located close to major transportation hubs and trade routes. Besides this, minimized use of pesticides, due to grower preferences for organic methods, together with planting of exotic crops, can lead to the appearance of exotic pests (Grasswitz, 2019). This represents a global threat to agricultural production, not only an isolated case in urban farming.

IPM could represent a sustainable production method. Even if pest management is considered for large farms, IPM can be applied in gardens. It is a practical and environmentally sensitive approach to pest management that relies on a combination of common-sense practices. This means using available pest control methods with the least possible hazard to humans and the environment, i.e., prevention, mechanical, and biological measures, using rational chemical pesticides (Šunjka et al., 2021). IPM can reduce pesticide use by up to 50% and successfully control harmful organisms (Pretty and Bharucha, 2015).

In the RS, peri-urban agriculture enables production according to ecological or even certified organic principles. Situated in the city vicinity and yet distant from the sources of pollution, this area provides suitable conditions for sustainable ecological food production for the urban population. To maintain the level of harmful organisms below the harm threshold, in this system, plant protection implies preventive, mechanical, and biological measures (Šunjka and Vuković, 2021), with the application of biopesticides or allowed chemical pesticides (Fenibo et al., 2021).

Biopesticides are PPPs based on living organisms and products synthesized by living organisms. Along with application in ecological and certified organic agriculture, they are intensively introduced into integrated as well as conventional agricultural production.

The advantages of biopesticides are numerous – natural origin, lower toxicity, impact only on the target pests, shorter PHI, and wide range of applications (Liu et al., 2021; Kumar et al., 2021; Fenibo et al., 2021; Šunjka and Mechora, 2022). These compounds have fast environmental degradation, resulting in decreased pesticide exposure and a negative impact (Šunjka and Mechora, 2022). Besides this, the application of biological agents in integrated plant protection programs enables the development of sustainable agriculture, reduces the need for chemical pesticides, and can help suppress resistant populations (Cawoy et al., 2011).

Currently, 14 biopesticides are registered for use in the RS and are available on the market. Those are biofungicides based on *Bacillus subtilis* Č13, *Bacillus subtilis* BS10, *Bacillus subtilis* Z3, *Bacillus amyloliquefaciens* Z3, *Bacillus amyloliquefaciens subtilis* Q713, *Bacillus amyloliquefaciens* MBI, *Trichoderma atroviride* SC1, *Pythium oligandrum*, essential oil of *Melaleuca alternifolia*, and orange essential oil and bioinsecticides – *Bauveria bassiana* ATCC 74040, *Bacillus turigiensis* subspec. kurstaki, *Cydia pomonella* granulovirus (CpGV-R5), *Cydia pomonella* granulovirus (CpGV-V22), azadirahtin, orange essential oil, and pyrethrin (Source: Ministry of Agriculture, Forestry and Water Management, Republic of Serbia, 2023a).

It is estimated that, by 2050, the importance of biopesticides in agricultural production will be equal to that of chemical pesticides (Olson, 2015). They are already included in IPM, along with those used in conventional agriculture. Besides biopesticides, within ecological agriculture, other plant protection agents are used, i.e., PPPs based on copper (copper-hydroxide, copper-oxide, copper-oxychloride, copper-sulfate, and Bordeaux mixture), sulfur, iron phosphate, paraffin oil, spinosad, and hydrogen peroxide. In total, there are 47 compounds registered in Serbia. Moreover, diatomaceous earth, kaolin clay, hydrogen peroxide, potassium bicarbonate, potassium permanganate, salts, and soaps are considered biopesticides. Along with formulated PPPs, self-made plant products for the control of different pests are largely used. Medicinal, aromatic, spice plants, vegetables, and weeds are sources of compounds with different pesticide activities. Although they have been known since ancient times, plant-based products have become actual with the increased interest in organic agriculture.

CONCLUSIONS

Although urban agriculture is considered a model of organically grown agricultural products using bio-intensive methods, a wide range of risks related to food safety cannot be underestimated. These risks originate from different anthropogenic factors (e.g., inappropriate use of pesticides, traffic, industry, and communal hygiene).

Monitoring carried out in the city of Novi Sad showed the presence of pesticides and heavy metals in soil and water used in urban agriculture. The presence of pesticide residues that have not been used for a long time was found, while in addition, residues of currently used pesticides were determined. Even if they do not exceed the MRL, they still could enter the food chain and have side effects on human health. Thus, before starting food production in urban areas, it is necessary to evaluate soil and water quality, considering the presence of various pollutants.

Urban agriculture's increasing importance in food security is a trend in developed and developing countries. However, food produced in urban agriculture cannot be a priori accepted as safe. Results of the analysis of pesticide residues in fruit from urban agriculture production sampled on the green market confirmed the necessity of their control.

Although organic production in Serbia is represented at less than 1%, it exists in the surrounding of cities. Results of the monitoring program of pesticide residues in organically produced vegetables in the peri-urban region of Novi Sad are like those obtained in the EU monitoring program. Whether it is about pesticides originating from the environment or inappropriate application, adequate and continuous monitoring programs are imperative.

To ensure this, the role of policymakers is crucial. In the RS, the Ministry of Agriculture, Forestry, and Water Management prescribed monitoring of pesticide residue analysis in agricultural products – 580 samples from conventional production, 395 from import, and 225 from organic production. The results of the monitoring program in 2022 are still not available.

The production of safe food includes plant protection, which is one of the most important measures. Plant protection in urban agriculture should be harmonized with existing conditions with maximum respect for the environment, using biological and other integrated measures. Finally, results presented in this chapter showed that promoting sustainable agricultural production and appropriate plant protection, with continuous monitoring of the pollutants, are prerequisites to producing healthy food in urban areas.

NOTES

1 European Commission. 2019. The European Green Deal. https://eur-lex.europa.eu/legal-content/EN/TXT/HTML/?uri=CELEX:52019DC0640&from=EN
2 European Commission, 2020. A Farm to Fork Strategy for a Fair, Healthy and Environmentally-Friendly Food System. Brussels: European Commission.
3 Statistical Office of the Republic of Serbia. Opštine i regioni u Republici Srbiji, 2022 [Municipalities and Regions in the Republic of Serbia, 2022]. Belgrade, 2022.
4 European Union, 2005. REGULATION (EC) NO 396/2005 OF THE EUROPEAN PARLIAMENT AND OF THE COUNCIL of 23 February 2005 on maximum residue levels of pesticides in or on food and feed of plant and animal origin and amending Council Directive 91/414/EEC. Official Journal of the European Union L 70/1.
5 Official Gazette of the Republic of Serbia, 91/2022. Regulation on maximum permitted quantities of residues of plant protection products in food and feed, 2022. Ministry of Agriculture, forestry and water management of the Republic of Serbia.
6 SANTE, European Commission Guidance Document on Pesticide Analytical Methods for Risk Assessment and Post-approval Control and Monitoring Purposes.

REFERENCES

Amato-Lourenco, L.F., Saiki, M., Saldiva, P.H.N., & Mauad, T., (2017), Influence of air pollution and soil contamination on the contents of polycyclic aromatic hydrocarbons (pahs) in vegetables grown in urban gardens of Sao Paulo, Brazil. *Frontiers in Environmental Science*, 5, 125–134, doi; 10.1016/j.envpol.2016.05.036

Ambrus, Á., & Yang, Y.Z., (2016), Global harmonization of maximum residue limits for pesticides. *Journal of Agricultural and Food Chemistry*, 64(1), 30–35, https://doi.org/10.1021/jf505347z

Anastassiades, M.S., Lehotay, J., Štajnbaher, D., & Schenck F.J., (2003), Fast and easy multiresidue method employing acetonitrile extraction/partitioning and "dispersive solid- phase extraction" for the determination of pesticide residues in produce. *Journal of AOAC International*, 86(2), 412–431, PMID: 12723926.

Antisari, L.V., Orsini, F., Marchetti, L., Vianello, G., & Gianquinto, G., (2015), Heavy metal accumulation in vegetables grown in urban gardens. *Agronomy for Sustainable Development*, 35, 1139–1147, doi: 10.1007/s13593-015-0308-z

Aubry, C., & Manouchehri, N., (2019), Urban agriculture and health: assessing risks and overseeing practices". *Field Actions Science Reports*, Special Issue 20, Urban Agriculture: Another Way to Feed Cities, 108-111, ISSN: 1867-139X.

Bao, X., (2020), New path of green development of urban agriculture in Beijing. *Qian Xian*, 10, 80–82.

Cawoy, H., Bettiol, W., Fickers, P., & Ongena, M., (2011), Bacillus-based biological control of plant diseases. In: Stoytcheva M (ed.); *Pesticides in the Modern World - Pesticides Use and Management*. Rijeka, Croatia: InTech, 274–302, doi: 10.5772/17184

Chen, T.B., Zheng, Y.M., Lei, M., Huang, Z.C., Wu, H.T., Chen, H., Fan, K-K., Yu, K., Wu, X., & Tian, Q.Z., (2005), Assessment of heavy metal pollution in surface soils of urban parks in Beijing, China. *Chemosphere*, 60, 542–551, doi: 10.1016/j.chemosphere.2004.12.072

Cooper, J., & Dobson, H., (2007), The benefits of pesticides to mankind and the environment. *Crop Protection*, 26, 1337–1348, doi:10.1016/j.cropro.2007.03.022

EFSA (European Food Safety Authority), (2018), Scientific opinion on pesticides in foods for infants and young children. *EFSA Journal*, 16(6), 5286.

EFSA (European Food Safety Authority), (2022), The 2020 European Union report on pesticide residues in food. *EFSA Journal*, 20(3), 7215.

European Commission, (2002), Directive 2002/63/EC of 11 July 2002 establishing Community methods of sampling for the official control of pesticide residues in and on products of plant and animal origin and repealing Directive 79/700/EEC. *Official Journal of the European Communities*, EN 16.7.2002. https://eur-lex.europa.eu/legal-content/EN/ALL/?uri=celex%3A32002L0063

European Commission, (2006), *Quality Control Procedures for Pesticide Residues Analysis*. Document No. SANCO/10232/2006, https://www.eurl-pesticides.eu/library/docs/allcrl/AqcGuidance_Sanco_2006_10232.pdf

European Commission, (2009), *Method Validation and Quality Control Procedures for Pesticide Residues Analysis in Food and Feed*. Document No. SANCO/10684/2009, https://www.eurl-pesticides.eu/library/docs/allcrl/AqcGuidance_Sanco_2009_10684.pdf

FAO, (2023), https://www.fao.org/news/story/en/item/1187738/icode/

FAO, Rikolto, & RUAF, (2022), *Urban and peri-urban agriculture sourcebook - From production to food systems*. Rome, FAO and Rikolto, https://doi.org/10.4060/cb9722en

Fenibo, E.O., Ijoma, G.N., & Matambo, T., (2021), Biopesticides in sustainable agriculture: A critical sustainable development driver governed by green chemistry principles. *Frontiers in Sustainable Food Systems*, 5, 619058, doi: 10.3389/fsufs.2021.619058

Filipović, V., Popović, V., & Subić, J., (2013), Organic agriculture and sustainable urban development: The Belgrade-Novi Sad metropolitan area case study. *In, Employment. Education and Entrepreneurship: Rural Entrepreneurship: Opportunities and Challenges;* Radović, M., Marković, D., Vojteski, K., Jovančević, D. (Eds.); Faculty of Business Economics and Entrepreneurship: Belgrade, Serbia, 337–353.

Fillion, J., Sauve, F., & Selwyn, J., (2000), Multiresidue method for the determination of residues of 251 pesticides in fruits and vegetables by gas chromatography/mass spectrometry and liquid chromatography with fluorescence detection. *Journal AOAC International*, 83(3), 698–713, PMID: 10868594

Gasperi, D., Pennisi, G., Rizzati, N., Magrefi, F., Bazzocchi, G., Mezzacapo, U., Centrone Stefani, M, Sanyé-Mengual, E., Orsini, F., & Gianquinto, G., (2016), Towards regenerated and productive vacant areas through urban horticulture: lessons from Bologna, Italy. *Sustainability*, 8, 1347, doi:10.3390/su8121347

Grasswitz, T.R., (2019), Integrated Pest Management (IPM) for small-scale farms in developed economies: challenges and opportunities. *Insects*, 10, 179, doi:10.3390/insects10060179

Hamzaoui-Essoussi, L., & Zahaf, M., (2012), Production and distribution of organic foods: assessing the added values. *InTech*, doi: 10.5772/52445

Handford, C.E., Elliott, C.T., & Campbell, K., (2015), A review of the global pesticide legislation and the scale of challenge in reaching the global harmonization of food safety standards. *Integrated Environmental Assessment and Management*, 11(4), 525–536, doi: 10.1002/ieam.1635

Hasan, R., Prodhan, M.D.H., Rahman, S.M.M., Khanom, R., & Ullah, A., (2017), Determination of organophosphorus insecticide residues in country bean collected from different markets of Dhaka. *Journal of Environmental & Analytical Toxicology*, 7, 4, doi: 10.4172/2161-0525.1000489

Keikotlhaile, M.B., & Spanoghe, P., (2011), Pesticide residues in fruits and vegetables. *InTech*, doi: 10.5772/13440

Kováts, N., Hubai, K., Sainnokhoi, TA., & Teke, G., (2021), Biomonitoring of polyaromatic hydrocarbon accumulation in rural gardens using lettuce plants. *Journal of Soils and Sediments*, 21, 106–117, https://doi.org/10.1007/s11368-020-02801-1

Kumar, J., Ramlal, A., Mallick, D., & Mishra, V., (2021), An overview of some biopesticides and their importance in plant protection for commercial acceptance. *Plants (Basel)*, 10(6), 1185, doi: 10.3390/plants10061185

Lazić, B., (2008), *Bašta Zelena Cele Godine*. KLM, Novi Sad.

Lazić, S., Šunjka, D., Grahovac, N., Vuković, S., & Jakšić, S., (2012), Determination of chlorpyrifos in water used for agricultural production. *Agriculture and Forestry*, 57(4), 17–25.

Lazić, S., Šunjka, D., Milovanović, I., Manojlović, M., Vuković, S., & Jovanov, P., (2015), Pesticide residues in agricultural soil of Vojvodina Province, *Serbia, 25th Annual Meeting SETAC Europe, 3-7 May 2015*, Barcelona, Spain, 431.

Lazić, S., Šunjka, D., Pucarević, M., Grahovac, N., Vuković, S., Indjić, D., & Jakšić, S., (2013), Monitoring of atrazine and its metabolites in groundwaters of the Republic of Serbia. *Chemical industry*, 67(3), 513–523, doi: 10.2298/HEMIND120508094L

Lehmann, E., Turrero, N., Kolia, M., Konaté, Y., & de Alencastro, L.F., (2017), Dietary risk assessment of pesticides from vegetables and drinking water in gardening areas in Burkina Faso. *Science of Total Environment*, 601-602, 1208–1216, DOI: 10.1016/j.scitotenv.2017.05.285

Liu, X., Cao, A., Yan, D., Ouyang, C., Wang, Q., & Li, Y., (2021), Overview of mechanisms and uses of biopesticides. *International Journal of Pest Management*, 67:1, 65–72, doi.org/10.1080/09670874.2019.1664789

Mandal, A., Sarkar, B., Mandal, S., Vithanage, M., Patra, A.K., & Manna, M.C., (2020), Impact of agrochemicals on soil health, *Chapter 7*, (Ed) *Majeti Narasimha Vara Prasad, Agrochemicals detection, treatment and remediation*, 161–187. Butterworth-Heinemann, ISBN 978-0-08-103017-2

Meftaul, I.M., Venkateswarlu, K., Dharmarajan, R., Annamalai, P., & Megharaj, M., (2020), Pesticides in the urban environment: A potential threat that knocks at the door. *Science of The Total Environment*, 711, 134612, doi: 10.1016/j.scitotenv.2019.134612

Ministry of agriculture, forestry and water management, Republic of Serbia, 2023. https://uzb.minpolj.gov.rs/wp-content/uploads/2022/10/Lista-registrovanih-SZB_1_10_2022.pdf

Ministry of agriculture, forestry and water management, Republic of Serbia, 2023a. https://uzb.minpolj.gov.rs/wpcontent/uploads/2023/01/Lista_sredstva_za_zastitu_bilja_za_organsku_proizvodnju_na_23dec2022.pdf

Montiel-Leon, J.M., Duy, S.V., Munoz, G., Verner, M.A., Hendawi, M.Y., Moya, H., Amyot, M., & Sauve, S., (2019), Occurrence of pesticides in fruits and vegetables from organic and conventional agriculture by QuEChERS extraction liquid chromatography tandem mass spectrometry. *Food Control*, 104, 74–82, doi: 10.1016/j.foodcont.2019.04.027

Olson, S., (2015), An analysis of the biopesticide market now and where is going. *Outlooks on Pest Management*, 26, 203–206, doi: 10.1564/v26_oct_04

Orsini, F., Pennisi, G., Michelon, N., Minelli, A., Bazzocchi, G., Sanyé-Mengual, E., & Gianquinto, G., (2020), Features and Functions of Multifunctional Urban Agriculture in the Global North: A Review. *Frontiers in Sustainable Food Systems*, 4, doi: 10.3389/fsufs.2020.562513

PAN, (2014), *Pesticide Action Network, Field Guide to Non Chemical Pest Management in Cowpea Production*. PAN, Hamburg.

Petit, C., Aubry, C., & Rémy-Hall, E., (2011), Agriculture and proximity to roads: How should farmers and retailers adapt? Examples from the Ile-de-France region. *Land Use Policy*, 28, 4, 867–876, doi:10.1016/j.landusepol.2011.03.001

Ping, H., Wang, B., Li, C., Li, Y., Ha, X., Jia, W., Li, B., & Ma, Z., (2022), Potential health risk of pesticide residues in greenhouse vegetables under modern urban agriculture: A case study in Beijing, China. *Journal of Food Composition and Analysis*, 105, 104222, doi: 10.1016/j.jfca.2021.104222

Pretty, J., & Bharucha, Z.P., (2015), Integrated pest management for sustainable intensification of agriculture in Asia and Africa. *Insects*, 6(1), 152–82, doi:10.3390/insects6010152

Prodhan, M.D.H., Papadakis, E.N., & Papadopoulou-Mourkidou, E., (2018), Variability of pesticide residues in eggplant units collected from a field trial and marketplaces in Greece. *Journal of the Science of Food and Agriculture*, 98, 2277–2284, doi: 10.1002/jsfa.8716

Pucarević, M., Stojić, N., Panin, B., Kecojević, I., & Bokan, N., (2015), Pesticide residues in organic products, XX Savetovanje o biotehnologiji Proceedings, 13-14. Mart 2015. *godine, Čačak, Srbija*, 20(22), 141–149.

Rémy, E., & Aubry, C., (2008), Le blé francilien à l'orée d'une profonde mutation: vers une partition de l'espace des risques? *Espaces et Sociétés*, 132-133, doi:10.3917/esp.132.0163

Saldanha, H., Sejerøe-Olsen, B., Ulberth, F., Emons, H., & Zeleny, R., (2012), Feasibility study for producing a carrot/potato matrix reference material for 11 selected pesticides at EU MRL level: Material
processing, homogeneity and stability assessment. *Food Chemistry*, 132(1), 567–573, doi: 10.1016/j.
foodchem.2011.10.071

Sanyé-Mengual, E., Specht, K., Grapsa, E., Orsini, F., & Gianquinto, G., (2019), How can innovation in
urban agriculture contribute to sustainability? Characterization and evaluation study from five Western
European cities. *Sustainability*, 11(15), 1–31, doi:10.3390/su11154221

Šunjka, D., & Mechora, Š., (2022), An alternative source of biopesticides and improvement in their formulation
- recent advances. *Plants (Basel)*, 11, 3172, doi: 10.3390/plants11223172

Šunjka, D., & Vuković, S., (2021a), Biopesticides in plant protection. In, *Agricultural studies on different subjects*. Arzu ÇIĞ (Ed.). IKSAD Publishing House, Golbasi (Turkey), ISBN: 978-625-8007-89-3

Šunjka, D., Lazić, S., Vuković, S., Alavanja, A., Nadj, Dj., & Mitrić, S., (2021), Residue and dissipa-
tion dynamic of spinetoram insecticide in pear fruits. *Plant protection science*, 57(4), 326–332, doi:
10.17221/154/2020-PPS

United Nations (UN), (2018), *Department of Economic and Social Affairs, Population Division. World
Urbanization Prospects: The 2018 Revision*. 2018. Report No.: ST/ESA/SER.A/420.

van de Merwe, J.P., Neale, P.A., Melvin, S.D., & Leusch, F.D.L., (2018), In vitro bioassays reveal that additives
are significant contributors to the toxicity of commercial household pesticides. *Aquatic Toxicology*, 199,
263–268, doi: 10.1016/j.aquatox.2018.03.033

Živanović Miljković, J., Popović, V., & Gajić, A., (2022), Land take processes and challenges for Urban agriculture: A spatial analysis for Novi Sad, Serbia. *Land*, 11, 769, https://doi.org/10.3390/land11060769

Znaor, D., & Landau, S., (2014), *Unlocking the Future: Sustainable Agriculture as a Path to Prosperity for the
Western Balkans*. Heinrich Böll Stiftung, Zagreb Office, Zagreb, ISBN 978-953-7723-08-8

14 The Vertical Farm
The Next-Generation Sustainable Urban Agriculture

Kheir Al-Kodmany

INTRODUCTION

The escalating problem of food insecurity has raised concerns among specialists who warn about the potential scarcity of sufficient food for the growing urban population. The relationship between the world population and the need for food is evident; as the former increases, the latter naturally follows suit. Concurrently, the stability of the food supply is jeopardized by many causes, including but not limited to climate change (De Bernardi and Azucar 2020), degradation of the environment, escalating food costs, and pandemics, exemplified by the recent COVID-19 outbreak (Levi and Robin, 2020). Hence, the global community will encounter a substantial predicament in the forthcoming decades, which involves the provision of sustenance to a growing populace while contending with limited natural resources. According to the Food and Agriculture Organization of the United Nations, the global population is projected to expand by around 2 billion individuals by the year 2050. Consequently, this will need a 70% augmentation in food production (FAO, 2023). In addition, urban sprawl is progressively encroaching into an increasing amount of farmland, diminishing its agricultural potential. Hence, it is imperative to prioritize the development of urban policies and innovations that may effectively facilitate optimal land utilization, enhance food production, and save water resources. Undoubtedly, the global community is currently confronted with the pressing issue of a food crisis, necessitating the implementation of efficacious measures (McCarthy et al., 2018).

The use of vertical farming can potentially address a portion of food insecurity. Vertical farming is a crop cultivation method that entails growing plants in vertically arranged layers or inclined surfaces, typically within controlled conditions such as skyscrapers or stacked containers. This novel methodology optimizes the utilization of spatial and material resources. Vertical farms are purposefully built to maximize resource efficiency. Advanced technologies, such as hydroponics, aeroponics, and aquaponics, are employed to optimize water usage and reduce the need for arable land. The observed level of efficiency plays a significant role in promoting sustainability. In contrast to conventional agricultural practices, vertical farming systems can work continuously throughout the year without being limited by seasonal fluctuations. The constant production of crops has the potential to mitigate the effects of seasonal food shortages and improve overall food accessibility. Vertical farming is especially well-suited for urban situations characterized by restricted availability of land. Integrating agricultural practices into urban environments diminishes reliance on remote rural areas for food production and transportation (Despommier, 2019).

This chapter aims to comprehensively analyze the overarching concept of vertical farming. It will delve into the underlying rationale and essential factors driving the adoption of vertical farming practices. Additionally, it will explore the fundamental techniques and systems employed in vertical agriculture. Furthermore, the advantages and disadvantages of vertical farming will be thoroughly examined, supplemented by relevant project examples. The intention is to offer a more detailed understanding of the complexities and hurdles faced by this particular project (Appendix).

DOI: 10.1201/9781003359425-14

THE PRIME DRIVERS OF THE VERTICAL FARM

Many factors drive the vertical farming project. However, the most salient ones include:

- Land efficiency and limited space in dense urban areas
- Climate change and its impact on agriculture
- Human health concerns and environmental degradation

LAND EFFICIENCY AND LIMITED SPACE IN DENSE URBAN AREAS

Urban agriculture refers to cultivating, processing, and distributing food inside urban environments. The potential advantages of this phenomenon encompass various aspects, such as promoting social integration, fostering ecological sustainability, and ensuring food sovereignty. Nevertheless, urban agriculture has notable challenges in identifying appropriate and cost-effective land. The presence of houses, roads, parks, and other forms of infrastructure in urban areas restricts the accessibility and appropriateness of land for agricultural purposes. In addition to its inherent costliness and susceptibility to taxation, regulations, and zoning ordinances that may impose limitations or prohibitions on agricultural operations, urban land often exhibits a propensity for elevated market values. Urban farmers may encounter competition from various land users, such as urban developers and commercial enterprises while seeking access to land. Urban farmers engaging in agricultural activities on disputed or unlawful land may face the risk of eviction or other legal consequences (Siregar et al., 2022).

Researchers have critiqued urban agriculture for decreasing density, entailing longer commutes, travel time, and more significant fuel costs and carbon emissions (Badami and Ramankutty, 2015). They explained that increased gas utilization from moving a small percentage of farmland into urban areas would create an extra 1.77 tons of CO_2 per household yearly. Therefore, urban agriculture suffers from finding space for farming due to scarcity and high land costs, the degradation and contamination of land, and air pollution. These challenges pose significant barriers and constraints to the development and expansion of urban agriculture (Song et al., 2015).

Vertical farming offers a solution by providing a controlled environment and maximizing agricultural work on a little lot. Harnessing the vertical dimension increases production by many folds. It can efficiently produce more food per unit area than traditional farms using vertical space. It significantly increases the yield per unit land area, as more plants can be grown on multiple stacked layers. For leafy greens, a vertical farm, for instance, can produce up to 600 times more wheat than a conventional outdoor field. The ability of vertical farming to grow crops all year-round multiplies production efficiency by 4 to 6 depending on the crop (Kozai et al., 2019).

CLIMATE CHANGE AND ITS IMPACT ON AGRICULTURE

Climate change is an emerging global phenomenon that affects the natural environment and human activities. It threatens food security and agriculture production in several ways. It alters the patterns of temperature, precipitation, droughts, floods, storms, and pests that affect crop growth and yield. For example, higher temperatures can reduce crop quality and quantity, increase water demand and evaporation, and cause heat stress to plants and animals. Changes in precipitation can lead to water scarcity or excess, soil erosion, nutrient leaching, and salinization. Droughts can reduce soil moisture and crop productivity, while floods can damage crops and infrastructure. Storms can cause wind damage, hail damage, and crop losses. Pests can spread diseases and reduce crop yields (De Bernardi and Azucar, 2020; Bhattacharya, 2019).

The vertical farm offers a solution using controlled-environment agriculture, independent of climate change. It provides stable and consistent crop yields because it does not depend on weather-altering conditions, seasons, animals, or pests, which can affect the quality and quantity of crops. Therefore, one of the main advantages of vertical farming is that it can reduce the vulnerability of crops to natural disasters such as droughts, floods, pests, and diseases. Climate-control technologies can also regulate temperature, humidity, carbon dioxide levels, and air circulation to create the ideal microclimate for each crop. Artificial lighting, sensors, cameras, and robots can monitor and adjust the growing parameters and perform seeding, harvesting, and packaging tasks. By avoiding the uncertainties and challenges of outdoor farming, vertical farms can ensure a consistent and reliable supply of fresh and nutritious food for urban populations. They can produce high-quality and reliable crops year-round, regardless of the external factors that affect conventional agriculture (Praveen and Sharma, 2019).

HUMAN HEALTH CONCERNS AND ENVIRONMENTAL DEGRADATION

Traditional farming can harm human health and the natural environment in various ways (Bernardi et al., 2020; Dhananjayan et al., 2020).

- Irrigation. Traditional farming relies heavily on irrigation to provide water for crops. Irrigation can deplete aquifers, river systems, and downstream groundwater. It can also cause waterlogging and soil salinization, reducing soil fertility and plant growth.
- Slash-and-burn. Traditional farming involves clearing forests and burning vegetation to create new cropland. Slash-and-burn can cause deforestation, soil erosion, biodiversity loss, greenhouse gas emissions, and air pollution. It can also expose people to respiratory diseases and fire hazards.
- Pesticides. Traditional farming uses pesticides to control pests and diseases that affect crops. They can contaminate food and water sources and pose health risks to farmworkers and consumers. Pesticides can also damage beneficial insects and wildlife.

Vertical farming may offer a viable solution to the above problems by providing several advantages over traditional agriculture, including (Al-Kodmany, 2018; Goodman and Minner, 2019):

- Environmentally friendly method. Unlike traditional agriculture, which relies on soil and natural conditions, vertical farming uses hydroponics or aeroponics to provide nutrients and water to plants in a controlled environment. It can lessen water pollution by eliminating the need for pesticides and fertilizers that can drain into groundwater or surface water. Vertical farming can also prevent soil erosion and deforestation caused by conventional farming methods.
- Lower water consumption. Vertical farming can use water more effectively than conventional farming. Because it can absorb and recycle water, limit evaporation from the plants or the soil, and reduce runoff, vertical farming can use up to 95% less water than conventional farming. As such, it can significantly reduce the water needed for crop production. This reduces the pressure on freshwater resources and land degradation from irrigation.
- Reduced land use. Vertical farming can preserve natural habitats and biodiversity by minimizing deforestation and land conversion. It can decrease the need for arable land, which is becoming scarce and degraded due to population growth, urbanization, and climate change.
- Reduced carbon footprint and waste. Vertical farming entails less waste and carbon emission because it produces food closer to markets and customers, cutting "food miles" (the distance food travels from where it is grown to where it is consumed) and storage costs. By extending shelf life and lowering spoilage, vertical farming can help reduce food waste.
- Improved food quality and safety. Vertical farming can produce fresh, nutritious, organic food free of contaminants and pathogens. One of the benefits of vertical farming is that

it can improve food nutrition by optimizing the growing conditions for each crop. It can increase the nutritional value of food by manipulating the environmental factors that affect plant growth and development. For example, it can adjust the temperature, humidity, lighting, and nutrient levels to suit the preferences of different crops. This can produce fresher, juicier, and more flavorful produce than conventional farming methods.
- Renewable energy. Vertical farming can also employ sustainable energy sources like solar or wind power to run its systems efficiently.

Therefore, some of the pros of vertical farming include efficient use of space, stable crop yields, protection from adverse outside conditions, less reliance on weather, reduced water usage, and fewer crop losses due to pests (Table 14.1).

VERTICAL FARMING METHODS

Researchers have advanced environmentally friendly techniques for producing food. The three primary vertical farming techniques—hydroponics, aeroponics, and aquaponics—are discussed in the following sections.

HYDROPONICS

In hydroponics, plants are grown in nutrient-rich water solutions rather than soil. There are different types of hydroponic systems as follows (Touliatos et al., 2016; Pascual et al., 2018):

- Wick system. The simplest and cheapest hydroponic system, where a wick draws the nutrient solution from a reservoir to the plant roots. It is suitable for small plants that do not need much water but have low control over the nutrient and oxygen levels.
- Deep water culture (DWC). It is a system where the plant roots are suspended in an extensive reservoir of nutrient solution aerated with an air pump and an air stone. It is simple to set up and maintain, providing high oxygen levels to the roots. However, it can be prone to algae growth and temperature fluctuations.
- Nutrient film technique (NFT). It is a system where a thin film of nutrient solution constantly flows over the plant roots in a sloped channel. It is efficient and productive, using less water and nutrients than other systems. However, it requires a reliable pump and electricity and can be vulnerable to power outages and clogging.
- Ebb and flow (or flood and drain). It is a system where the plant roots are periodically flooded with nutrient solution and then drained back to the reservoir. It allows for greater control over the nutrient and moisture levels and can accommodate different growing media types. However, it also requires a reliable pump and electricity, and it can be affected by power outages and leaks.
- Drip system. It is a system where the nutrient solution is dripped onto the plant roots or growing media through emitters or tubes. It is precise and versatile, as it can deliver different nutrients to plants. However, it can also be complex and expensive to set up and maintain and prone to clogging and salt buildup.

AQUAPONICS

Aquaponics is a way of cultivating vegetation and fish together in a symbiotic system, where the fish waste provides nutrients for the vegetation, and the vegetation filters the water for the fish. Different aquaponic systems exist depending on the design, size, location, and purpose. Some common types are (Khandaker and Kotzen, 2018):

TABLE 14.1
The Advantages of Vertical Farming Using the Sustainability Framework

#	Benefit	Environmental	Social	Economic
1	Decreases food miles (travel distances)	Decreases air pollution	Enhances air quality, which improves the environment and people's health. People receive "fresher" local food	Decreases energy consumption, packaging, and fuel to transport food
2	Reduces water consumption for food production using high-tech irrigation methods and recycling systems	Reduces surface water runoff of traditional farms	Makes potable water available to more people	Reduces costs
3	Recycles organic waste	Saves the environment by reducing needed landfills	Improves food quality and, subsequently, consumers' health	Turns waste into an asset
4	Generates local jobs	The proximity of the workplace to home will reduce employees' travel and ecological footprint	Creates a local community of workers and connections with farmers	Supports the domestic economy and local employment
5	Reduces the use of fertilizers, herbicides, and pesticides	Improves the environmental well-being	Improves food quality and, subsequently, consumers' health	Minimizes costs
6	Improves productivity	Needs less space	Reduces laborious work and saves time to do productive and socially rewarding activities	Offers greater yields
7	Avoids crop losses due to floods, droughts, hurricanes, overexposure to the sun, and inclement weather	Reduces environmental damage and requires cleanups of farms after damage	Improves food security	Avoids economic loss
8	Controls product/produce regardless of seasons	Produces regarding season	Increases accessibility year-round and improves response to population demand	Fuels economic activities year-round
9	Uses renewable energy	Reduces fossil fuel	Improves air quality	Reduces costs
10	Brings nature closer to the city	Increases biodiversity	Enhances the health and psychological well-being	Generates local jobs
11	Promotes science and green technology	Green technology reduces harm to the urban and natural environments	Encourages seeking higher education and modern skills	Offers new jobs in bioengineering, biochemistry, biotechnology, construction, and research and development
12	Decreases traditional farming activities and practices	Preserves the natural ecological system	Improves the health of citizens	Saves money required to correct environmental damage
13	Repurposes dilapidated buildings	Enhances the environment. Removes eye sores and stigma from neighborhoods	Creates opportunities for social interaction	Revives economy

- Media-based systems. These systems use solid media, such as gravel or clay pebbles, to support the plant roots and act as a biofilter for fish waste. The medium is usually flooded and drained periodically with a pump or a siphon, allowing the plants to absorb nutrients and oxygen. The media also acts as a biological filter, breaking fish waste into nitrates.
- DWC systems (also known as floating raft aquaponics systems). These systems use large tanks or ponds to grow the plants on polystyrene boards or rafts that float on top of the water. The plant roots are submerged in the water, which is aerated and circulated by pumps or air stones. The water also passes through a biofilter before returning to the fish tank.
- NFT systems. These systems use narrow pipes or channels to deliver a thin film of nutrient-rich water to the plant roots. The plants are usually grown in small pots or baskets that fit into holes in the pipes or channels. The water is recirculated back to the fish tank after passing through the plants.
- Vertical systems. These systems use vertical towers or walls to grow plants in a small footprint. The water trickles through the plants after being pumped from the fish tank to the top of the building or wall. After that, the water is gathered at the bottom and transferred back to the fish tank.

Aquaponic systems can also be classified into three main categories based on the ratio of fish feed to plant area:

- High-density systems. These systems have a high ratio of fish feed to plant area, producing more fish than plants. They require more filtration and aeration to maintain water quality and avoid ammonia buildup. They are suitable for commercial fish production or hobbyists who want to harvest fish regularly.
- Low-density systems. These systems have a low ratio of fish to plant, producing more plants than fish. They require less filtration and aeration as the plants absorb most nutrients and oxygenate the water. They are suitable for growing leafy greens, herbs, or ornamental plants.
- Balanced systems. These systems have a balanced fish-to-plant area ratio, producing equal fish and plants. They require moderate filtration and aeration to maintain water quality and nutrient balance. Balanced systems are suitable for growing various plants and fish for personal consumption or small-scale marketing.

AEROPONICS

Aeroponics is a way of growing vegetation without soil, using a mist of nutrient solution to deliver water and minerals to the roots. Aeroponic systems consist of a reservoir, a pump, a spray nozzle, and a chamber or enclosure that holds the plants and exposes their roots to the mist (Al-Kodmany, 2018). There are three types of aeroponics systems (Eldridge et al., 2020).

- High-pressure aeroponics systems. In these systems, mist is generated by high-pressure pumps. Overall, these systems are more efficient and can produce smaller droplets better absorbed by the roots. However, they are also more expensive and complex to set up and maintain.
- Low-pressure aeroponics systems. A simpler alternative, low-pressure system is the more basic type of aeroponics. They are cheaper and easier to use but are less effective than high-pressure techniques. They use regular water pumps and sprinklers to create the mist.
- Commercial aeroponics systems. These are state-of-the-art aeroponics today. They are designed for large-scale production and can handle thousands of plants simultaneously. They use advanced technology and automation to optimize the growth conditions and yield of the crops. They are costly and require professional installation and maintenance.

VERTICAL FARMING SUPPORTIVE TECHNOLOGIES

LED LIGHTING

It is a crucial technology for vertical farming, and it offers several advantages over traditional light sources for vertical farming, as follows (Kalantari et al., 2017):

- It can be customized to emit specific wavelengths of light that match the photosynthetic needs of different crops to enhance plant growth, quality, and yield.
- It can be controlled to adjust the intensity and duration of light according to each crop's growth stage and cycle to optimize vertical farming systems' energy use and productivity.
- It can be placed close to the plants without causing heat damage or stress to reduce vertical farming facilities' space and cooling requirements.

SENSORS AND AUTOMATION

Sensors are essential for creating a perfect environment for vertical farming. They can help automate vertical farming by monitoring and controlling the growing environment. These devices monitor and control various aspects of the vertical farming system, such as temperature, humidity, pH, CO_2 levels, nutrient levels, and water flow. Sensors enable growers to collect data on various aspects of vertical farming and use them to improve their decision-making and optimize production. They would also allow growers to remotely observe and supervise their vertical farms without human intervention. Some of the main types of sensors used in vertical farming are (Chuah et al., 2019; Saad et al., 2021):

- Climate sensors. They measure the vertical farm's air temperature, humidity, CO_2 levels, and airspeed. These parameters affect plant growth, development, and risk of diseases and pests. By adjusting the climate conditions according to the plant's needs, growers can optimize crop yield and quality.
- Irrigation water sensors. These sensors measure the pH (acidity) and EC (electrical conductivity) of the irrigation water, which reflects the nutrient composition and availability of the plants. The pH and EC values can vary depending on the type of hydroponic system, the water source, and the fertilizer used. By maintaining optimal pH and EC levels in the irrigation water, growers can ensure optimal plant growth and prevent nutrient deficiencies or toxicities.
- Light sensors. These sensors measure the light level and spectrum as perceived by the plants. The light level and range affect the photosynthesis rate, morphology, flowering, and coloration of the plants. The light level and spectrum perceived by the plants can deviate from the light level and scope installed depending on factors such as plant density, canopy structure, and the reflectance of materials. By measuring the light level and spectrum as perceived by the plants; growers can adjust the LED lighting modules to provide optimal light conditions for each crop stage.
- Phenotyping sensors. These sensors measure the plant morphology, growth, color, biomass, etc. These parameters reflect the growth results of the plants under different environmental conditions. By measuring these parameters, growers can evaluate the performance of different cultivars, treatments, or management practices. They can also identify problems such as diseases or pests early and take corrective actions accordingly.

MACHINE LEARNING AND DATA ANALYTICS

Data analytics collects, processes, analyzes, and visualizes data to gain insights and support decision-making. Machine learning enables computers to learn from data and make predictions or decisions without being explicitly programmed. It can help vertical farmers optimize various aspects of their production, such as crop selection, lighting, temperature, humidity, irrigation, nutrient delivery, harvesting, and packaging. Machine learning can also help detect and prevent crop diseases, pests, and environmental stress. Similarly, data analytics can help vertical farmers monitor and control their systems, improve efficiency and profitability, and enhance product quality and safety. They can help vertical farmers understand customers' preferences and demands and tailor their products accordingly. Consequently, vertical farmers can achieve higher yields, lower costs, lower environmental impact, and higher customer satisfaction using machine learning and data analytics (Siregar et al., 2022).

VERTICAL FARM PROJECTS

Vertical farming is sprouting rapidly. The following section highlights significant projects in different parts of the world (Table 14.2).

TABLE 14.2
Examples of Vertical Farming Projects around the Globe

Continent	Name	Location
North America	AeroFarms	Newark, New Jersey
	Gotham Greens	Brooklyn, New York
	Green Spirit Farms	New Buffalo, Michigan
	The Plant	Chicago, Illinois
	FarmedHere	Chicago, Illinois
	Vertical Harvest	Jackson, Wyoming
	Plenty	San Francisco, California
	Terrasphere	Vancouver, B.C., Canada
Europe	Infarm	Bedford, UK
	Nordic Harvest	Copenhagen, Denmark
	Agri-science	Norwich, UK
	Skyberries	Vienna, Austria
	Agricool	Paris, France
	PlantLab	Den Bosch, Holland
Asia	Pasona Headquarters	Pasona Headquarters
	Kameoka Plant	Kameoka, Kyoto, Japan
	Keihanna Techno Farm	Keihanna, Kyoto, Japan
	VertiVegies	Singapore
	Sky Greens	Singapore
	YesHealth iFarm	Taoyuan, Taiwan
	Rural Development Administration	Suwon, South Korea
	Nuvege	Kyoto, Japan
Middle East	ECO1	Dubai, UAE
	Madar Farms	Dubai, UAE
	Mowreq	Riyadh, Saudi Arabia
	Green Box	Kuwait, Kuwait
Australia	Stacked Farms	Gold Coast, QLD
	Feed the Bush	Melbourne
	Eden Towers	Perth
Africa	Agrotonomy	Kenya, Equatorial Guinea, Nigeria, and Cape Verde
	Kibera Vertical Farm	Kenya
	Malabo Montpellier Panel	Equatorial Guinea
	Green Studios	Egypt
South America	AgroUrbana	Chile

The Plant, Chicago, Illinois, US

The Plant is a unique example of a net-zero vertical farm in a historic building in Chicago's old stockyards. The former meatpacking plant and Peer Foods warehouse, built in the 1920s, covers an area of 8,686 m² (93,500 ft²) and has four floors. The Plant aims to be a hub for food businesses, research labs, and educational and training programs for vertical farming. A $1.5 million funding from the Illinois Department of Commerce and Economic Opportunity assisted in transforming the building into a vertical farm and food company incubator. The project began in 2010 and was completed in 2016 (Birkby, 2016; Al-Kodmany, 2018).

The Plant uses an innovative anaerobic digestion system to convert food waste into biogas that powers, heats, and cools the facility. The system can process 27 tons of food waste per day and 11,000 tons per year, generating methane that is burned to produce electricity and heat. The Plant produces various food items, such as greens, mushrooms, bread, and kombucha tea. It integrates a tilapia farm, beer brewery, kombucha brewery, communal kitchen, an aquaponics system, and green energy production. The project aims to illustrate how sustainable food production and economic development can be achieved through on-site gardening, the incubation of small craft food businesses, brewing beer and kombucha, and other activities. The project utilizes materials that would otherwise be wasted through a closed-loop system by connecting one output to another's input (Al-Kodmany, 2020). The main features of the closed-loop system as follows:

- An anaerobic digester at the heart of the system turns organic materials into biogas piped into a turbine generator to make electricity for plant growth light.
- The plants make oxygen for the kombucha tea brewery, and the kombucha tea brewery produces CO_2 for the shrub.
- Waste from the fish feeds the vegetation, and the vegetation cleanses the water for the fish.
- More fish waste goes to the digester along with plants and wastes from outside sources and spent grain from the brewery.
- Spent barley from the brewery feeds the fish.
- Sludge from the digester becomes algae duckweed that also feeds the fish.
- Along with electricity, the turbine makes steam piped to the commercial kitchen, brewery, and the entire building for heating and cooling.
- In a nutshell, we get kombucha tea, fresh vegetables, fish, beer, and food from the kitchen, all with no waste.

Sky Greens, Singapore

Singapore is a small island nation with a dense population that relies heavily on food imports. Due to the limited availability and high land cost, only about 250 acres of the island are used for agriculture, producing just 7% of the food its residents need. The rest of the food demand is met by importing food from various countries, which adds to transportation costs and environmental impact. Singapore has been exploring and investing in vertical farming to address these challenges. Vertical farming offers several benefits for Singapore, such as reducing the dependence on food imports, enhancing food security and quality, saving water and energy, and creating more green jobs. Although Singapore's situation is unique, it also serves as an example for other cities facing similar land scarcity issues, population growth, and climate change (Al-Kodmany, 2018; Wong et al., 2020).

Sky Greens, Singapore's first commercial urban vertical farm for tropical vegetables, operates with minimal energy, water, and land consumption to produce nutritious and tasty greens. The farm, which started in 2010, has a three-story (9 m/30 ft) structure that employs "A-Go-Gro (AGG) Vertical Farming" in transparent greenhouses to cultivate various types of leafy vegetables throughout the year with much higher yields than conventional farming methods. The farm can harvest 1 ton of fresh greens every two days and offers a range of tropical vegetables, such as Chinese cabbage, Spinach, Lettuce, Xiao Bai Cai, Bayam, Kang Kong, Cai Xin, Gai Lan, and Nai Bai. The farm has successfully delivered high-quality food at reasonable prices and plans to increase its production and diversify its vegetable offerings (Al-Kodmany, 2018; Wong et al., 2020).

The AGG system is a vertical farming technology that uses aluminum A-frames to support rotating troughs of plants. The A-frames can reach up to 9 m (30 ft) in height and have 38 levels of troughs with different growing media, such as soil or hydroponics. The system occupies only 5.6 m^2 (60 ft2) of floor space, which makes it ten times more productive than traditional farming. The troughs rotate around the A-frames three times a day to ensure even sunlight exposure for the plants. This also reduces or eliminates the need for artificial lighting in some parts of the building. The rotation is driven by a patented low-carbon hydraulic system that holds the plant trays. The hydraulic system is a modern adaptation of an ancient technique; it is a closed loop that uses gravity and consumes very little energy. Each 9 m (30 ft) tower uses only 60 W of power; thus, the monthly electricity cost for the farm is only about \$360 (\$3/tower) (Wong et al., 2020).

KEIHANNA TECHNO FARM, KEIHANNA, KYOTO, JAPAN

Spread Company's Keihanna Technopolis facility is one of its most advanced projects. Completed in 2018, it is a highly automated vertical farm that uses Techno FarmTM, a cutting-edge food production system. It can grow leaf lettuce varieties without pesticides and produce 30,000 heads per day. The facility has a robotic arm that plants the trays of vegetation, which are stacked vertically. The irrigation and harvesting processes are also done mainly by robotic arms. The crops undergo photosynthesis continuously for 24 hours under specialized LED lights of white and purple colors. The facility also saves water by recycling 90% of it. Automation reduces labor costs by 50%. Spread strives to maintain high hygiene standards in the cultivation environment and aims to obtain the international food safety certification "FSSC22000" (Despommier, 2019).

CHALLENGES AND OPPORTUNITIES

Vertical farming has some advantages, such as saving space, reducing water use, and increasing crop yield. However, it faces some challenges (Despommier, 2013; Al-Kodmany, 2018; Despommier, 2019).

- High initial costs. Vertical farming requires a lot of investment to set up the facilities, equipment, and systems needed to grow crops indoors. Land prices in urban areas are also very high, and finding suitable locations for vertical farms is challenging. Some vertical farms use existing structures, such as shipping containers or abandoned buildings, to reduce costs, but these may not be optimal for growing conditions.
- High operational costs. Vertical farming consumes much energy, especially artificial lighting, climate control, and water circulation. Energy costs account for 40% to 50% of the total production costs of vertical farming. Labor costs are also high, as vertical farming requires highly skilled laborers and staff to run and maintain the advanced systems. Other operational costs include nutrients, seeds, growing mediums, and pest control.
- Limited crop variety. Vertical farming can only grow a few crops that are suitable for indoor conditions and have a high economic value. Most crops grown in vertical farms are leafy greens, herbs, and microgreens, which have a low caloric density and do not provide enough food security for the growing population. Vertical farming cannot grow crops that need pollination, such as fruits and nuts, or crops that require a lot of space, such as grains and tubers.
- Environmental impact. Vertical farming may reduce some environmental problems associated with conventional agriculture, such as land degradation, water pollution, and pesticide use. However, it also creates new ecological challenges, such as greenhouse gas emissions

from energy consumption, waste generation from packaging, and loss of biodiversity from monoculture. Vertical farming also depends on external inputs, such as electricity and water, which may not be reliable or sustainable in some regions.
- Social impact. Vertical farming may negatively impact rural communities that depend on agriculture. It may reduce the demand for traditional farming jobs and skills, leaving many farmers unemployed or underemployed.

Nevertheless, new technologies enable faster, more efficient, and more sustainable crop production. Technological advancements offer unique opportunities to enhance the prospect of the vertical farm. Some of these technologies are (Chin and Audah, 2017; bin Ismail and Thamrin, 2017; Krishnan, & Swarna, 2020):

- Artificial intelligence (AI). AI boosts the capabilities of various systems used in vertical farms, such as lighting, irrigation, nutrient delivery, pest control, and harvesting. AI can also analyze data from sensors and cameras to optimize environmental conditions and crop quality.
- Internet of things (IoT). Sensors are the backbone of vertical farming technologies that track heat, illumination, humidity, CO_2 levels, pH, EC, and other factors. IoT devices and cloud platforms can communicate with sensors for real-time feedback and control.
- Autonomous robotics. Robots can perform tasks such as seeding, transplanting, pruning, harvesting, and packaging in vertical farms. Robots can also move plants between different levels or locations within the farm to ensure optimal growth.
- Agritech analytics. Data-driven solutions can help vertical farmers monitor and improve their operations. Agritech analytics can provide insights into crop performance, yield prediction, resource consumption, market demand, and profitability.

CONCLUSIONS

Vertical farming, although not a universally applicable solution for all agricultural issues, undoubtedly presents a beneficial augmentation to the current food system, with the potential to contribute substantially toward a future characterized by enhanced resilience and equity. Recognizing this phenomenon's potential benefits and constraints is essential for comprehending its significance in the broader framework of worldwide food production. Vertical farming is a potentially effective response to the increasing food requirements, especially in urban regions, offering a range of advantages regarding the environment, economy, and society. There are several noteworthy benefits associated with this approach, such as the preservation of water and land resources, the mitigation of greenhouse gas emissions and food miles through localized production, the enhancement of food quality and safety through controlled environments, the generation of new markets and employment prospects, and the improvement of overall food security and resilience.

Notwithstanding these advantages, vertical farming encounters obstacles, including big upfront expenses, energy usage, and technical complexities. Financial obstacles can impede the general acceptance and implementation of controlled environments, while the high energy consumption associated with such environments may counterbalance certain environmental benefits. Technical challenges in the field of plant cultivation encompass several aspects, such as the optimization of growing conditions, the implementation of automated systems, and the mitigation of potential issues associated with disease control. It is crucial to emphasize the necessity of a collaborative endeavor in research and innovation to fully exploit the potential of vertical farming and address its inherent constraints. The persistent pursuit of scientific investigation has the potential to yield significant advancements in energy efficiency, infrastructure cost-effectiveness, and enhancement of farming practices. The integration of automation and precision agriculture techniques in vertical farming systems has the potential to optimize operations and improve overall efficiency.

In conclusion, while vertical farming is not a panacea, its integration into the broader food system offers substantial benefits and represents a step toward a more sustainable and resilient future. Recognizing its potential contributions and actively addressing its limitations through ongoing research and innovation will be pivotal in maximizing the positive impact of vertical farming on global agriculture (O'Sullivan et al., 2019; Riccaboni et al., 2021).

REFERENCES

Al-Kodmany, K. (2018). The vertical farm: A review of developments and implications for the vertical city. *Buildings*, *8*(2), 24.

Al-Kodmany, K. (2020). The vertical farm: Exploring applications for peri-urban areas. In *Smart Village Technology: Concepts and Developments*, edited by Srikanta Patnaik, Siddhartha Sen, and Magdi S. Mahmoud, 203–232. Cham: Springer.

Badami, M. G., & Ramankutty, N. (2015). Urban agriculture and food security: A critique based on an assessment of urban land constraints. *Global Food Security*, *4*, 8–15.

Bhattacharya, A. (2019). Global climate change and its impact on agriculture. *Changing Climate and Resource Use Efficiency in Plants*; Academic Press: Cambridge, MA, pp. 1–50.

Birkby, J. (2016). Vertical farming. *ATTRA Sustainable Agriculture*, *2*, 1–12.

Chin, Y. S., & Audah, L. (2017). Vertical farming monitoring system using the internet of things (IoT). In *AIP conference proceedings* (Vol. 1883, No. 1, p. 020021). AIP Publishing LLC.

Chuah, Y. D., Lee, J. V., Tan, S. S., & Ng, C. K. (2019). Implementation of smart monitoring system in vertical farming. In *IOP Conference Series: Earth and Environmental Science* (Vol. 268, No. 1, p. 012083). Philadelphia, PA: IOP Publishing.

De Bernardi, P., Azucar, D., (2020). The food system grand challenge: A climate smart and sustainable food system for a healthy Europe. *Innovation in Food Ecosystems: Entrepreneurship for a Sustainable Future*, 1–25.

Despommier, D. (2013). Farming up the city: The rise of urban vertical farms. *Trends Biotechnol*, *31*(7), 388–389.

Despommier, D. (2019). Vertical farms, building a viable indoor farming model for cities. *Field Actions Science Reports. The Journal of Field Actions*, (Special Issue 20), 68–73.

Dhananjayan, V., Jayakumar, S., & Ravichandran, B. (2020). Conventional methods of pesticide application in agricultural field and fate of the pesticides in the environment and human health. *Controlled Release of Pesticides for Sustainable Agriculture*, 1–39.

Eldridge, B. M., Manzoni, L. R., Graham, C. A., Rodgers, B., Farmer, J. R., & Dodd, A. N. (2020). Getting to the roots of aeroponic indoor farming. *New Phytologist*, *228*(4), 1183–1192.

FAO (2023). *The Food and Agriculture Organization of the United Nations*, https://www.fao.org/home/en (accessed, November 30, 2023))

Goodman, W., & Minner, J. (2019). Will the urban agricultural revolution be vertical and soilless? A case study of controlled environment agriculture in New York City. *Land Use Policy*, *83*, 160–173.

bin Ismail, M. I. H., & Thamrin, N. M. (2017). IoT implementation for indoor vertical farming watering system. In *2017 International Conference on Electrical, Electronics and System Engineering (ICEESE)* (pp. 89–94). New York: IEEE.

Kalantari, F., Mohd Tahir, O., Mahmoudi Lahijani, A., & Kalantari, S. (2017). A review of vertical farming technology: A guide for implementation of building integrated agriculture in cities. In *Advanced Engineering Forum* (Vol. 24, pp. 76–91). Trans Tech Publications Ltd.

Khandaker, M., & Kotzen, B. (2018). The potential for combining living wall and vertical farming systems with aquaponics with special emphasis on substrates. *Aquaculture Research*, *49*(4), 1454–1468.

Kozai, T., Niu, G., & Takagaki, M. (Eds.). (2019). *Plant Factory: An Indoor Vertical Farming System for Efficient Quality Food Production*. Academic Press.

Krishnan, A., & Swarna, S. (2020). Robotics, IoT, and AI in the automation of agricultural industry: A review. In *2020 IEEE Bangalore Humanitarian Technology Conference (B-HTC)* (pp. 1–6). New York: IEEE.

Levi, E., & Robin, T. (2020). COVID-19 did not cause food insecurity in Indigenous communities but it will make it worse. *Yellowhead Institute website*, 29. https://yellowheadinstitute.org/2020/04/29/covid19-food-insecurity/

McCarthy, U., Uysal, I., Badia-Melis, R., Mercier, S., O'Donnell, C., & Ktenioudaki, A. (2018). Global food security: Issues, challenges and technological solutions. *Trends in Food Science & Technology*, *77*, 11–20.

O'sullivan, C. A., Bonnett, G. D., McIntyre, C. L., Hochman, Z., & Wasson, A. P. (2019). Strategies to improve the productivity, product diversity and profitability of urban agriculture. *Agricultural Systems*, *174*, 133–144.

Pascual, M. P., Lorenzo, G. A., & Gabriel, A. G. (2018). Vertical farming using hydroponic system: Toward a sustainable onion production in Nueva Ecija, Philippines. *Open Journal of Ecology*, *8*(01), 25.

Praveen, B., & Sharma, P. (2019). A review of literature on climate change and its impacts on agriculture productivity. *Journal of Public Affairs*, *19*(4), e1960.

Riccaboni, A., Neri, E., Trovarelli, F., & Pulselli, R. M. (2021). Sustainability-oriented research and innovation in 'farm to fork'value chains. *Current Opinion in Food Science*, *42*, 102–112.

Saad, M. H. M., Hamdan, N. M., & Sarker, M. R. (2021). State of the art of urban smart vertical farming automation system: Advanced topologies, issues and recommendations. *Electronics*, *10*(12), 1422.

Siregar, R. R. A., Seminar, K. B., Wahjuni, S., & Santosa, E. (2022). Vertical farming perspectives in support of precision agriculture using artificial intelligence: A review. *Computers*, *11*(9), 135.

Song, W., Pijanowski, B. C., & Tayyebi, A. (2015). Urban expansion and its consumption of high-quality farmland in Beijing, China. *Ecological Indicators*, *54*, 60–70.

Touliatos, D., Dodd, I. C., & McAinsh, M. (2016). Vertical farming increases lettuce yield per unit area compared to conventional horizontal hydroponics. *Food and Energy Security*, *5*(3), 184–191.

Wong, C., Wood, J., & Paturi, S. (2020). Vertical farming: An assessment of Singapore City. *Etropic: Electronic Journal of Studies in the Tropics*, *19*, 228–248.

APPENDIX: VERTICAL FARMING BY DIGITS

9 billion	Expected global population by 2050, according to the United Nations.
70%	More food that will be needed to meet minimum human nutritional requirements by 2050, compared to 2013.
38%	Global land surface used for agriculture, according to the Food and Agriculture Organization.
80%	Share of the earth's population that will reside in urban centers by 2050.
$1.4 billion	VC funding in vertical farming in 2022, a record amount, according to PitchBook.
10–20 times	Vertical farming yield per acre compared to open-field crops.
25%	How much the vertical farming market is expected to grow annually over the next decade, according to a 2021 Morgan Stanley report.
12,000 years ago	Organized plant agriculture begins; irrigation is the first environmental control.
30 CE	The first greenhouses emerge.
1940s–1980s	The Green Revolution begins. The period sees a major increase in the production of food grains, thanks to new technologies such as high-yielding varieties of wheat and rice and widespread use of chemical fertilizers, particularly, benefitting countries like Mexico and India.
1952	The first phytotrons—large, centralized, controlled-environment growing facilities—are launched around the world.
1984	The Netherlands pioneers the concept of stacked layers to grow crops in an industrial factory.
1989–1990	Kewpie Corporation—the Japanese packaged food company behind the eponymous mayonnaise—creates its TS Farm method, which uses aeroponics, fluorescent lighting, and A-frame design.
2004	AeroFarms, an indoor vertical farming company, is founded in Newark, New Jersey.
2005	Widespread conversion to LE begins in vertical farms.
2010	*The Vertical Farm: Feeding the World in the 21st Century* is published, but the book's futuristic visions of high-rise indoor urban agriculture prompt pushback from agricultural engineers and controlled environment agriculture experts, who raise questions about energy usage.
2014	Plenty, an indoor plant-based company based in San Francisco, is founded.
2015	New York City-based Bowery Farming comes along.

Source: Michelle Cheng: Vertical farming: Controlled worlds, *QUARTZ*, February 8, 2023, https://qz.com/emails/quartz-obsession/1850087379/vertical-farming-controlled-worlds

Index

Note: **Bold** page numbers refer to tables; *italic* page numbers refer to figures and page numbers followed by "n" denote endnotes.

Abdulkadir, A. 52, 53
Abdu, N. 49
acoustical ground effect 201
aeroponics 223, 228
Africa
    agriculture 14–15
    water supply systems 16
A-Go-Gro (AGG) vertical farming 231–232
agricultural production
    contemporary 211
    conventional 218
    RS *see* Republic of Serbia (RS)
    sustainability of 209
    threat to 217
Agriculture and Livestock Policy (1997) 136
agri-food system transitions in Kerala
    and landscape-level changes 65–66
    and urban home gardening *see* urban home gardening,
        in Kerala
agritech analytics 233
agro-urbanism 164
air quality 200–201
Akbar (Mughal Emperor) 13
Akoto-Danso, E. K. 50
allotment gardening 29
"allotment gardens" (Germany) 197
Ana, T. 116
ancient civilizations 17
    China 11–12
    Egypt 10
    India 10
    Japan 11
anthropogenic pollutants 209
aquaponics 223, 226–228
Arab Engineering Bureau 173
Arabian Nights 12
arable land 88, 223, 225
    in Singapore 110
    UA 114
Aravindakshan, S. 121
Architectural and Urban Design Program 5
Arthaśāstra 12–13
artificial intelligence (AI) 233
Artmann, M. 118
Ashoka (Indian ruler) 12
Assurnasirpal II (Assyrian king) 15
Asuero, R. P. 29
atmosphere 48
autonomous robotics 233
Ayllon, M. J. Z. 32

backyard gardens *see* urban home gardens
balanced aquaponics systems 228
Bellwood-Howard, I. 52, 54

Belmont Forum for the Sustainable Urbanization Global
        Initiative 166
Binns, T. 142
biodiversity 48–49
biopesticide 218
Bohn, K. 206
Bougnom, B. P. 52
Boussini, H. 52
Brinkmann, K. 54
brown agenda 1, 23, 32
building-based farming 197
building-bound space 197–198
building construction and design
    constructability and substrate weight 205
    rooftops 204–205
    vertical farming 205
building facades 197–198; *see also* built environment,
        urban farming on
built environment, urban farming on
    aesthetics enhancement 202–203
    air quality improvement 200–201
    building construction and design *see* building
        construction and design
    GHG emissions 203–204
    internal building temperature, moderation of 198–199
    lower energy costs 202
    noise pollution levels reduction 201–202
    water pollution 204
Burkina Faso *see* urban and peri-urban agriculture (UPA)
Bussotti, F. 201

Camps-Calvet, M. 30
carbon emissions (CEs) 115
Cauwenbergh, N. van 117
Chagomoka, T. 51
China
    agriculture in the Yangtze River region 11–12
    vegetation 15
    water supply systems 16
Chongyi Shangbao Terraces 16
circular cities concept 24–25
Cissé, O. 54, 55
climate change 48
    challenges hindering the development of UPA in
        MENA region 157–159
    and its impact on agriculture 224–225
    MENA region *see* UPA in MENA region
climate sensors 229
commercial aeroponics systems 228
community gardening 29–30
"Community gardens" (UK) 197
community-supported agriculture (CSA) 30
continuous productive urban landscapes (CPULs) 205–206
Cool Farm Tool 121, 128

Court of Palms 15
cultivable lands 114

Danso, G. 1
Dao, J. 49, 52
Dar es Salaam (Tanzania) see UPA in Dar es Salaam and
        Morogoro Municipality
Dasgupta, N. 9
data
    analytics and machine learning 229–230
    UA in India 182–183
deep water culture (DWC)
    aquaponics 228
    hydroponics 226
demographic determinants
    UA practices: age, by 183–184, 184
    gender, by 184–185, 185
dietary diversity 24
Di Leo, N. 48, 54
Diogo, R. V. C. 50, 52
Doha Living Lab at QU 173
domestication syndrome 10
Dongus, S. 143
dooryard gardens see urban home gardens
Dossa, L. H. 50, 51
drip system 72, 226
drylands nexus see food-water-energy nexus
Dugué, P. 53, 54
Dunnet, N. 202
Dutt, D. 80

early agricultural practices
    Africa 14–15
    China 15
    Egypt 13–14
    Ethiopia 14
    India 14
    Mesopotamia 13
    Syria 15
ebb and flow (or flood and drain) 226
ecological systems
    advantages of urban and peri-UA 3
    global warming issue 17
economic impact
    UPA in Dar es Salaam and Morogoro Municipality
        140–144
    vertical farming 232
ecosystem services (EES) 21
edible city concept 21, 24
Egypt
    civilizations 10
    gardens in 12
    vegetation 13–14
"Egyptian Switchers Community" 157
El Bilali, H. 154
Elzen, B. 77
Ethiopia, vegetation in 14

Facebook-based (FB) agriculture groups 67–68
farmers
    Singapore see traditional farmers (Singapore)
    UPA in Dar es Salaam and Morogoro Municipality
        137–140
farm-to-fork strategy 209

fertilizers 209
flowerbeds or flower meadows 203
Foeken, D. 146
Food and Agriculture Organization (FAO) 211
food deserts 24
food insecurity
    MENA region see UPA in MENA region
    vertical farming see vertical farming
food miles 23
food production 24
food security 40, 114–115, 118–119
    challenges hindering the development of UPA in
        MENA region 157–159
    India 126, 129
    refugees, UPA advantages for 156
    Singapore 85
    UA in MENA region 154–155
    UPA see urban and peri-urban agriculture (UPA)
food urbanism 164, 166–167, 172
food-water-energy nexus
    Qatar see Qatar
    urban agriculture and 164–165
Food–Water–Energy Nexus Master Plan 173
formal settlements 190, 194n1
Fuller, G. 15
fungicide residues 213

Garden of Eden 12, 13
gardens, ancient times 12–13
Gaza Urban and Peri-Urban Agriculture Platform
        (GUPAP) 158
Gṛhārāma 13
Ghosh, S. 116
Gilgamesh Epic 12
Global Farm 173
global population 113–114
Global South see urban home gardening in Kerala
global warming 17
Golden, S. 144
Gomgnimbou, A. P. K. 48, 52
Good Agricultural Practice (GAP) Program 94
Govind, M. 20, 27
Graefe, S. 44, 52
green agenda 1, 23, 32
Green Deal goals 209
green economy 209
greenhouse gas (GHG) emissions 48, 114, 202
green jobs 194
green spaces 2
green walls (GWs) 198, 202–204
ground dip 201
"Grow Your Own" in Cairo governorate 157

Halloran, A. 136, 138, 140
Hanging Gardens of Babylon 15
"harapitani/halapita" 12
Häring, V. 49
Hayashi, K. 52
heavy metals
    in fruits 214–216
    in soil 212–214
    in water 214
Helmand civilization 10
herbicides 209, 213, 214

high-density aquaponics systems 228
high-pressure aeroponics systems 228
high-tech farming 94–95, 105–107, 110
home-garden food production 24
Hoover, B. 26
house gardens *see* urban home gardens
house-lot gardens *see* urban home gardens
hydraulic civilizations 15; *see also* irrigation
hydroponics 223, 226

India
    Ashokan period 12–13
    civilization 10
    UA *see* UA in India
    vegetation 14
    water supply systems 16
industrialization 197
Indus Valley civilization 10
informal settlements *190*, 194n1
insecticides 209
integrated pest management (IPM) 218
integrated rooftop greenhouses (i-RTGs) 25
internet of things (IoT) 233
intra-urban agriculture 210
irrigation 225
    technologies 15–16
    water sensors 229

Jabal al-Qal'ah district project 159
Janapadas 10
Japan
    ancient civilizations 11
    Keihanna Techno Farm 232
Jomon culture 11
Jomon hunter-gatherers 11
Jomon period 11

Kaboré, W.-T. T. 48
Kamel, I. M. 154
Karshika Vipani Online Organic Agricultural Market 68
Keihanna Techno Farm (Keihanna, Kyoto, Japan) 232
Kerala
    agri-food regime and landscape-level changes 65–66
    geography and population 64
    urban home gardening *see* urban home gardening, in Kerala
Kerala State Biodiversity Board 79
Kerala State Planning Board 79
Khabiya, P. 194
Kiba, D. I. 48, 49
"Kilimo cha Kufa na Kupona" (Agriculture for Life and Death) in 1974/75 135
Kingsbury, N. 202
Kiran, A. 194
kitchen gardens *see* urban home gardens
Kohsaka, R. 31
Korbéogo, G. 52, 54
Kranji Countryside Association (KCA) 99, 100
Krusche, P. 198

land management 49
landscape urbanism 179
land use dynamics 114
Levasseur, V. 52

light sensors 229
limit of detection (LOD) 213
livelihoods *see* urban and peri-urban agriculture (UPA)
Livestock Policy 2006 136
Lompo, D. J. 49
low-density aquaponics systems 228
low-pressure aeroponics systems 228

Maani, O. 159
machine learning 229–230
Maćkiewicz, B. 29
Magid, J. 136, 138, 140
Mahajanpadas 10
Mahalingam, A. 194
Manka'abusi, D. 48
Markard, J. 77
Martellozzo, F. 154
Matsuzawa, T. 31
maximum residue levels (MRLs) 211, 217
media-based aquaponic systems 228
Mesopotamia
    gardens 12
    vegetation 13
Middle East and North Africa (MENA) region *see* UPA in MENA region
Milan Urban Food Policy Pact (MUFPP) 41
Ming Dynasty (1368–1644) 16
mini-polyhouse 72
mixed method approach 182
Mkwela, H. S. 136
Mlozi, M. R. S. 140
MNEX International Workshops 176
Mntambo, B. 194
monitoring program of pesticide residues (Novi Sad) 212
    green market in Novi Sad *215*
    and heavy metals in fruits 214–216
    and heavy metals in soil 212–214, *213*, **214**
    in vegetables from ecological agriculture *216*, 216–217
    in water 214
Morogoro Municipality (Tanzania) *see* UPA in Dar es Salaam and Morogoro Municipality
Moustier, P. 141
movable and soil-independent systems 197
MUFFP Framework for Action 41
multifunctionality
    meaning 20
    of urban agriculture 21–22
multi-level perspectives (MLPs)
    agri-food system transitions in Kerala 65
    anchoring 63
    fit-and-conform and stretch-and-transform patterns 64
    landscapes, regimes, and niches 63
Mwalukasa, M. 146

Nagananda, K. 194
Naik, G. 194
nattuchantha (weekly market in Thrissur) 68
nature-based solutions (NbS) 3, 20
    edible city concept 21
    to urban sustainability 21–22
Nature Ecology & Evolution 15
Near East 153
Neolithic Revolution 10
New Urban Agenda (NUA) 41

niches
    anchoring of innovations 77–78
    expectations and visions, people's networks, and
        learning 78–79
    policy actors 78
    scientists 78
    socio-cultural factors 79
    urban home gardening 77 *see also* urban home
        gardening, in Kerala
Nicholls, E. 29
Niger *see* urban and peri-urban agriculture (UPA)
Nile's water channels 16
"Nin-Gishzida" 12
North Africa *see* UPA in MENA region
nutrient film technique (NFT)
    aquaponics 228
    hydroponics 226
nutrients and materials 49–50

Olivier, D. W. 193
Onam (harvest festival of Kerala) 68
on-ground vegetation 206
Open Data Kit (ODK) platform 182
organochlorine insecticides 213
organoponics/organopónicos 31
Orsini, F. 40, 41, 44, 51
Ouédraogo, A. 49
Ozainne, S. 11

Paganini, N. 116
Palace at Mari (1800 B.C.) 15
Palace at Ugarit 15
Panchanathan, B. 194
pandemic-induced disruptions 114, 122, 129
Patenković, A. 31
peri-urban agriculture 210
pest control 209
pesticide residues
    in fruits and vegetables 212
    monitoring program of pesticide residues *see*
        monitoring program of pesticide residues (Novi
        Sad)
    plant protection and 211
pesticides 209, 225
phenotyping sensors 229
"Pingree Potato Patches" (Detroit) 197
Pinheiro, A. 20, 27
plant protection
    and pesticide residues 211
    in urban agriculture 217–218
plant protection products (PPPs) 211
Plant, The (project) (Chicago, Illinois, US) 231
Pleistocene period (20,000–16,000 B.C.) 10
policies
    current policies in UPA 135–137
    MENA region 160
    to promote or support UPA activities 157–158
    restricting UPA activities 136
    statements acknowledge UPA activities 136
    UPA in Tanzania, promoted 135
political ecology 3–4, 85
Poonacha, P. 194
"P-Patch" (Seattle) 197

practices and policies; *see also* policies
    early agricultural practices *see* early agricultural
        practices
    UA in India *see* UA in India
    urban home gardening in Kerala 68–72, *69–72*
pramodavana 13
Preferred Reporting Items for Systematic Reviews 41
pre-harvest interval (PHI) 209
Proksch, G. 205
psychological benefits 156
pyrethroid insecticides 215

Qatar 165
    dynamic adaptive master plan and urban interventions
        173–175
    Environmental Science Centre 179
    food–water–energy nexus 165
    permaculture garden and edible campus to university
        living lab, from 171–175
    Qatar University Master Plan for food urbanism *172*
    QU Living Lab *see* Qatar University Living Lab
Qatar University Living Lab 5, 166
    composting system creation *177*
    permaculture garden and edible campus to, from
        171–175
    planting edible permaculture garden *177*
    policy recommendations summary **178–179**
    research to design to policy, from 175–180
    stakeholders of *174*
Qing Dynasty (1644–1911) 16
QuEChERS-based method 214

rain shelters 72
Rao, N. 194
RedMart 103
regenerative urbanism 165
Republic of Serbia (RS)
    monitoring program of pesticide residues (case studies)
        see monitoring program of pesticide residues
        (Novi Sad)
    peri-urban agriculture 218
    pesticide residues *see* pesticide residues
    plant protection *see* plant protection
    urban agriculture 210
resilience, social-ecological 25
resource availability, UA in India
    fertilizers and pesticides 187–188
    formal and informal settlements, source of information
        in *190*, 194n1
    income and other economic resources 190–191
    plants grown along the roof and sides of houses *189*
    seeds 187
    social networks 189–190
    soil 187
    sunlight, exposure to 188
    training, education, and peer learning 188–189
Rigveda 10
road verge gardening 32
Robineau, O. 53, 54
rodenticides 209
Roessler, R. 50
rooftop gardens 31, 204–205
Roy, P. 194

sack gardening 25, 30
Sahel *see* urban and peri-urban agriculture (UPA)
Sangare, S. K. 48
Sargon II (722–704 B.C.) (King) 12
Sartison, K. 118
Sawio, C. J. 143
Schlecht, E. 49
Schmutz, U. 23
Science Advances 14
Sennacherib (704–68 B.C.) 12, 15
sensors 229
"The Severn Project" (Bristol) 197
"Shagara Roofs" in El Abour City 157
Sharma, S. 194
short-supply food chains (SSFCs) 23–24
Siasa ni Kilimo (Politics is Agriculture) in 1972 135
Simatele, D. M. 142
Singapore
    change in forested, cultivated, and urban land cover
        (1915–2018) 87
    context map of the location in Southeast Asia 86
    family business 110
    geography 86–88
    impacts on dietary or other habits 109
    labor 110
    macro social network: government, landlord, market,
        and others 101–104, *102*
    meso social network: hired laborers, other farmers,
        vendors/distributors, and customers 97–101, *98*
    micro-social network: individual, family 95–97, *97*
    motivation to practice farming 108–109
    other income sources 109
    political ecology 85
    Sky Greens 231–232
    traditional farmers *see* traditional farmers (Singapore)
    traditional farms *see* traditional farms (Singapore)
Singapore Agro-Food Enterprises Federation Limited
    (SAFEF) 99, 102
Singapore Green Plan 2030, '30 by 30' Grow Local target
    3, 85, 88
Singapore Land Authority (SLA) 88
Sky Greens (Singapore) 231–232
slash-and-burn 225
small-scale social-ecological systems 28
small-scale urban agriculture systems 20, 27, *28*
smart farming techniques 7
social-ecological resilience 25
socio-economic factors role in UA in India *see* UA in India
soil-bound spaces 197
soil fertility 49
soil-less cultivation technologies 72
Song Dynasty (960–1279) 16
Songgu Irrigation Scheme 16
S-shaped trajectory 11–12
stakeholders
    action through cross-sectoral dialogue, moving
        175–176
    Qatar University Living Lab *174*
Stenchly, K. 48, 49, 53
sustainability; *see also* sustainability transitions
    urban farming and sustainable cities 205–206
sustainability transitions; *see also* urban home gardening
        in Kerala
    government actors 64

lock-ins, regime's resistance 79–80
    MLPs 63–64
    UA in research 64
sustainable agriculture **227**
Sustainable Development Goals (SDGs) 26–27, 41, **74–76**,
        154
sustainable intensification (SI) 114
    CEs 115
    sustainability claims 115–116
    UA for SI *see* UA for SI
Swai, O. 194
Syria, vegetation in 15

Tankari, M. R. 51
Tanzania
    population 134
    UPA, policies promoted 135 *see also* UPA in Dar es
        Salaam and Morogoro Municipality
tapovana 13
13th FYP working group (Kerala) 66–67
Thys, E. 50, 52
Tongjiyan Irrigation System 16
Town and Country Planning Regulations (1992) 136
traditional farmers (Singapore) 109–110
    benefits and barriers 90, 104–108
    farm practices 88
    interview 90–93
    livelihood alternatives 108–109
    perspectives, documenting 88–93
    skills-building 106
    social networks 88–89, *90*
traditional farming 225
traditional farms (Singapore) 93–94
    GAP program 94
    high-tech operations 94–95
    SFA 94
    summary of farm practices based on form and function
        93–94
Tripp, A. M. 135
Truffer, B. 77
Turba Farm 173
2013 Agricultural Policy 136

UA for SI 116–117, 128–129; *see also* UA in India
    cultural dimension 118
    economic dimension 117
    environmental dimension 117
    expert assessment 121
    India *see* UA in India
    literature search and review 119–120
    political dimension 118–119
    public documents and databases, analysis of 120
    social dimension 117–118
    survey and primary data collection 120
    sustainability index construction and statistical
        analysis 121
    theoretical framework of sustainability 119
UA in India 128–129, 181–182, 193–194
    age, practices by 183–184, *184*
    attitudes and perceptions of practices 191–193
    carbon emissions 125
    collecting data 182
    contextual determinants 185–191
    crop irrigation water productivity 124

UA in India (*cont.*)
    crop nitrogen use productivity 125
    cultural significance 126
    data analyzing 183
    demographic determinants 183–185
    depicting socio-economic determinants of practices
        183
    determinants shaped practices 183
    economic dimensions 125–126
    environmental efficiency 125
    environmental impact in cities 124–125
    gender, practices by 184–185, *185*
    GVO 123, 128–129
    inherent determinants 191–193
    observations 183
    research approach 182
    resource availability 187–190
    respondents' perception regarding benefits *192*
    schemes, efficiency in implementing 126–127
    societal impact 126
    space 185–187, *186*
    status, demand, and supply dynamics in India 121–123
    subjective knowledge of practices 193
    sustainability and potential for SI in UASs, assessment
        of 127–128
Ulm, K. 32
al-'Umari, Hisham 159
UPA in Dar es Salaam and Morogoro Municipality
    activities types 138–139, **139**
    challenges and opportunities 146–147
    current policies in 135–137
    defined 137
    economic benefits 141
    economic impact 140–144
    economic savings on food 142–143
    extra income provided **142**
    farmers characterization and 137–140
    job creation 143–144
    location and land ownership 139–140
    motivation to start 140
    physical exercise **145**
    policies promoted UPA in Tanzania 135
    policy statements acknowledge activities 136
    restricting UPA activities, policies 136
    saving money spent on vegetables **142**
    social impact 144–146
    social network **144**
    socio-demographic characteristics of respondents 138
    women in 145–146
UPA in MENA region 4, 159–160
    actions 160
    advantages 154–156
    Amman (JORDAN) 159
    challenges hindering development 157–159
    GUPAP 158
    land and space availability 157
    real estate development 157
    recognition, lack of 157
    refugees, advantages for 156
    strategies and policies promote or support UPA
        activities 157–158
urban agriculture (UA) 1–2, 9–10, 17, 62, 114, 208
    allotment gardening 29
    believes and practices 12–13

characteristic of 210
for circular cities 24–25
community gardening 29–30
CSA 30
defined 1
early settlements 10–12
economic benefits 26
enhanced food production and dietary diversity 24
food miles 23
global sustainable development initiatives 41
green and brown agendas, bridging 23
India *see* UA in India
multifunctionality *see* multifunctionality
multi-scale approach 166
organoponics or organopónicos 31
plant protection in 217–218
Qatar *see* Qatar
Republic of Serbia 210
research and teaching, integrating into 165–170
road verge gardening 32
rooftop gardens 31
sack gardening 30
SDGs 26–27
SI *see* UA for SI
Singapore *see* Singapore
small-scale urban agriculture systems pattern 27, *28*
social and cultural benefits of 25–26
social-ecological resilience 25
SSFCs 23–24
sustainability potential 20
timeline of projects, initiatives, and events on UA
    (2012–2020) 167–170
urban beekeeping 31
urban home gardening 28–29
vegetation types 13–15
water management systems 15–16
urban and peri-urban agriculture (UPA) 3, 4, 40, 55–56,
    134–136, 151–153
    anthropogenic pollutants 209
    atmosphere and climate change 48
    biodiversity 48–49
    cycles of nutrients and materials 49–50
    Dar es Salaam and Morogoro Municipality *see* UPA in
        Dar es Salaam and Morogoro Municipality
    defined 41, 151, 208
    documents included in review 42–44
    economic dimension 52–53, **53**
    economic impact 140–144
    environmental dimension 44–50
    India 134
    land management and soil fertility 49
    plant protection and nutrition 209
    political dimension 53–55
    selected documents addressing environmental issues
        45–47
    social dimension 50–52
    social, economic, and policy impacts of 134–147
    UPA in MENA region *see* UPA in MENA region
    water resources 48
Urban and Peri-urban Agriculture as Green Infrastructure
    (UPAGrI) project 182, 194
urban farming
    growing spaces, typology of 197–198
    impact on built environment 198–204

and sustainable cities 205–206
urban gardening 52, 157
urban green infrastructure (UGI) 21
urban heat island (UHI) effect 199
urban home gardening 3, 63
    Kerala *see* urban home gardening in Kerala
    motivation for setting up 28–29
    scientific production methods 29
    size of 29
urban home gardening in Kerala
    and agri-food system transitions **67**, 80
    anchoring of niche innovations 77–78
    annual production of vegetables harvested *73*
    emerging niche, as a 77
    Facebook-based agriculture groups 67–68
    government interventions for promotion 66–67
    network anchoring 77
    niche development, enablers of 78–79
    normative anchoring 77
    people's movement, as 67–68
    sustainability contributions 73, **74–76**
    technologies and practices 68–72, *69–72*
urban home gardens 28–29; *see also* urban home
        gardening
urbanization 1, 114
    meaning 40
    of poverty 151
urban land
    CPUL 205–206
    cultural impact of UA 118
    in India 122
    in Singapore *87*
urban-scale landscape concept 206
urban sustainability 1–2

Van Veenhuizen, R. 1
Vedic civilization 10
Vegetable Development Programme (VDP) 66, *68*
Vegetable Initiative for Urban Clusters (VIUC) 66
vertical aquaponics systems 228

vertical farming 6, 165, 205, 223, 233–234
    advantages using sustainability framework **227**
    aeroponics 223, 228
    aquaponics 223, 226–228
    challenges and opportunities 232–233
    climate change and its impact on agriculture 224–225
    human health concerns and environmental degradation
        225–226
    hydroponics 223, 226
    land efficiency and limited space in dense urban areas
        224
    LED lighting 229
    machine learning and data analytics 229–230
    projects **230**, 231–232
    sensors and automation 229
vertical gardens 206
"Victory gardens" 197
Vigouroux, Y. 14, 15
Viljoen, A. 206
Viśvavarman period 13
volatile organic compounds (VOCs) 200

Wang, S.-J. 52
water-bound spaces 198
water pollution 204
water resources 48
Weinreb, A. 40
West Africa *see* urban and peri-urban agriculture (UPA)
wick system 72, 226
women in UPA 145–146
World Population Prospects 2017 report 152

Xinghua Duotian Irrigation and Drainage System 16

Yayoi era 11
Yayoi people 11
"Your Roof Is Your Paradise" in Beheira governorate 157

Zail, D. B. 118
Zhang, X. 116

Printed in the United States
by Baker & Taylor Publisher Services

Printed in the United States
by Baker & Taylor Publisher Services